COMPLEX QUANTUM SYSTEMS

Analysis of Large Coulomb Systems

LECTURE NOTES SERIES
Institute for Mathematical Sciences, National University of Singapore

ISSN: 1793-0758

Published

Vol. 13 Econometric Forecasting and High-Frequency Data Analysis
edited by Roberto S Mariano & Yiu-Kuen Tse

Vol. 14 Computational Prospects of Infinity — Part I: Tutorials
edited by Chitat Chong, Qi Feng, Theodore A Slaman, W Hugh Woodin & Yue Yang

Vol. 15 Computational Prospects of Infinity — Part II: Presented Talks
edited by Chitat Chong, Qi Feng, Theodore A Slaman, W Hugh Woodin & Yue Yang

Vol. 16 Mathematical Understanding of Infectious Disease Dynamics
edited by Stefan Ma & Yingcun Xia

Vol. 17 Interface Problems and Methods in Biological and Physical Flows
edited by Boo Cheong Khoo, Zhilin Li & Ping Lin

Vol. 18 Random Matrix Theory and Its Applications: Multivariate Statistics and Wireless Communications
edited by Zhidong Bai, Yang Chen & Ying-Chang Liang

Vol. 19 Braids: Introductory Lectures on Braids, Configurations and Their Applications
edited by A Jon Berrick, Frederick R Cohen, Elizabeth Hanbury, Yan-Loi Wong & Jie Wu

Vol. 20 Mathematical Horizons for Quantum Physics
edited by H Araki, B-G Englert, L-C Kwek & J Suzuki

Vol. 21 Environmental Hazards: The Fluid Dynamics and Geophysics of Extreme Events
edited by H K Moffatt & E Shuckburgh

Vol. 22 Multiscale Modeling and Analysis for Materials Simulation
edited by Weizhu Bao & Qiang Du

Vol. 23 Geometry, Topology and Dynamics of Character Varieties
edited by W Goldman, C Series & S P Tan

Vol. 24 Complex Quantum Systems: Analysis of Large Coulomb Systems
edited by Heinz Siedentop

*For the complete list of titles in this series, please go to
http://www.worldscibooks.com/series/LNIMSNUS

Lecture Notes Series, Institute for Mathematical Sciences,
National University of Singapore
Vol. 24

COMPLEX QUANTUM SYSTEMS

Analysis of Large Coulomb Systems

Editor

Heinz Siedentop
Ludwig-Maximilians-Universität München, Germany

World Scientific

NEW JERSEY • LONDON • SINGAPORE • BEIJING • SHANGHAI • HONG KONG • TAIPEI • CHENNAI

Published by

World Scientific Publishing Co. Pte. Ltd.
5 Toh Tuck Link, Singapore 596224
USA office: 27 Warren Street, Suite 401-402, Hackensack, NJ 07601
UK office: 57 Shelton Street, Covent Garden, London WC2H 9HE

British Library Cataloguing-in-Publication Data
A catalogue record for this book is available from the British Library.

Lecture Notes Series, Institute for Mathematical Sciences, National University of Singapore — Vol. 24
COMPLEX QUANTUM SYSTEMS
Analysis of Large Coulomb Systems

ISBN 978-981-4460-14-9

CONTENTS

Foreword vii

Preface ix

Stability of Matter 1
Rafael D. Benguria and Benjamín A. Loewe

Mathematical Density and Density Matrix Functional Theory (DFT and DMFT) 61
Volker Bach

On the Dynamics of a Fermi Gas in a Random Medium with Dynamical Hartree–Fock Interactions 121
Thomas Chen

On the Minimization of Hamiltonians over Pure Gaussian States 151
Jan Dereziński, Marcin Napiórkowski, and Jan Philip Solovej

Variational Approach to Electronic Structure Calculations on Second-Order Reduced Density Matrices and the N-Representability Problem 163
Maho Nakata, Mituhiro Fukuda, and Katsuki Fujisawa

Fermionic Quantum Many-Body Systems: A Quantum Information Approach 195
Christina V. Kraus

Hydrogen-Like Atoms in Relativistic QED 219
Martin Könenberg, Oliver Matte, and Edgardo Stockmeyer

FOREWORD

The Institute for Mathematical Sciences at the National University of Singapore was established on 1 July 2000. Its mission is to foster mathematical research, both fundamental and multidisciplinary, particularly research that links mathematics to other disciplines, to nurture the growth of mathematical expertise among research scientists, to train talent for research in the mathematical sciences, and to serve as a platform for research interaction between the scientific community in Singapore and the wider international community.

The Institute organizes thematic programs which last from one month to six months. The theme or themes of a program will be chosen from areas at the forefront of current research in the mathematical sciences and their applications.

Generally, for each program there will be tutorial lectures followed by workshops at research level. Notes on these lectures are usually made available to the participants for their immediate benefit during the program. The main objective of the Institute's Lecture Notes Series is to bring these lectures to a wider audience. Occasionally, the Series may also include the proceedings of workshops and expository lectures organized by the Institute.

We hope that through the regular publication of these lecture notes the Institute will achieve, in part, its objective of promoting research in the mathematical sciences and their applications.

January 2013

Chi Tat Chong
Wing Keung To
Series Editors

PREFACE

Quantum theory is the success story of the past eighty years. Starting with the hydrogen atom, an example of any textbook in the field, it is now believed that quantum mechanics can describe the most intricate and complex structures of matter. Accordingly, many-body physics has developed as an indispensable tool of theoretical physics. As usual, in physics, the language used to communicate those discoveries is mathematics. This however, contrasts with the fact that essential parts of the arguments are merely on a heuristic level. For mathematicians, this created not only a formidable challenge to structure those developments and to understand the logical core but also a bonanza of new possibilities of developing so far unknown beautiful mathematics.

The purpose of the IMS program on "Complex Quantum Systems" in 2010 was exactly to focus on these new developments in analysis with particular input from the emerging field of quantum information. The contributions collected in this book are based and inspired by this program. The organization of the material is as follows: for the newcomers in the field we start with two review articles:

I: Rafael Benguria and Benjamín Loewe give an overview over the — by now — classical results of stability of matter that go back to Dyson and Lenard and to Lieb and Thirring. The particular tools like the Lieb–Thirring inequality and simple correlation inequalities are presented in a pedagogical way that emphasizes clarity over generality.

II: Correlation inequalities are also in the focus of the second review by Volker Bach. However, the focus is on more recent developments, in particular, the correlation inequalities by Bach and by Graf and Solovej that allow to find the asymptotic behavior of large Coulomb systems very accurately. Recently, Bach, Knörr, and Menge were able to derive those correlation inequalities from the N-representability conditions of the one-particle density matrix giving some added impetus from the theoretical side to the use of those conditions in the context of atomic and molecular structure calculations. All of this can be found in this chapter.

III: The problem of deriving macroscopic or mesoscopic equations from microscopic ones is a long-standing task. Thomas Chen reviews progress in the field for an interacting Fermi gas with random potentials.

IV: For systems of large number of particles, a description in Fock space is often convenient. In addition, it is often helpful to introduce new creation and annihilation operators (new quasi-particles) yielding a new effective quadratic Hamiltonian. Dereziński, Napiórkowski, and Solovej discuss when such a procedure can be justified.

V: That the N-representability can be used in atomic physics was already hinted above. Nakata, Fukuda, and Fujisawa give an overview of the method covering the theory behind it and its methodology. They also give a brief history of the subject and report new computational results. Typically, the results obtained by the reduced-density-matrix method are comparable to those of the CCSD(T), which is a rather sophisticated traditional approach in quantum chemistry.

VI: Quantum information is a new and fast developing field of theoretical physics. The question of entanglement is intimately related with correlations. It is therefore not astonishing that this field has new things to say about large correlated systems. Christina Kraus is reviewing the development of matrix product and other states originating from it.

VII: Quantum electro-dynamics is considered to be *the* theory of charged particles interacting with the radiation field. In the absence of any consistent mathematical foundation beyond perturbation theory, models of the photon field interacting with a fixed number of particle have been considered. Recently this has been done also with realistic relativistic models like the no-pair operator. Könenberg, Matte, and Stockmeyer — who are the world leaders in the subject — present their results in the context of the hydrogen atom.

Finally, I would like to use the opportunity to express my appreciation: Thanks go to all who contributed to the program on "Complex Quantum Systems" in particular Matthias Christandl, Andreas Winter, and Georg-Berthold Englert who actually brought the researchers in quantum information and mathematical physics together.

Special thanks go to Professor Louis Chen, Director of the IMS, and his team who created a unique environment which made all this possible. Support of the IMS and the Center for Quantum Technologies is gratefully acknowledged.

November 2012

Heinz Siedentop
Ludwig-Maximilians-Universität München
Germany
Editor

STABILITY OF MATTER

Rafael D. Benguria
Departamento de Física, Pontificia Universidad Católica de Chile
Casilla 306, Santiago 22, Chile
rbenguri@fis.puc.cl

Benjamín A. Loewe
Departamento de Física, Pontificia Universidad Católica de Chile
Casilla 306, Santiago 22, Chile
baloewe@puc.cl

These are extended notes based on a series of lectures on the Stability of Matter given by one of us (RB) as part of the workshop on "Complex Quantum Systems" held at the National University of Singapore, February 17–March 27, 2010.

Contents

1 Introduction: *The stability of quantum systems: A historical overview* 2
2 Stability of Matter: *The classical proof of Lieb and Thirring* 5
2.1 Stability of the hydrogen atom in non-relativistic quantum mechanics 6
2.2 Stability of a system of N electrons in non-relativistic quantum mechanics 10
2.3 Stability of a many particle system via Thomas–Fermi theory 13
2.4 Bibliographical remarks 17
3 Lieb–Thirring Inequalities 18
3.1 Use of commutation methods to prove the Lieb–Thirring inequality for $\gamma = 3/2$ in dimension 1 20
3.2 The Eden–Foias bound ([46]) 22
3.3 Bibliographical remarks 24
4 Electrostatic Inequalities 25
5 The Maximum Number of Electrons an Atom Can Bind 29

5.1 The maximum number of electrons for a one center case in the Thomas–Fermi model 32
5.2 Bound on $N_c(Z)$ for the TFW model in the atomic case 33
6 The Stability of Matter for a Relativistic Toy Model 35
6.1 Bibliographical remarks 41
7 A New Lieb–Oxford Bound with Gradient Corrections 41
Acknowledgments 46
Appendix: A Short History of the Atom 46
References 52

Mais ce n'est pas tout; la Physique ne nous donne pas
seulement l'occasion de résoudre des problèmes;
elle nous aide à en trouver les moyens, et cela de deux manières.
Elle nous fait présentir la solution;
elle nous suggère des raisonaments.
Conference of H. Poincaré at the ICM, in Zürich, 1897, see [122], p. 340.

1. Introduction: *The stability of quantum systems: A historical overview*

Stability properties of two or more particle systems interacting among themselves via one over distance potentials (i.e., via Coulomb or Kepler interactions) have been a central and recurrent theme in Physics since the second half of the 17th century. After the introduction of the universal law of gravitation by Newton in 1687 [114], many people studied the stability of the solar system. These studies not only shed light on the physical properties of a multiple body system interacting via Kepler potentials but, in fact, produced fundamental advances in mathematics. On the one hand it triggered the introduction of perturbation theory in classical mechanics in terms of action and angle variables. On the other hand it triggered a whole realm of new mathematics centered around dynamical systems, which ultimately produced the Kolmogorov–Arnold–Moser theory (see, e.g., the classical book of Siegel and Moser [138]).

Stability also played a fundamental role in atomic physics during the early 20th century. At the beginning of the century several people (e.g., Thomson [153], Nagaoka [111], Rutherford [131, 132]) introduced different models of atoms (basically neutral systems with positive and negative charge distributions interacting via a Coulomb potential). The early models of Nagaoka and later of Rutherford have a strong influence from their gravitational counterpart. Soon after their introduction, they realized that (at least in the framework of classical physics) these systems were unstable. In the Rurtherford model, the accelerated electron losing energy via electromagnetic radiation, would fall into the nucleus in a very short time. It was partly because of the (classical) instability of the Rutherford atom that Bohr introduced his model [24, 25, 26, 27], that gave birth to the *old quantum mechanics* in 1913. At the beginning of the *old quantum mechanics*, Jeans observed: *There seems to be no difficulty about the supposition that at very small distances the law of force is different from the inverse square. On the contrary, there would be a very real difficulty in supposing the law* $1/r^2$ *held down to zero values of* r. *For the force between two charges at zero distance would be infinite; we should have charges of opposite sign continuously rushing together and, when once together, no force would be adequate to separate them... Thus the matter in the universe would tend to shrink into nothing or to diminish indefinitely in size. The observed permanence of matter precludes any such hypothesis... We should be wrong in regarding a molecule as a cluster of electrons and positive charges. A more likely suggestion, put forward by Larmor and others is that the molecule may consist, in part at least, of rings of electrons in rapid orbital motion* (see, [75], p. 168). The old quantum mechanics was successful in explaining the observed spectrum of the Hydrogen atom, but failed in explaining the spectra of more complicated atomic structures. Following the de Broglie's [39] suggestion that matter had also a wavelike behavior, Schrödinger in 1926 [134] introduced his (now classical) wave equation for the electron. The solution of the time independent Schrödinger equation for the Hydrogen atom (i.e., with a potential $V(x) = -Z/|x|$) reproduced the known spectra of this system. It is easy to see, either by exact computations or by using Heisenberg's *uncertainty principle* (with the help of Hardy's or better Sobolev's inequality) that the spectrum of the Hydrogen atom is bounded below (and thus, this one particle system is *stable*). At this point it should be emphasized that stability of Hydrogen does not follow from the classical formulation of the Heisenberg uncertainty relation, but only from the Hardy or Sobolev formulations. For a more complex system one not only needs to prove a similar lower bound on the ground state energy, but more important, one needs to prove that the energy per particle of the system is bounded below. This last property is what one understands as the *Stability of Matter*. Speculations about matter's stability go back to the early days of quantum mechanics. In 1931, Paul Ehrenfest observed that someone holding a piece of metal or stone should be *astonished that this quantity of matter should occupy so large a volume. Admittedly,* Ehrenfest remarked, *the molecules are packed tightly together and likewise the atoms within each molecule. But why are the atoms themselves so big?* For an answer, he summoned a principle first stated by Wolfgang Pauli: No two particles of the same kind can occupy the same quantum state at the same time. Pauli's principle means that electrons cannot all fall into the lowest energy, smallest orbital around an atomic nucleus but have to fill successively larger orbitals. *That is why atoms are so unnecessarily big, and why metal and stone are so bulky,* Ehrenfest concluded. It took forty years after the Schrödinger equation to have the first proof of the Stability of Matter in the context of Nonrelativistic Quantum Mechanics. This was accomplished by Freeman Dyson and Andrew Lenard [45, 81]. What Dyson and Lenard proved was that for a system of fermions interacting via Coulomb forces the ground state energy per particle of the system is bounded from below by a constant. The constant in their proof was unnecessarily large. In 1975, Lieb and Thirring [103] provided a simpler proof, based on the fact that the Thomas–Fermi energy is a good approximation to the Quantum Mechanical energy [99], and that in the Thomas–Fermi model there are no molecules [150]. The lower bound to the energy per particle in the proof of Lieb and Thirring is much more realistic (of the order of a few Rydbergs per particle). After the proof of Lieb and Thirring there has been a huge literature on the subject, with extensions to many different physical situations. There are several reviews (e.g., [84, 93, 96, 108, 137]) on the subject. In these notes we start by reviewing the now classical proof of the stability of non-relativistic matter by Lieb and Thirring. In Section 3 we review some basic facts about Lieb–Thirring inequalities. In Section 4 we review some basic facts about electrostatic inequalities. In Section 5 we review some problems related with the maximum ionization of an atom. In Section 6 we give the details of the proof of stability of a relativistic toy model. Finally, in Section 7, we review some new results on Lieb–Oxford type inequalities. At the end of these notes (in the Appendix) we present a brief essay on the history of atoms. One of us (RB) has also given related series of lectures on the Stability of Matter in the conference "Spectral Days – 2010", held in Santiago de Chile in September 2010 [10], and also as a short course in the Faculty of Mathematics at the Ludwig Maximilians Universität, München, Germany, at the end of May, 2011.

2. Stability of Matter: *The classical proof of Lieb and Thirring*

In this section we will introduce some notation and definitions and we will review the classical proof of Elliott Lieb and Walter Thirring of the stability of matter in nonrelativistic quantum mechanics.

In *Nonrelativistic Quantum Mechanics* we consider a system of N electrons and K fixed nuclei, described by the Hamiltonian,

$$H = -\frac{\hbar^2}{2m}\sum_{i=1}^{N}\Delta_i - e^2\sum_{i=1}^{N}V(x_i) + e^2\sum_{1\le i<j\le N}\frac{1}{|x_i - x_j|} + U, \tag{2.1}$$

acting on $L^2(\mathbb{R}^{3N}; C^{2^N})$ (bosons) or on the space $\bigwedge^N L^2(\mathbb{R}^3, C^2)$ (fermions). Here,

$$V(x) = \sum_{j=1}^{k}\frac{z_j}{|x - R_j|} \tag{2.2}$$

is (minus) the electrostatic potential created by the k fixed nuclei in $x \in \mathbb{R}^3$. Here, the nuclei are located at the positions $R_1, R_2, \ldots, R_k$, and have electric charge $ez_1, ez_2, \ldots, ez_k$ respectively, where e is the absolute value of the charge of the electron. Moreover, m is the mass of the electron. Finally,

$$U = e^2\sum_{1\le i<j\le k}\frac{z_i\, z_j}{|R_i - R_j|} \tag{2.3}$$

denotes the interaction energy between the nuclei. We denote as

$$E_\psi = \frac{(\psi, H\psi)}{(\psi, \psi)}, \tag{2.4}$$

the expectation value of the Hamiltonian in the state $\psi \in \bigwedge^N L^2(\mathbb{R}^3, C^2)$, and

$$E(N, k, \underline{Z}, \underline{R}) = \inf_{\psi} E_\psi. \tag{2.5}$$

Here, $\underline{Z} = (z_1, z_2, \ldots, z_N)$, etc. Now, if we minimize over all possible configurations of the nuclei we get the ground state energy of the system of k nuclei and N electrons, which we denote by,

$$E(N, k, \underline{Z}) = \inf_{\underline{R}} E(N, k, \underline{Z}, \underline{R}). \tag{2.6}$$

There are two notions of **Stability**:

(A) Stability of the first type

$E(N, k, \underline{Z}, \underline{R})$ is finite for all N, k, $\underline{Z}$, $\underline{R}$.

The stability of the first type amounts to the fact that the Hamiltonian given by (2.1) is bounded below.

(B) Stability of the second type

$E(N, k, \underline{Z}) \geq -A(z)(N+k)$, assuming that each $z_i \leq z$. In Nature, $z_j \leq 92$.

The stability of the second kind amounts to the fact that in the configuration of least energy, the energy per particle is bounded below by a constant (i.e., the constant $A(z)$).

2.1. *Stability of the hydrogen atom in non-relativistic quantum mechanics*

As we discussed in the Introduction, the stability of atoms was one of the reasons that gave rise to quantum mechanics at the beginning of the 20th century. By *stability of atoms* one means the fact that the energy of the ground state of an atom is finite. Let us first consider the Hamiltonian of the Hydrogen atom,

$$H = -\frac{1}{2}\Delta - \frac{Z}{|x|}, \tag{2.7}$$

acting on $L^2(\mathbb{R}^3)$. Here, we have chosen Hartree Atomic Units (au), i.e., units in which $\hbar = 1$, $m = 1$, $|e| = 1$. Let E_0 the bottom of the spectrum of H. Of course, one knows (by exact calculations) that $E_0 = -Z^2/2$ [Ha] $= -Z^2$ [Ry] (in Hartree Atomic Units, the unit of energy is a Hartree, which is equivalent to $= 27,2$ [eV], or, in other words, 2 Rydbergs). However, we can pretend for a moment that we do not know this exact value and we will use functional analytic inequalities to estimate E_0 from below. We will use three different inequalities (Hardy's, Sobolev's and the Coulomb's Uncertainty Principle), yielding different lower bounds. Just a few years before the Schrödinger equation, Hardy [61] proved the following inequality

$$\int_{\mathbb{R}^3} |\nabla\psi|^2 \, dx > \frac{1}{4}\int_{\mathbb{R}^3} \frac{\psi^2}{|x|^2} \, dx, \tag{2.8}$$

for all $\psi \in H^1(\mathbb{R}^3)$. Using the Rayleigh–Ritz principle, one has

$$E_0 = \inf \frac{(\psi, H\psi)}{(\psi,\psi)} = \inf \frac{(1/2)\int_{\mathbb{R}^3} |\nabla\psi|^2\,dx - Z\int_{\mathbb{R}^3} \psi^2|x|^{-1}\,dx}{\int_{\mathbb{R}^3}\psi^2\,dx}, \tag{2.9}$$

where the infimum is taken over all $\psi \in H^1(\mathbb{R}^3)$. Using Hardy's inequality (2.8) in (2.9), we obtain,

$$E_0 = \inf \frac{\int_{\mathbb{R}^3} \psi^2(a^2/8 - a\,Z)\,dx}{\int_{\mathbb{R}^3}\psi^2\,dx}, \tag{2.10}$$

where we have set $a = 1/|x|$. Finally, using the estimate $a^2/8 - aZ \geq -2\,Z^2$, one gets from (2.10) that

$$E_0 \geq -2\,Z^2\ [\mathrm{Ha}]. \tag{2.11}$$

This estimate is way too low (it is four times the actual value), but it is simple to obtain and it proves that the bottom of the spectrum of the Hydrogen atom is bounded below, and that the atom is *stable*. One can get an improved lower bound if one uses Sobolev's inequality [142], i.e.,

$$\int_{\mathbb{R}^3} (\nabla\psi)^2\,dx \geq K_s \left(\int_{\mathbb{R}^3} \psi^6\,dx\right)^{1/3}, \tag{2.12}$$

with $K_s = 3\left(\frac{\pi}{2}\right)^{4/3} \approx 5,478\ldots$. Using this in (2.9) we get,

$$(\psi, H\psi) \geq (K_s/2)\left(\int \rho^3\,dx\right)^{1/3} - \int \frac{Z}{|x|}\rho\,dx \equiv h(\rho), \tag{2.13}$$

where we have set $\rho = \psi^2$. From the previous equation we have that for all $\psi \in L^2(\mathbb{R}^3)$, normalized (i.e., such that $\int \psi^2\,dx = 1$),

$$(\psi, H\psi) \geq \min_{\rho} h(\rho),$$

where

$$h(\rho) \equiv (K_s/2)\left(\int \rho^3\,dx\right)^{1/3} - Z\int \frac{\rho}{|x|}\,dx, \tag{2.14}$$

with $\rho \geq 0$, and $\int \rho\,dx = 1$. It is straightforward to find the function ρ that minimizes the functional $h(\rho)$ subject to these constraints. It is given by,

$$\hat{\rho}(x) = C\left(\frac{1}{|x|} - \frac{1}{R}\right)^{1/2},$$

for $|x| \leq R$, and $\hat{\rho}(x) = 0$, for $|x| \geq R$, where $R = 3/(2Z)$. From here, it follows that,

$$\min_\rho h(\rho) = h(\hat{\rho}) = -\frac{2}{3}Z^2 \qquad [\text{Ha}],$$

and, therefore,

$$E_0 \equiv \inf_\psi \frac{(\psi, H\psi)}{(\psi,\psi)} \geq -\frac{2}{3}Z^2 \qquad [\text{Ha}], \tag{2.15}$$

which is a better approximation to the actual value $E_0 = -Z^2/2$ [Ha].

There is a somewhat different way to estimate the ground state energy of the Hydrogen atom. The new way, although gives a slightly worse bound, can be easily generalized to the case of N electrons. In fact, let us consider *Hölder's inequality*,

$$|\int f(x)\, g(x)\, dx| \leq \left(\int |f(x)|^p\, dx\right)^{1/p} \left(\int |g(x)|^q\, dx\right)^{1/q}, \tag{2.16}$$

with p and q such that $(1/p) + (1/q) = 1$, and $p \geq 1$. Using (2.16) with $f = \rho$, $g = \rho^{2/3}$, $p = 3$, and $q = 3/2$, one gets,

$$\int \rho^{5/3}\, dx \leq \left(\int \rho(x)^3\, dx\right)^{1/3} \left(\int \rho(x)\, dx\right)^{2/3}. \tag{2.17}$$

Now, using Sobolev's inequality together with (2.17) we have,

$$\int (\nabla\psi^2)\, dx \geq K_s \int \rho_\psi(x)^{5/3}\, dx. \tag{2.18}$$

Here, $\rho_\psi \equiv \psi^2$, is such that $\int \rho_\psi\, dx = 1$.

Remark: Of course, one can study directly the minimization problem:

$$K_1 \equiv \min_\psi \frac{\int (\nabla\psi)^2\, dx}{\int \psi(x)^{10/3}\, dx} \qquad \text{subject to } \int \psi^2\, dx = 1. \tag{2.19}$$

We leave this minimization problem as an exercise to the reader. Numerically one finds $K_1 \approx 9,578\ldots$. For reasons that will be clear below, it is convenient at this point to introduce the constant (c here stand for *classical*)

$$K_c \equiv \frac{3}{5}(6\pi^2)^{2/3} \approx 9,116\ldots \tag{2.20}$$

Numerically, $K_1 > K_c > K_s$. Using this bound, i.e.,

$$\int (\nabla\psi)^2\, dx \geq K_c \int \psi^{10/3}\, dx = K_c \int \rho^{5/3}\, dx, \tag{2.21}$$

in order to estimate from below the expectation value of the kinetic energy of the electron one gets,

$$(\psi, H\psi) \geq h_c(\rho),$$

with,

$$h_c(\rho) = (K_c/2) \int \rho^{5/3}\, dx - Z \int \frac{\rho}{|x|}\, dx.$$

One is lead to the minimization problem,

$$\min_\rho \{h_c(\rho) \mid \rho(x) \geq 0, \quad \int \rho(x)\, dx\}. \tag{2.22}$$

The minimizer of $h_c(\rho)$ is given by

$$\rho_c(x) = \left(\left(\frac{6}{5} \frac{Z}{K_c} \right) \left(\frac{1}{|x|} - \frac{1}{R} \right) \right)^{3/2}, \tag{2.23}$$

for $|x| \leq R$ while $\rho_c(x) = 0$ for $|x| > R$. The radius R is determined by the condition $\int \rho(x)\, dx = 1$ and it turns our to be

$$R = \frac{K_c}{2\, Z} \left(\frac{2}{\pi} \right)^{4/3}.$$

Also, the value of $\min h_c(\rho) = h(\rho_c)$ is given by $-3^{1/3} Z^2/2$ [Ha]. So, finally, using this variational principle one gets,

$$E_0 \geq -\frac{3^{1/3}}{2} Z^2 \qquad [\text{Ha}].$$

So far, we have discussed (as in [84]) the use of Hardy's and Sobolev's inequalities in order to prove the *stability* of the Hydrogen atom. This discussion has some relevance in particular to make the connection with the N-particle case. Of course the sharp (exact) lower bound (for the Hydrogen atom) is obtained using the *Coulomb Uncertainty Principle*:

Theorem 2.1: Coulomb Uncertainty Principle (see, e.g., [94] and [108], p. 14). *Let $\psi \in L^2(\mathbb{R}^3)$, such that $\nabla\psi \in L^2(\mathbb{R}^3)$, then,*

$$\int_{\mathbb{R}^3} \frac{1}{|x|} \psi^2\, dx \leq \|\nabla\psi\|_2 \|\psi\|_2. \tag{2.24}$$

Equality is attained in (2.24) if and only if $\psi(x) = A \exp -c|x|$, for any $A \in \mathbb{R}$, and $c > 0$.

Proof: There are many ways of proving this sharp result. Perhaps the simplest proof (see [108], p. 14) uses commutator techniques. Here we will use rearrangements (see the Bibliographical Remarks i) at the end of this chapter for references on rearrangements). If we denote by ψ^* the *symmetric decreasing rearrangement* of ψ (i.e., a radially symmetric, non increasing function equi-measurable to ψ), one has that $\|\psi\|_2 = \|\psi^*\|_2$ (in general L^p norms stay the same under rearrangements) and $\|\nabla\psi\|_2 \geq \|\nabla\psi^*\|_2$. Moreover, since $1/|x|$ is symmetric decreasing, one also has,

$$\int_{\mathbb{R}^3} \frac{1}{|x|}\psi(x)^2\, dx \leq \int_{\mathbb{R}^3} \frac{1}{|x|}\psi(x)^{*2}\, dx.$$

Because of these three facts, it is enough to prove (2.24) for radially symmetric functions. For radially symmetric functions, (2.24) follows by integration by parts and Schwarz's inequality. In fact. if $\psi(x) = \psi(r)$, with $r = |x|$, we have,

$$\begin{aligned}&\textstyle\int_{\mathbb{R}^3} |x|^{-1}\psi(x)^2\, dx = 4\pi \int_0^\infty \psi^2(r)\, r\, dr = -4\pi \int_0^\infty \psi(r)\, \psi'(r)\, r^2\, dr \\ &\leq \left(\textstyle\int_0^\infty \psi^2(r)\, 4\pi r^2\, dr\right)^{1/2} \left(\textstyle\int_0^\infty \psi'^2(r)\, 4\pi r^2\, dr\right)^{1/2} = \|\psi\|_2 \|\nabla\psi\|_2. \qquad (2.25)\end{aligned}$$

One has equality, if and only if $\psi'(r) = -c\psi(r)$, i.e., if $\psi(r) = A \exp -cr$.□

Remark 2.2: i) It follows immediately from (2.24) that $E_0 = -Z^2/2$ [Ha] (and the infimum in (2.9) is actually attained at $\psi(x) = A \exp -Z|x|$. ii) *Coulomb Uncertainty Principles* play an important role in most of the proofs of Stability of Matter, since they provide a way of controlling the Coulomb singularities with the help of the kinetic energy term (see, e.g., Theorem 6.2, below, for an example of the use of a *Coulomb Uncertainty Principle* in a simple relativistic toy model).

2.2. *Stability of a system of N electrons in non-relativistic quantum mechanics*

Consider now a system of N noninteracting fermions in a cubic box of volume V, with Dirichlet boundary conditions. (In the rest of this chapter we will switch to units in which $\hbar^2/(2m) = 1$ and $e = 1$). It is well known that for large values of N the ground state energy of this system, which we will denote by T_V, (the Hamiltonian of this system in these units is just $-\sum_{i=1}^N \Delta_i$, acting on $\bigwedge^N L^2(\mathbb{R}^3, C^2)$), is approximately given by

$$T_V \approx q^{-2/3} K_c V \rho^{5/3} \qquad (2.26)$$

where $\rho = N/V$ and q is the number of spin states (as usual, $q = 2$ for electrons). In other words, T_V is proportional to $N^{5/3}$ for fermions. On the other hand, T_V is proportional to N in the case of bosons. As we have seen above (see equation (2.21)), the expectation value of the kinetic energy of one electron satisfies,

$$T_\psi \geq K_c \int \rho_\psi(x)^{5/3}\, dx, \tag{2.27}$$

and (2.26) suggests that for a system of N electrons,

$$T_\psi \approx q^{-2/3} K_c \int \rho_\psi(x)^{5/3}\, dx. \tag{2.28}$$

In some sense this intuition is correct. In 1975, Lieb and Thirring [103] proved that the right side of (2.28) with a smaller constant is in fact a lower bound to the expectation of the kinetic energy of a system of N fermions (see, (2.32)) below.

A system of N electrons is described by the wavefunction

$$\psi(x_1, x_2, \ldots, x_N; \sigma_1, \sigma_2, \ldots, \sigma_N) \tag{2.29}$$

with $x_i \in \mathbb{R}^3$, $\sigma_i \in \{1, 2, \ldots, q\}$. Because of the *Pauli Principle*, the function ψ must be antisymmetric in the pairs (x_i, σ_i). The norm of ψ is given by

$$(\psi, \psi) \equiv \sum_{\sigma_i=1}^{q} \int |\psi(x_1, x_2, \ldots, x_N; \sigma_1, \sigma_2, \ldots, \sigma_N)|^2\, dx_1\, dx_2\, \ldots\, dx_N.$$

Also, the expectation value of the kinetic energy of the fermions is given by

$$T_\psi = \sum_{i=1}^{N} \sum_{\sigma_i=1}^{q} \int |\nabla_i \psi(x_1, x_2, \ldots, x_N; \sigma_1, \sigma_2, \ldots, \sigma_N)|^2\, dx_1\, dx_2\, \ldots\, dx_N. \tag{2.30}$$

Recall that the kinetic energy of the system of N fermions is given by $-\sum_{i=1}^{N} \Delta_i$, in the units we are using in this section. The *single particle density* of this system of N fermions is defined, as usual, as

$$\rho_\psi(x) = N \sum_{\sigma_i=1}^{q} \int |\psi(x, x_2, \ldots, x_N; \sigma_1, \sigma_2, \ldots, \sigma_N)|^2\, dx_2\, \ldots\, dx_N \tag{2.31}$$

and, it is normalized in such a way that $\int \rho_\psi(x)\, dx = N$. In 1975, Lieb and Thirring [103] proved that if ψ is normalized (so that $(\psi, \psi) = 1$), and the single particle density is given by (2.31), then one has

$$T_\psi \geq (4\pi)^{-2/3} q^{-2/3} K_c \int \rho_\psi(x)^{5/3}\, dx. \tag{2.32}$$

The numerical value of $(4\pi)^{-2/3}$ is approximately 0.185 (this constant was later improved to 0.277).

In order to prove this theorem, Lieb and Thirring devised a new type of functional inequalities that we describe in Chapter 2 below. The study of these inequalities, together with their extensions in many directions, has developed greatly in the last 35 years. The whole subject is now generically known under the name *Lieb Thirring inequalities*. For the problem at hand we only need the following special case: Consider the Schrödinger operator $H = -\Delta + V$ acting on $L^2(\mathbb{R}^3)$, where $V(x) \leq 0$. Let $e_1 \leq e_2 \leq \cdots \leq 0$ denote the negative eigenvalues of H. Then one has

$$\sum_j |e_j| \leq \frac{4\pi}{15\pi^2} \int |V(x)|^{5/2}\, dx. \tag{2.33}$$

Notice that if we were to use the semiclassical approximation, one would approximate

$$\begin{aligned} \textstyle\sum_j |e_j| &\approx \tfrac{1}{(2\pi)^3} \textstyle\int (p^2 + V)_-\, dx\, dp \\ &= \tfrac{1}{(2\pi)^3} \textstyle\int \left[4\pi \int_0^{|V(x)|^{1/2}} (p^2 + V)_- p^2\, dp \right] dx \end{aligned}$$

where $f(x)_- \equiv \max(0, -f(x))$ denotes the negative part of the function f. The integral on p can be done explicitly, and one gets the semiclassical estimate

$$\sum_j |e_j| \approx \frac{1}{15\pi^2} \int |V(x)|^{5/2}\, dx,$$

which differs from the upper bound (2.33) by the factor (4π).

Central to the discussion here is the physical intuition that in average a single electron "feels" the (electrostatic) potential generated by the nuclei plus the average potential generated by the whole cloud of electrons. In fact, each single electron will "feel" the Thomas–Fermi potential associated to the single particle density. It is this physical intuition plus an appropriate Lieb–Thirring inequality that allows to find a lower bound on the kinetic energy of the N electrons in terms of a functional of the single particle ρ_ψ. With that in mind consider the single particle Hamiltonian for the i–th electron, given by

$$h_i = p_i^2 - g\rho_\psi^{2/3}, \tag{2.34}$$

where g is a coupling constant to be determined. Also, consider the following

N-particle Hamiltonian,

$$\hat{H} = \sum_{i=1}^{N} h_i. \tag{2.35}$$

Then, the ground state energy of this N-particle Hamiltonian, E_0 say, satisfies

$$E_0 \leq (\psi, \hat{H}\psi) = T - g \int_{\mathbb{R}^3} \rho_\psi(x)^{5/3}\, dx. \tag{2.36}$$

On the other hand (by Pauli's principle) E_0 is greater than or equal to the sum of negative eigenvalues of $\hat{H}$, that can be estimated using the Lieb–Thirring inequality (2.33). Thus, one gets,

$$E_0 \geq -\frac{4}{15\pi} g^{5/2} \int_{\mathbb{R}^3} \rho_\psi(x)^{5/3}\, dx. \tag{2.37}$$

Combining (2.36) and ((2.37), one finally gets,

$$T \geq (g - \frac{4}{15\pi} g^{5/2}) \int_{\mathbb{R}^3} \rho_\psi(x)^{5/3}\, dx. \tag{2.38}$$

Maximizing over g, one obtains,

$$T \geq \frac{3}{5}\left(\frac{3\pi}{2}\right)^{2/3} \int_{\mathbb{R}^3} \rho_\psi(x)^{5/3}\, dx, \tag{2.39}$$

which is the desired lower bound on the kinetic energy of the system in terms of a functional of the single particle density.

2.3. *Stability of a many particle system via Thomas–Fermi theory*

In the previous paragraphs we have seen how to use the uncertainty principle (either using Hardy's inequality, Sobolev's inequality, or better the Coulomb Uncertainty Principle) to show that in non-relativistic quantum mechanics the energy of an atom with one electron is bounded below (and that therefore there is no collapse). For atoms with many electrons, the proof of the stability of matter (understood as above) is more difficult, and it was accomplished by F. Dyson and A. Lenard. In 1975, Lieb and Thirring [103] gave a different proof, with a major improvement on the lower bound for the energy per particle. The proof of Lieb and Thirring relied on two physical facts: i) That the energy of the Thomas–Fermi model of atoms is a good approximation to the groundstate energy of the non-relativistic Hamiltonian of the system of N electrons, and ii) that in the Thomas–Fermi model there are no molecules (Teller's Theorem). As pointed above, in order to use this physical intuition, Lieb and Thirring used what it is nowadays known as *Lieb–Thirring inequalities* to find a lower bound to the quantum mechanical energy in terms of the Thomas Fermi model.

The Thomas–Fermi model (introduced independently by Thomas [151] and Fermi [55]) is defined through the energy functional [82]

$$\xi(\rho) = q^{-2/3} K_c \int_{\mathbb{R}^3} \rho^{5/3}\, dx - \int_{\mathbb{R}^3} V(x)\rho(x)\, dx + \frac{1}{2}\int_{\mathbb{R}^3}\int_{\mathbb{R}^3} \frac{\rho(x)\rho(y)}{|x-y|}\, dx\, dy + U \tag{2.40}$$

where $\rho(x) \geq 0$ denotes the electronic density. On the other hand,

$$V(x) = \sum_{i=1}^{k} \frac{z_i}{|x - R_i|},$$

denotes the Coulomb potential created by the k fixed nuclei of charge $z_i > 0$, located at R_i, $i = 1, \ldots, k$. Finally,

$$U = \sum_{1 \leq i < j \leq k} \frac{z_i\, z_j}{|R_i - R_j|}$$

denotes the repulsion energy between the nuclei. The term $\int \rho^{5/3}$ in (2.40) represents the kinetic energy of the electrons, and takes into account the Fermi statistics of them. The energy of the system of λ electrons and k nuclei in this model is given by

$$E_{TF}(\lambda) = \inf\{\xi(\rho) \mid \rho \geq 0, \int \rho\, dx = \lambda\}. \tag{2.41}$$

The corresponding Euler equation, i.e., the *Thomas–Fermi equation*, is given by,

$$\frac{5}{3} K_c q^{-2/3} \rho^{5/3} \equiv \max(\phi(x) - \mu, 0), \tag{2.42}$$

with

$$\phi(x) = V(x) - \int \rho(y) \frac{1}{|x-y|}\, dy. \tag{2.43}$$

Here, μ, the *chemical potential*, is a Lagrange multiplier that is introduced to take into account the constraint $\int \rho(x)\, dx = \lambda$. The equation (2.42) has a solution if and only if $\lambda \leq \sum_{i=1}^{k} z_i$. The Thomas–Fermi energy of an isolated *neutral atom* of nuclear charge Z (i.e., with $k = 1$ and $\lambda = Z$) is found numerically to be

$$E_{TF}(Z) = -2.21 q^{2/3} \frac{1}{K_c} Z^{7/3}. \tag{2.44}$$

One of the main inputs in the proof of stability by Lieb and Thirring is Teller's *no binding* theorem [150] that asserts that "there are no molecules in the Thomas–Fermi model". If there are at least two nuclei, decompose

$$V(x) = \sum_{i=1}^{k} z_i \frac{1}{|x - R_i|} = V^1 + V^2,$$

with

$$V^1(x) = \sum_{i=1}^{m} z_i \frac{1}{|x - R_i|},$$

with $m < k$. Let $E_{TF}^1(\lambda)$ be the TF energy for a system of λ electrons in the presence of the nuclei $1, \ldots, m$ and respectively $E_{TF}^2(\lambda)$ for the nuclei $m+1, \ldots, k$. Given λ, let $\lambda_1 \geq 0$ and $\lambda_2 \equiv \lambda - \lambda_1 \geq 0$ be chosen in such a way that $E_{TF}^1(\lambda_1) + E_{TF}^2(\lambda_2)$ is minimized. Then, Teller's theorem says that

$$E_{TF}(\lambda) \geq E_{TF}^1(\lambda_1) + E_{TF}^2(\lambda_2), \tag{2.45}$$

i.e., there is *no binding*.

Using the *no binding* theorem of Teller one has that for all $\lambda \geq 0$ and $Z \equiv \sum_{i=1}^{k} z_i$,

$$E_{TF}(\lambda) \geq E_{TF}(Z) \geq -2.21 q^{2/3} \frac{1}{K_c} \sum_{j=1}^{k} z_j^{7/3}. \tag{2.46}$$

The first inequality in (2.46) follows from the fact that the Thomas–Fermi energy is a decreasing function on the number of electrons and that the absolute minimum of the energy functional without a restriction on the number of electrons is attained at neutrality (i.e., when $\lambda = Z$). The second inequality in (2.46) follows from (2.44) and Teller's Theorem.

Using (2.46), Lieb and Thirring realized that taking all the nuclear charges $z_i = 1$, and replacing K_c by γ, one has that for all $\rho(x) > 0$ with $\int \rho(x)\, dx < \infty$ and $\int \rho(x)^{5/3}\, dx < \infty$,

$$\begin{aligned} \textstyle\sum_{1 \leq i < j \leq N} \frac{1}{|x_i - x_j|} &\geq -\tfrac{1}{2} \textstyle\int \rho(x) \frac{1}{|x-y|} \rho(y)\, dx\, dy \\ &\quad + \textstyle\int \rho(y) V_X(y)\, dy - 2.21 \frac{N}{\gamma} - \gamma \int \rho(x)^{5/3}\, dx \end{aligned} \tag{2.47}$$

where $V_X(y) \equiv \sum_{j=1}^N 1/|y - x_j|$. Taking the expectation value of the equation (2.47) with respect to a normalized antisymmetric function ψ, and setting $\rho(x) = \rho_\psi(x)$ (i.e., the *single particle density* associated to ψ), one readily gets,

$$\left(\psi, \sum_{1\le i<j\le N} \frac{1}{|x_i-x_j|}\psi\right) \ge \frac{1}{2}\int \rho_\psi(x)\frac{1}{|x-y|}\rho_\psi(y)\,dx\,dy \\ - 2.21\frac{N}{\gamma} - \gamma\int \rho_\psi(x)^{5/3}\,dx. \tag{2.48}$$

To control the expectation value of the kinetic energy we use (2.32). Collecting terms we have,

$$E_\psi^Q \ge \alpha\int \rho_\psi^{5/3} - \int V(x)\rho_\psi(x)\,dx \\ +\frac{1}{2}\int \rho_\psi(x)\frac{1}{|x-y|}\rho_\psi(y)\,dx\,dy + U - 2.21\frac{N}{\gamma}, \tag{2.49}$$

with $\alpha \equiv (4\pi q)^{-2/3}K_c - \gamma$. So far, γ was arbitrary. From now on we restrict its value so that $\alpha > 0$, and later one optimizes over γ. From Teller's no binding theorem, one has,

$$E_\psi^Q \ge -2.21\left\{\frac{N}{\gamma} + \frac{1}{\alpha}\sum_{j=1}^k z_j^{7/3}\right\}.$$

One can optimize the right side of this equation with respect to γ. The optimal γ is given by

$$\gamma = (4\pi q)^{-2/3}K_c\left[1 + \left(\sum_{j=1}^k z_j^{7/3}/N\right)^{1/2}\right]^{-1},$$

and thus one gets,

$$E_N^Q \ge -2.21\frac{(4\pi q)^{2/3}N}{K_c}\left(1 + a^{1/2}\right)^2,$$

with $a \equiv \sum_{j=1}^k z_j^{7/3}/N$. Completing squares, one has $(1 + a^{1/2})^2 \le 2 + 2a$; therefore,

$$E_N^Q \ge -4.42(4\pi q)^{2/3}\frac{1}{K_c}\left(N + \sum_{j=1}^k z_j^{7/3}\right). \tag{2.50}$$

Thus, if the nuclear charges z_j are bounded above by some fixed z, E_N^Q is bounded below by a constant times the *Total Number* of particles, $N + k$.

If $z_j = 1$ (i.e., one has a bunch of Hydrogen atoms) and $N = k$ (i.e., we are in the case of neutrality) (2.50) implies

$$E_N^Q \geq -22.24N \qquad [\text{Ry}].$$

In the previous formulas, the q dependence is kept in purpose in order to say something about *bosons*. In fact, if $q = N$, the requirement of antisymmetry on ψ is no restriction at all. In this case $E_N^Q = \inf \operatorname{spec} H_N$ over all $L^2(\mathbb{R}^{3N})$. Hence,

$$E_N^Q \geq -\frac{2.21}{K_c}(4\pi)^{2/3} N^{5/3} \left[1 + \left(\sum_{j=1}^{k} \frac{z_j^{7/3}}{N}\right)^{1/2}\right]^2. \tag{2.51}$$

2.4. *Bibliographical remarks*

i) Rearrangements of functions were introduced by G. Hardy and J. E. Littlewood. Their results are contained in the classical book, G. H. Hardy, J. E. Littlewood, and G. Pólya, *Inequalities*, 2nd edn., Cambridge University Press, 1952. The fact that the L^2 norm of the gradient of a function decreases under rearrangements was proven by Faber and Krahn [52, 76, 77]. A more modern proof as well as many results on rearrangements and their applications to PDE's can be found in [149]. The reader may want to see also the article by E. H. Lieb, *Existence and uniqueness of the minimizing solution of Choquard's nonlinear equation*, Studies in Appl. Math. **57**, 93–105 (1976/77), for an alternative proof of the fact that the L^2 norm of the gradient decreases under rearrangements using heat kernel techniques. An excellent expository review on rearrangements of functions (with a good bibliography) can be found in G. Talenti, *Inequalities in rearrangement invariant function spaces*, in *nonlinear analysis, function spaces and applications, Vol. 5 (Prague, 1994)*, 177–230, Prometheus, Prague, 1994. (available at the website: http://www.emis.de/proceedings/Praha94/). See also, Chapter 3, *Rearrangement Inequalities* of [94], and also Lecture 2 of my recent lecture notes [11].

ii) Since the work of Dyson and Lenard and later of Lieb and Thirring, many new alternative proofs of the stability of matter have appeared. In particular, many of the recent proofs do not use the Thomas–Fermi model as a physical and mathematical tool. Instead, the use of electrostatic inequalities initiated by Onsager in the thirties is perhaps the most common method nowadays. See for example [108, 96, 137] and the references therein.

3. Lieb–Thirring Inequalities

The Lieb–Thirring inequalities (LT for short) originated in the proof of the stability of matter of Elliott Lieb and Walter Thirring [103] in 1975, and have given place to a vast mathematical literature since then. Consider a Schrödinger operator

$$H = -\Delta + V(x), \tag{3.1}$$

acting on $L^2(\mathbb{R}^n)$, and denote $e_1 \leq e_2 \leq \cdots < 0$ its negative eigenvalues. Then, the Lieb–Thirring inequalities amount to the fact that

$$\sum_j |e_j|^\gamma \leq L_{\gamma,n} \int_{\mathbb{R}^n} V_-(x)^{\gamma+n/2}\, dx \tag{3.2}$$

where $V_-(x) = \max(-V(x), 0)$ is the negative part of $V(x)$, for suitable values of γ. The range of possible values of γ depend on n. For $n = 1$, LT inequalities hold for all $\gamma \geq 1/2$ (the case $\gamma = 1/2$ was proved in [158]). For $n = 2$, they hold for all $\gamma > 0$, and for dimensions larger or equal to 3 they hold for all $\gamma \geq 0$. Notice that the case $\gamma = 0$ in (3.2) yields the number of negative eigenvalues of the operator H. The LT in this special case (i.e., the LT inequality with $\gamma = 0$, $n \geq 3$) was obtained independently by Cwickel [34], Lieb, [86] and Rosenbljum [127], and it is usually referred to as the CLR bound (see also [32] for an alternative proof of the CLR bound). Associated with (3.2) is the semiclassical estimate for $\sum_j |e_j|^\gamma$ which in fact motivates it. Using the *Planck dictum* (with $\hbar = 1$),

$$\sum_{j\geq 1} |e_j|^\gamma \approx (2\pi)^{-n} \int_{\mathbb{R}^n\times\mathbb{R}^n} (p^2 + V(x))_-^\gamma\, dp\, dx = L^c_{\gamma,n} \int_{\mathbb{R}^n} V_-(x)^{\gamma+n/2}\, dx, \tag{3.3}$$

where

$$L^c_{\gamma,n} = \left(\frac{1}{4\pi}\right)^{n/2} \frac{\Gamma(\gamma+1)}{\Gamma(\gamma+1+(n/2))}. \tag{3.4}$$

By considering $V(x) = \lambda W(x)$ with W sufficiently smooth and letting the coupling constant $\lambda \to \infty$ (which is equivalent to letting $\hbar \to 0$ in the usual semiclassical picture) one concludes that

$$L_{\gamma,n} \geq L^c_{\gamma,n}. \tag{3.5}$$

Since the original paper of Lieb and Thirring there have been many results concerning the sharp values of the constants $L_{\gamma,n}$. In 1978, Aizenman and

Lieb [1] proved that

$$R_{\gamma,n} \equiv \frac{L_{\gamma,n}}{L^c_{\gamma,n}}, \tag{3.6}$$

is a monotonically non-increasing function of γ, and thus, if there is a value of γ, say $\hat{\gamma}$ for which $R_{\hat{\gamma},n} = 1$, then,

$$L_{\gamma,n} = L^c_{\gamma,n},$$

for all $\gamma \geq \hat{\gamma}$. Using the Buslaev–Faddeev–Zakharov trace formula [29, 53], Lieb and Thirring [104] proved that

$$L_{3/2,1} = L^c_{3/2,1} = \frac{3}{16}$$

(see also the proof of this fact below, using commutation methods). Hence, $L_{\gamma,1} = L^c_{\gamma,1}$ for all $\gamma \geq 3/2$. The corresponding result

$$L_{\gamma,n} = L^c_{\gamma,n},$$

for $\gamma \geq 3/2$ and all $n \geq 2$ was established by Laptev and Weidl [80] (see also [19] for an alternative proof). The fact that

$$L_{1/2,1} = 2\, L^c_{1/2,1} = \frac{1}{2}, \tag{3.7}$$

was established in [70]. It is still an open problem to determine the sharp values of $L_{\gamma,n}$ for $1/2 < \gamma < 3/2$ in one dimension and for $\gamma < 3/2$ in higher dimensions. However, there is a conjecture due to Lieb and Thirring [104], which says that $L_{\gamma,1}$, for $\gamma \leq 3/2$, should be given by

$$L^1_{\gamma,1} = \sup \frac{|e_1|^\gamma}{\int_{\mathbb{R}} V_-(x)^{\gamma+1/2}\, dx} \tag{3.8}$$

where the supremum is taken over all potentials V such that $V_- \in L^{\gamma+1/2}(\mathbb{R})$. This maximization problem had been solved by Keller [78], who found

$$L^1_{\gamma,1} = \frac{1}{\sqrt{\pi}(\gamma - \frac{1}{2})} \frac{\Gamma(\gamma+1)}{\Gamma(\gamma+\frac{1}{2})} \left(\frac{\gamma - \frac{1}{2}}{\gamma + \frac{1}{2}}\right)^{\gamma+(1/2)},$$

using the direct method of the calculus of variations (see [20] for an alternative method of determining $L^1_{\gamma,1}$). This conjecture has only been proved in the case $\gamma = 1/2$ [70], where $L^1_{1/2,1} = 1/2$.

Moreover, for higher dimensions ($n > 1$), it was conjectured by Lieb and Thirring (see [104], pp. 286–287), that there is a critical value of γ, $\gamma_{c,n}$ say, such that

$$L_{\gamma,n} = L^C_{\gamma,n}, \qquad \gamma \geq \gamma_{c,n},$$

whereas

$$L_{\gamma,n} = L^1_{\gamma,n}, \qquad \gamma \leq \gamma_{c,n}.$$

Here, $\gamma_{c,n}$ is defined to be that γ for which $L^C_{\gamma,n} = L^1_{\gamma,n}$; the uniqueness of this γ_c is part of the conjecture. Moreover, $\gamma_{c,1} = 3/2$ (as we have discussed in the paragraph above), $\gamma_{c,2} \approx 1.2$, $\gamma_{c,3} \approx .86$, and the smallest n such that $\gamma_{c,n} = 0$ is $n = 8$.

In their original proof of the LT inequalities [103] (i.e., in the special case $\gamma = 1$ and $n = 3$) Lieb and Thirring used the Birman–Schwinger kernel together with the counting function $N_{-\alpha}(V)$ (i.e., the number of negative eigenvalues of $H = -\Delta + V$ acting on $L^2(\mathbb{R}^3)$ which lie below $-\alpha$, $\alpha > 0$) to derive it. See also the review articles [108, 137] and the book [96] for details. In what follows I will present two different alternative methods that have been used to obtain LT inequalities: i) The *commutation method*, and ii) The Eden–Foias method.

3.1. *Use of commutation methods to prove the Lieb–Thirring inequality for $\gamma = 3/2$ in dimension 1*

Commutation methods were introduced by Jacobi in 1837 [74], and later by Darboux [38] to derive spectral properties of Sturm–Liouville operators, and since then, they have had a long history. Its modern appearance seems to be due to Crum [33]. For a rigorous discussion of the use of commutation methods we refer to [40] and [58]. Here we will present the proof of (3.7), using commutation methods as given in [19]. Let $-\lambda_1$ be the lowest eigenvalue of the one dimensional Schrödinger operator

$$H = -\frac{d^2}{dx^2} + V(x),$$

acting on $L^2(\mathbb{R})$. Here we assume for simplicity that $V(x)$ has compact support, which lies in $[-a, a]$. It is well known that the lowest eigenvalue is not degenerate and the corresponding eigenfunction ϕ_1 can be chosen to be strictly positive. Moreover, outside the support of the potential we have,

$$\phi_1(x) = \begin{cases} \text{const.}e^{-\sqrt{\lambda_1}\,x}, & \text{if } x > a, \\ \text{const.}e^{+\sqrt{\lambda_1}\,x}, & \text{if } x < -a. \end{cases} \tag{3.9}$$

Thus the logarithmic derivative,

$$F(x) = \frac{\phi_1'(x)}{\phi_1(x)}, \tag{3.10}$$

is defined and satisfies the Riccati equation,

$$F' + F^2 = V + \lambda_1, \tag{3.11}$$

together with the conditions

$$F(x) = \begin{cases} -\sqrt{\lambda_1}\, x, & \text{if } x > a, \\ +\sqrt{\lambda_1}\, x, & \text{if } x < -a. \end{cases} \tag{3.12}$$

A simple computation shows that the Hamiltonian H can be written as

$$H = D^* D - \lambda_1, \tag{3.13}$$

where

$$D = \frac{d}{dx} - F, \tag{3.14}$$

and

$$D^* = -\frac{d}{dx} - F. \tag{3.15}$$

It is a general fact [40, 58] that the operators $D^*\, D$ and $D\, D^*$ acting on $L^2(\mathbb{R})$ have the same spectrum with the possible exception of the zero eigenvalue. Note that $D^*\, D$ has a zero eigenvalue which corresponds to the ground state of H. The operator $D\, D^*$ does not have a zero eigenvalue. This follows from the fact that the corresponding eigenfunction ψ satisfies

$$\psi' = -F\psi,$$

and hence $\psi(x) = c/\phi_1(x)$ (where c is a constant) which grows exponentially and is not normalizable. Thus, the new Schrödinger operator,

$$\tilde{H} = D\, D^* - \lambda_1 = -\frac{d^2}{dx^2} - F' + F^2 - \lambda_1 = -\frac{d^2}{dx^2} + V - 2F', \tag{3.16}$$

has, except for the eigenvalue $-\lambda_1$, precisely the same eigenvalues as H. Also, notice that the potential $V - 2F'$ is smooth and has compact support in the same interval as the potential V. Next, we compute using the Riccati equation (3.11)

$$\int (V - 2F')^2\, dx = \int V^2\, dx + 4 \int (\lambda_1 - F^2) F'\, dx. \tag{3.17}$$

The last term can be computed explicitly using (3.12) and we obtain,

$$\int (V - 2F')^2\, dx = \int V^2\, dx - \frac{16}{3} \lambda_1^{3/2}. \tag{3.18}$$

Thus,

$$\sum_{k=1}^{L} \lambda_k^{3/2} - \frac{3}{16}\int V^2\,dx = \sum_{k=2}^{L} \lambda_k^{3/2} - \frac{3}{16}\int (V - 2F')^2\,dx \tag{3.19}$$

and the Schrödinger operator with the potential $V - 2F'$ has precisely the eigenvalues $-\lambda_2, \ldots, -\lambda_L$. Continuing this process we remove one eigenvalue after another. After the last one is removed a manifestly negative quantity is left over, and this proves Theorem 1 in the scalar case.

3.2. *The Eden–Foias bound* ([46])

A completely different approach to derive LT inequalities was introduced by Eden and Foias in 1991, [46]. This method yields the best (not sharp) LT constants to date in the $n = \gamma = 1$ case. Recently it has been extended also to cover the n dimensional case with $\gamma = 1$ (see, [42], where the best constants up to date for the case $n \geq 1$ and $\gamma = 1$ are obtained).

Lemma 3.1: An L^∞ Bound for Functions $f \in L^2(\mathbb{R})$ and $f' \in L^2(\mathbb{R})$. *If $f \in L^2(\mathbb{R})$ and $f' \in L^2(\mathbb{R})$, then,*

$$f(x)^2 \leq \|f\|_2\|f'\|_2. \tag{3.20}$$

Proof: Using the fundamental theorem of calculus we write,

$$f(x)^2 = \int_{-\infty}^{x} \tfrac{d}{ds} f^2(s)\,ds = 2\int_{-\infty}^{x} f(s)f'(s)\,ds$$
$$\leq \alpha \int_{-\infty}^{x} f^2(s)\,ds + \tfrac{1}{\alpha}\int_{-\infty}^{x} f'^2(s)\,ds,$$

for any constant $\alpha > 0$. Analogously, we have,

$$f(x)^2 \leq \beta \int_x^\infty f^2(s)\,ds + \frac{1}{\beta}\int_x^\infty f'^2(s)\,ds,$$

for any $\beta > 0$. We choose $\beta = \alpha$, and add the last two equations to get,

$$2f(x)^2 \leq \alpha\|f\|_2^2 + \frac{1}{\alpha}\|f'\|_2^2,$$

and the result follows optimizing the right side in α, i.e., choosing $\alpha = \|f'\|_2/\|f\|_2$. □

Now, consider an orthonormal set $\{u_n\}_{n=1}^N$ of functions in $L^2(\mathbb{R})$, and introduce the function

$$K(x,y) = \sum_{n=1}^{N} u_n(x)u_n(y). \tag{3.21}$$

In the sequel we use the result of Lemma 3.1, with the function $f_y(x) \equiv K(x,y)$, i.e., we consider $K(x,y)$ as a function of x with y as a parameter. Thus, setting $f_y(x) = K(x,y)$, we compute first,

$$\|f_y\|_2^2 = \sum_{n,m=1}^{N} \int_{\mathbb{R}} u_n(x)u_n(y)u_m(x)u_m(y)\,dx = \sum_{n=1}^{N} u_n(y)^2, \tag{3.22}$$

where the last equality follows from the orthonormality of the set $\{u_n\}_{n=1}^N$. Analogously, one can compute,

$$\|f_y'\|^2 = \sum_{n,m=1}^{N} u_n(y)u_m(y) \int_{\mathbb{R}} u_n'(x)u_m'(x)\,dx. \tag{3.23}$$

Using (3.20) with the function $f_y(x)$, squaring, using (3.22) and (3.23), and choosing $x = y$ on the right side of the result, we finally get,

$$\left[\sum_{n=1}^{N} u_n(y)^2\right]^4 \leq \sum_{n=1}^{N} u_n(y)^2 \sum_{n,m=1}^{N} u_n(y)u_m(y) \int_{\mathbb{R}} u_n'(x)u_m'(x)\,dx. \tag{3.24}$$

Simplifying one gets,

$$\left[\sum_{n=1}^{N} u_n(y)^2\right]^3 \leq \sum_{n,m=1}^{N} u_n(y)u_m(y) \int_{\mathbb{R}} u_n'(x)u_m'(x)\,dx.$$

Finally, integrating the above inequality in y over the whole line, one obtains,

$$\int_{\mathbb{R}} \left[\sum_{n=1}^{N} u_n(y)^2\right]^3 dy \leq \int_{\mathbb{R}} \sum_{n=1} u_n'(y)^2\,dy, \tag{3.25}$$

where we have used again the orthonormality of the set $\{u_n\}_{n=1}^N$.

Now, one can use the result embodied in (3.25) to derive a Lieb–Thirring inequality with exponent $\gamma = 1$ in one dimension. In fact, consider the Schrödinger operator,

$$H = -\frac{d^2}{dx^2} - V(x), \tag{3.26}$$

acting on $L^2(\mathbb{R})$. Here the potential $V(x) \geq 0$ goes to *zero* at $\pm\infty$. Let $-\lambda_1 < -\lambda_2 < \cdots$ the eigenvalues of H, where $\lambda_n \geq 0$, all n. Also, denote by u_n the corresponding (normalized) eigenfunctions. Thus, we have,

$$-u_n'' - Vu_n = -\lambda_n u_n. \tag{3.27}$$

Since H is self adjoint, the set $\{u_n\}_{n=1}^N$ is an orthonormal set (where N is the number of eigenvalues; without loss of generality we can always assume that V is of compact support, so the number of negative eigenvalues (i.e., N) is finite. Multiplying (3.27) by u_n and integrating in x over the line $\mathbb{R}$, one has,

$$\lambda_n = \int_{\mathbb{R}} V u_n^2 \, dx - \int_{\mathbb{R}} (u_n')^2 \, dx. \tag{3.28}$$

Summing (3.28) over n, we get,

$$\sum_{n=1}^N \lambda_n = \int_{\mathbb{R}} V \left(\sum_n u_n^2 \right) dx - \int_{\mathbb{R}} \left(\sum_n {u'}_n^2 \right) dx. \tag{3.29}$$

Now, for any positive θ, we have the point–wise bound,

$$V(x)\theta \leq \theta^3 + \left(\frac{2}{3\sqrt{3}} \right) V(x)^{3/2}. \tag{3.30}$$

From (3.25), (3.29) and (3.30) we finally conclude that

$$\sum_{n=1}^N \lambda_n \leq \frac{2}{3\sqrt{3}} \int_{\mathbb{R}} V(x)^{3/2} \, dx. \tag{3.31}$$

The constant $c = 1/(3\sqrt{3}) \approx 0.3849$ is the best constant to date, but is almost sixty percent larger than the conjectured value $L_{1,1}^1 = 4/(3\pi\sqrt{3}) \approx 0.2450$.

3.3. *Bibliographical remarks*

i) The function $K(x, y)$ used in the proof of the Eden–Foias bound (i.e., equation (3.21) above) appears frequently in the proof of many results concerning orthogonal polynomials, in particular in connection with the Christoffel–Darboux formula (see, e.g., [67], p. X).

Lieb–Thirring inequalities are not only important in connection to the mathematics of the stability of matter. They have also connections to interesting geometrical problems. Just to mention one particular problem in that direction, Benguria and Loss [20] proved that the following geometrical problem for ovals in the plane is related to the Lieb–Thirring conjecture in the case $n = 1$, $\gamma = 1$: Denote by C a closed curve in the plane, of length 2π, with positive curvature κ, and let

$$H(C) \equiv -\frac{d^2}{ds^2} + \kappa^2 \tag{3.32}$$

acting on $L^2(C)$ with periodic boundary conditions. Let $\lambda_1(C)$ denote the lowest eigenvalue of $H(C)$. Certainly, $\lambda_1(C)$ depends on the geometry of the curve C. In [20] it is proven that a particular case of the Lieb–Thirring conjecture for $\gamma = 1$ and $n = 1$ is equivalent to proving that

$$\lambda_1(C) \geq 1, \tag{3.33}$$

with equality if and only if C belongs to a one-parameter family of ovals which include the circle (in fact the one-parameter family of curves is characterized by a curvature given by $\kappa(s) = 1/(a^2 \cos^2(s) + a^{-2} \sin^2(s))$[20]). It is a simple matter to see that if C is a circle of length 2π, the lowest eigenvalue of $H(C)$ is precisely 1. The fact that there is degeneracy of the conjectured minimizers makes the problem much harder.

The conjecture (3.33) is still open. Concerning (nonoptimal) lower bounds, Benguria and Loss proved [20]

$$\lambda_1(C) \geq 1/2, \tag{3.34}$$

and more recently Linde [93] found the best lower bound to date,

$$\lambda_1(C) > \left(1 + \frac{\pi}{\pi + 8}\right)^{-2} \approx 0.60847. \tag{3.35}$$

4. Electrostatic Inequalities

A key role in the different proofs of the stability of matter is played by the nature of the Coulomb potential and its mathematical properties. In fact, the nature of the Coulomb potential (in electromagnetism) or the Kepler potential in gravitation, has played a central role since the publication of Newton's Principia in 1687 [114]. It is in the Principia that Newton proved his celebrated theorem that *outside a radially symmetric distribution of mass its gravitational potential is given by the same expression one would obtain if the total mass were concentrated at the center of the distribution.* For a discussion of Newton's theorem, and several extensions, see [114], Section XII, of Book I, Theorems 30, 31, and 32. Further analytic properties of the Coulomb (or Kepler) potential were later studied (among others) by Lagrange, Laplace and Poisson in the late eighteenth and early nineteenth century, and in particular by George Green in his famous essay of 1828 [59] where he introduced the term "potential", as we use it today.

The use of electrostatic inequalities in the context of the Stability of Matter were introduced by Onsager (see the original reference [116], or Onsager's collected works [117]) who proved the following lemma.

Lemma 4.1: (Onsager's Lemma) *Consider N unit point charges located at the different points $x_1, x_2, \ldots, x_N$ in $\mathbb{R}^3$. For each $1 \leq i \leq N$, let μ_{x_i} be*

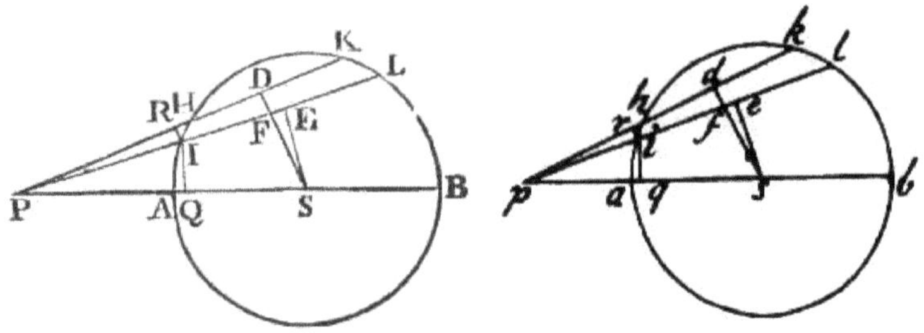

Fig. 1. Original figures used in the proof of Newton's theorem.

a non-negative, bounded function that is radially symmetric about the point x_i *and whose total integral* $\int_{\mathbb{R}^3} \mu_{x_i}(x)\,dx = 1$. *Then, for any non-negative integral function* ρ *we have that*

$$\sum_{1\le i<j\le N} \frac{1}{|x_i - x_j|} \ge -D(\rho,\rho) + 2\sum_{i=1}^{N} D(\rho,\mu_{x_i}) - \sum_{i=1}^{N} D(\mu_{x_i},\mu_{x_i}), \quad (4.1)$$

where

$$D(f,g) = \frac{1}{2}\int_{\mathbb{R}^3\times\mathbb{R}^3} f(x)\frac{1}{|x-y|}g(y)\,dx\,dy. \quad (4.2)$$

Proof: Onsager's Lemma is a consequence of two important properties of $D(f,g)$, namely, i) $D(f,g)$ is a positive definite quadratic form (for the proof of this fact, see for example [96], Theorem 5.1, p. 90), and ii)

$$\frac{1}{2}\sum_{1\le i<j\le N} \frac{1}{|x_i - x_j|} \ge \sum_{1\le i<j\le N} D(\mu_{x_i},\mu_{x_j}),$$

which is a consequence of Newton's theorem. In fact, just consider $D(\rho - \sum_{i=1}^N \mu_{x_i}, \rho - \sum_{j=1}^N \mu_{x_j}) \ge 0$ by i), expand, and use ii). □

Onsager's Lemma was used in the early proof of Dyson and Lenard [45, 81], and several of its different extensions have been used in the more recent results around the stability of matter.

In the sequel we need some notation. Consider a set of K nuclei of identical nuclear charge $Z > 0$, located at different points, $R_1, R_2, \ldots, R_K$ in $\mathbb{R}^3$. Given the distribution of points R_i, we introduce the nearest neighbor, or Voronoi, cells [156] (see also the review [96]), $\{\Gamma_j\}_{j=1}^K$, defined by

$$\Gamma_j = \{x \mid |x - R_j| \le |x - R_k|\}. \quad (4.3)$$

The boundary of Γ_j, $\partial\Gamma_j$, consists of a finite number of planes. We also define the distance

$$D_j = \mathrm{dist}(R_j, \partial\Gamma_j) = \frac{1}{2}\min\{|R_k - R_j| \mid k \neq j\}. \tag{4.4}$$

Finally, we denote by B_j the ball of radius D_j centered at R_j, $j = 1, \ldots, K$. In fact, D_j is the distance from R_j to the boundary $\partial\Gamma_j$.

With this notation we can state another basic electrostatic inequality due to Lieb and Yau, [105, 106] (a weaker earlier version was found by W. Hughes in [69], Section I, pp. 8-ff). Define the piecewise function $\Phi(x)$ on $\mathbb{R}^3$ with the aid of the Voronoi cells mentioned above. In the cell Γ_j, $\Phi(x)$ equals the electrostatic potential generated by all the nuclei except for the nucleus situated in Γ_j itself, i.e., for $x \in \Gamma_j$,

$$\Phi(x) = \sum_{\substack{i=1\\ i\neq j}}^{K} \frac{Z}{|x - R_i|}. \tag{4.5}$$

Then, one has,

Theorem 4.2: (Basic Electrostatic Inequality of Lieb and Yau)

$$D(\rho,\rho) - \int_{\mathbb{R}^2} \Phi(x)\rho(x)\,\mathrm{d}x + U \geq \frac{Z^2}{8}\sum_{j=1}^{K}\frac{1}{D_j} \tag{4.6}$$

with

$$U = \sum_{1\leq k<\ell\leq K} \frac{Z^2}{|R_k - R_\ell|}.$$

The Lieb–Yau electrostatic inequality also hold for signed measures (charge distributions) (see [96], eq. (5.2.6)). For a rigorous, detailed proof of this Theorem see [96]. Here we will sketch a heuristic proof of this theorem, first in the case of two centers ($K = 2$) and then we will say a few comments on the general proof. So consider the $K = 2$ case. One may view the left side of (4.6) as a quadratic functional on ρ. If one minimizes this functional over measures, one derives the appropriate Euler equation for the minimizer, i.e.,

$$\int_{\mathbb{R}^3} \frac{\rho(y)}{|x - y|}\,dy - \Phi(x) = 0. \tag{4.7}$$

Now, the potential $\Phi(x)$ is harmonic in the interior of the Voronoi cells (i.e., away from the boundaries). Thus, the laplacian of Φ will be a measure, supported on the boundaries between the Voronoi cells. In particular, in the $K = 2$ case, it will have support on the bisector plane of the segment joining R_1 and R_2. Notice that in this case $D_1 = D_2$, and we will call D this common quantity. Thus, from (4.7) we see that instead of a minimizing density $\rho(x)$ we have a minimizing measure supported on the bisector plane. In the sequel we will use cartesian coordinates such that the segment joining R_1 and R_2 lies along the x-axis, and the bisector plane is the $x = 0$ plane. We will then denote by $\sigma(y, z)$ the density (in the bisector plane) of the minimizing measure alluded to above. According to (4.7), Φ is the electric potential created by σ. Since we know the expression for the potential, we can then use the standard Gauss' law of Electricity and Magnetism to compute the density σ. In fact, using Gauss' law, we easily get that

$$\sigma(y,z) = \frac{1}{4\pi}\left[(-\nabla\Phi))\cdot\hat{\imath}\,\Big|_{x=0^+} - (-\nabla\Phi))\cdot\hat{\imath}\,\Big|_{x=0^-}\right] \tag{4.8}$$

where $\hat{\imath}$ is the normal to the bisector plane. Notice that while Φ is continuous everywhere, the normal (to the bisector plane) derivatives of Φ have a jump. Evaluating (4.8), using the expression for Φ one gets,

$$4.9\sigma(y,z) = \frac{Z}{4\pi}\frac{D}{(D^2+y^2+z^2)^{3/2}}. \tag{4.9}$$

Using this expression for the minimizing charge density, and the Euler equation (4.7), we get the minimum possible value for the left side of (4.6), which is given (in this case) by,

$$LS = -\frac{1}{2}\int_{-\infty}^{\infty}\int_{-\infty}^{+\infty}\sigma(y,z)\Phi(0,y,z)\,dy\,dz + U. \tag{4.10}$$

Here, $U = 1/|R_2 - R_1| = 1/2D$. Moreover, $\Phi(0, y, z) = Z/\sqrt{D^2 + y^2 + z^2}$. Finally, using polar coordinates to parametrize points in the bisector plane, we can express the left side as,

$$LS = -\frac{1}{2}\int_0^{2\pi}\int_0^{+\infty}\frac{Z^2}{4\pi}\frac{D}{(D^2+\rho^2)^2}\rho\,d\rho\,d\theta + \frac{Z^2}{2D} = \frac{Z^2}{4D}, \tag{4.11}$$

which is precisely the right side of (4.6) in this case. In the general case, with K nuclei, the structure of the Voronoi cells is slightly more complicated, and a similar procedure yields the desired estimate (this time however, the estimate we get in general is not sharp, like in the $K = 2$ case). For the rigorous proof in the general case see, e.g., [96].

Before we close this section, we would like to state an inequality that has played an important role in different modern proof of stability, which is a consequence of the Lieb–Yau inequality (for the proof see [96], Theorem 5.4).

Theorem 4.3: (Baxter's inequality [7, 96]) *Let V_C be the Coulomb potential of N electrons with coordinates $x_i \in \mathbb{R}^3$ and K nuclei with common nuclear charge Z and with coordinates $R_j \in \mathbb{R}^3$, i.e.,*

$$V_C = \sum_{1\le i<j\le N} \frac{1}{|x_i - x_j|} - \sum_{i=1}^{N}\sum_{j=1}^{K} \frac{Z}{|x_i - R_j|} + \sum_{1\le k<\ell\le K} \frac{Z^2}{|R_k - R_\ell|}. \quad (4.12)$$

As before let Γ_j denote the Voronoi cell belonging to R_j and let $\mathcal{D}(x)$ denote the distance from x to the nearest nucleus (in other words, if $x \in \Gamma_j$, then $\mathcal{D}(x) = |x - R_j|$). Also define D_j as before. Then, one has,

$$V_C \ge -(2Z+1)\sum_{i=1}^{N} \frac{1}{\mathcal{D}(x_i)} + \frac{Z^2}{8}\sum_{j=1}^{M} \frac{1}{D_j}. \quad (4.13)$$

Remarks:
i) The importance of (4.13) is that allows to obtain lower bounds on the Hamiltonian of N interacting electrons in terms of the sum of independent one particle hamiltonians This type of bounds has played an important role in obtaining energy estimates for many–body hamiltonians. Bounds of this sort go back to the original work of Hertel, Lieb and Thirring [63].
ii) The result of Baxter's [7] did not include the second term on the right side of (4.13).
iii) See [147] and [137] for a proof of the Stability of Matter using Baxter's inequality and the Lieb–Thirring Inequalities.
iv) Baxter's inequality also plays a crucial role in Lieb's proof of the $(2Z + K)$ [88, 89] bound on the maximum number of electrons a molecule of K centers, with total nuclear charge Z can bind (see the next section).

5. The Maximum Number of Electrons an Atom Can Bind

The ability of atoms and molecules to catch or give electrons is one of the fundamental pillars of chemistry and biology, and it has somewhat attracted attention since the late 18th century with the birth of modern chemistry. Given a neutral atom, one can extract one electron by providing enough energy. The energy needed to remove one electron is called its *ionization energy* (e.g., the energy needed to extract the electron of the Hydrogen atom in its ground state is approximately 13.6 eV). One can continue this process of removing electrons by providing increasing amounts of energy (a fact that seems intuitively obvious but yet to be proven, see however [28]), up to stripping the atom completely of all its electrons. On the other hand, the question up to what extent one can add electrons to a neutral atom is not yet completely understood. After the introduction of the Schrödinger equation in 1926, Hans Bethe was the first to study the possibility of having the negative Hydrogen ion H^-. Using the variational principle of E. Hylleraas, he proved the existence of a bound state for H^- [23], and computed quite accurately the ionization energy of the first electron to be 17 kcal/mol (i.e., approximately 0.74 eV, whereas its present value is approximately 0.75 eV). With the discovery of Rupert Wildt that the presence of H^- in the solar atmosphere is the main cause of its opacity (in the visible, but specially in the infrared) [161] there was a renewed interest in the spectral properties of the H^- anion (see, in particular, the review articles [31], [71], and the monograph [62], pp. 404–ff). A rigorous proof that H^- has only one bound state (i.e., the ground state) and no singly excited states had to wait approximately fifty years, until the work of Robert Hill [65, 66]. For a review of the literature on the Hydrogen anion up to 1996, we refer the reader to the article of Rau [126].

In the last half a century there has been a big effort in trying to determine the maximum number of electrons an atomic nucleus or a molecule can bind. These efforts come from three different fronts. From one side, in experimental physics, there has been an intensive search for *dianions* (i.e., doubly ionized atoms and molecules). On another front, there has been an intensive numerical search for the possibility of stable atomic dianions [68] giving strong evidence that stable atomic dianions do not exist. Finally, in mathematical physics there has been an interesting quest to solve this problem. Based on the knowledge coming from these three fronts, we expect that an atom of atomic number Z can bind at most $Z+1$ electrons, while a molecule of K nuclei of total nuclear charge Z can bind at most $Z+K$ electrons.

There are two main ingredients involved in this question: the fact that the maximum number of electrons an atom of nuclear charge Z can bind is at least Z (i.e., that neutral atoms do exist) is very much related to the mathematical properties of the Coulomb interaction between charged particles. In particular, a crucial role is played by Newton's theorem. On the other hand, the expected fact that at most $Z+1$ electrons can be bound has to do with Pauli's Exclusion Principle (i.e., more precisely with the fact that electrons obey Fermi statistics).

We have seen above that the first rigorous results (of Bethe and Hill) on negative ions dealt with H^-. Concerning the more general situation of N electrons and K (usually fixed) nuclei interacting via Coulomb potentials the first results were obtained by Zhislin (see [162, 163]) who proved that below neutrality (i.e., when the total number of electrons is strictly less than the total nuclear charge) the corresponding Hamiltonian in non–relativistic quantum mechanics has an infinite number of bound states, whereas at neutrality or above it, the number of possible bound states is at most finite. Upper bounds on the number of bound states for bosonic matter above neutrality were obtained later in [3, 130]. At the beginning of the 80's, Ruskai and Sigal [128, 129, 139, 140], using the IMS localization formula and appropriate partitions of unity obtained the first actual upper bounds on the maximum number of electrons an atom or molecule can bind. In 1983, Benguria and Lieb [17] proved that the Pauli principle is crucial when considering the problem of the maximum number of electrons an atom can bind. In fact, they proved that $N_c(Z) - Z \geq cZ$ as $Z \to \infty$, where c is obtained by solving the Hartree equation (which is equation (5.11) below with $\gamma = 0$). Here and in the sequel we denote by $N_c(Z)$ the maximum number of electrons a nucleus of charge Z can bind. Then, Baumgartner [6] solved numerically the Hartree equation to find $c = 0,21$. Later, Solovej [144] obtained an upper bound which showed that $N_c(Z) = 1.21\,Z$ is the appropriate asymptotic formula for large Z. In 1984, Lieb obtained the simple upper bound $N_c(Z) < 2Z+K$ independently of statistics [88, 89], and Lieb, Sigal, Simon and Thirring proved that fermionic matter is asymptotically neutral (i.e., $N_c(Z)/Z \to 1$ as Z goes to infinity [97, 98]). In 1990, Fefferman and Seco [54] obtained a correction term to this asymptotic neutrality, namely they proved that

$$N_c(Z) \leq Z + cZ^{1-\alpha},$$

for some constant c, with $\alpha = 9/56$. The proof of this result was later simplified by Seco, Sigal and Sovolej [135] who established a connection between the ionization energy and the excess charge $N_c(Z) - Z$, and estimated asymptotically the ionization energy. Recently P.-T. Nam [112] proved that the maximum number N_c of non-relativistic electrons that a nucleus of charge Z can bind is less than $1.22\,Z + 3\,Z^{1/3}$, which improves Lieb's upper bound $N_c < 2Z + 1$ when $Z \geq 6$.

The conjecture we mentioned at the beginning to the effect that the excess charge $N_c(Z) - Z \leq 1$ for an atom is still open. However, for semiclassical models (including the Thomas–Fermi model and its extensions, the Hartree–Fock theory, and others) there are sharper results. It was proven by Lieb and Simon ([99, 100]) that $N_c(Z) = Z$ for the Thomas–Fermi model, whereas for the gradient correction (i.e., for the Thomas–Fermi–Weizsäcker model) Benguria and Lieb proved that $N_c(Z) - Z \leq 1$. In 1991, Solovej [145] proved that $N_c(Z) - Z \leq c$ for some constant c for a reduced Hartree–Fock model. Finally, Solovej in 2003 [146] proved a similar bound for the full Hartree–Fock model. We refer the reader to the recent monograph of Lieb and Seiringer [96], chapter 12, for a more complete up to date summary on the maximum ionization.

In the rest of this section we will present two simple arguments used to find upper bounds on the number of electrons in simple models. First we will show how Newton's theorem implies that $N_c(Z) \leq Z$ for the Thomas–Fermi model in the atomic case (i.e., in the radial case). Second we sill show how in the Thomas–Fermi–Weizsäcker model $N_c(Z) \leq 2Z$.

5.1. *The maximum number of electrons for a one center case in the Thomas–Fermi model*

The Thomas–Fermi (TF for short) model is defined through the energy functional,

$$\mathcal{E}(\rho) = \frac{3}{5}\gamma \int_{\mathbb{R}^3} \rho^{5/3}\, dx - \int_{\mathbb{R}^3} V(x)\rho(x)\, dx + D(\rho, \rho), \tag{5.1}$$

where γ is a positive constant. The following minimization problem was rigorously studied in [100] (see also the review article [87]):

$$\min\{\mathcal{E}(\rho) \mid \rho \in L^1(\mathbb{R}^3) \cap L^{5/3}(\mathbb{R}^3), \rho(x) \geq 0, \int_{\mathbb{R}^3} \rho(x)\, dx = N\}. \tag{5.2}$$

Here, we will show how Newton's theorem implies $N_c(Z) \leq Z$. When estimating $N_c(Z)$, all we have to do is to consider (5.2) without the constraint on the number of particles. We prove this in the radially symmetric case, $V(x) = Z/|x|$. We write,

$$\mathcal{E}(\rho) = \frac{3}{5}\gamma \int_{\mathbb{R}^3} \rho^{5/3}\, dx - P(\rho), \tag{5.3}$$

with

$$P(\rho) = \int_{\mathbb{R}^3} V(x)\rho(x)\, dx - D(\rho, \rho). \tag{5.4}$$

In the sequel we show that $N_c \leq Z$. Assume, by contradiction, that the density ρ that minimizes $\mathcal{E}(\rho)$ is such that $N_C = \int_{\mathbb{R}^3} \rho(x)\, dx > Z$. Then, we can always find an R such that $\int_{|x|\le R}\rho(x)\,dx = Z$. Then, define $\rho' = \rho$ if $|x| \le R$ and $\rho' = 0$ otherwise. Clearly $\int_{\mathbb{R}^3}\rho'(x)^{5/3}\,dx < \int_{\mathbb{R}^3}\rho(x)^{5/3}\,dx$. We will show that $P(\rho') \ge P(\rho)$, hence $\mathcal{E}(\rho') < \mathcal{E}(\rho)$ which is a contradiction. From (5.4) we have,

$$P(\rho') - P(\rho) = \int_{\mathbb{R}^3} V(\rho' - \rho)\,dx - D(\rho',\rho') + D(\rho,\rho). \tag{5.5}$$

If we set $\Delta(x) = \rho(x) - \rho'(x)$, we have that $\Delta(x) = 0$ inside $B_0(R)$, the ball of radius R centered at 0, while $\Delta \ge 0$ in the complement of the ball. In terms of Δ we have

$$P(\rho')-P(\rho) = \int_{\mathbb{R}^3} V(x)\Delta(x)\,dx + D(\Delta,\rho+\rho') = \int_{B_R(0)^c}\Delta(x)\left[\varphi(x) - \frac{Z}{|x|}\right] \tag{5.6}$$

where,

$$\varphi(x) = \frac{1}{2}(\rho+\rho') * \frac{1}{|x|}. \tag{5.7}$$

By Newton's theorem we have that

$$\varphi(x) \ge \frac{Z}{|x|} \tag{5.8}$$

in $B_R(0)^c$, hence it follows from (5.6) that $P(\rho') - P(\rho) \ge 0$, and we are done.

5.2. *Bound on $N_c(Z)$ for the TFW model in the atomic case*

The Thomas–Fermi–Weizsäcker (TFW for short) model is a gradient correction to the standard Thomas–Fermi model, and it is defined through the energy functional,

$$\mathcal{E}_{T\mathcal{F}W}(\psi) = A\int_{\mathbb{R}^3}(\nabla\psi)^2 dx + \frac{3}{5}\gamma\int_{\mathbb{R}^3}\psi^{10/3}dx - \int_{\mathbb{R}^3} V(x)\psi^2(x)\,dx + D(\psi^2,\psi^2). \tag{5.9}$$

Here A and γ are positive constants. The following minimization problem was rigorously studied in [9, 14]:

$$\min\{\mathcal{E}_{T\mathcal{F}W}(\psi) \mid \psi \in H_1(\mathbb{R}^3)\cap L^{10/3}(\mathbb{R}^3), \int_{\mathbb{R}^3}\psi^2\,dx = N\}. \tag{5.10}$$

Consider the TFW equation for the most negative ion, i.e., the Euler equation associated with the variational principle (5.10) without constraints (i.e., without requiring $\int_{\mathbb{R}^3}\psi^2\,dx = N$). This equation is given by [9, 14]

$$-A\Delta\psi + \gamma\psi^{7/3} - \varphi(x)\psi = 0, \tag{5.11}$$

where $\varphi(x)$ is the total Coulomb potential (due to the nucleus and the electrons) given by,

$$\varphi(x) = \frac{Z}{|x|} - \int_{\mathbb{R}^3} \rho(y) \frac{1}{|x-y|}\, dy. \tag{5.12}$$

Here $\rho(x) = \psi^2(x)$ is the electronic density, and $N = \int_{\mathbb{R}} \rho(x)\, dx$ is the (maximum) number of electrons. Here we are interested in estimating N from above. In order to do that we multiply (5.11) by $|x|$ and integrate over $\mathbb{R}^3$. Hence we get,

$$-A \int_{\mathbb{R}^3} \psi(x)|x|\Delta\psi\, dx + \gamma \int_{\mathbb{R}^3} |x|\rho^{5/3}\, dx = \int_{\mathbb{R}^3} |x|\varphi(x)\psi^2(x)\, dx. \tag{5.13}$$

In the sequel we will consider the two sides of (5.13) separately. Using (5.13) and (5.12) we have

$$\int_{\mathbb{R}^3} |x|\varphi(x)\psi^2\, dx = ZN - \frac{1}{2}\int_{\mathbb{R}^3\times\mathbb{R}^3} \rho(x)\rho(y)\frac{|x|+|y|}{|x-y|}\, dx\, dy. \tag{5.14}$$

The last equality follows by symmetrizing in x and y. Now, using the triangle inequality we have,

$$\frac{|x|+|y|}{|x-y|} \geq 1,$$

and thus, it follows from (5.14) that

$$\int_{\mathbb{R}^3} |x|\varphi(x)\psi^2(x)\, dx \leq ZN - \frac{1}{2}N^2. \tag{5.15}$$

Concerning the first term on the left side of (5.13), since ψ is radially symmetric, if we write $\psi(x) = u(r)/r$, with $r = |x|$ one has as usual $\Delta\psi = u''/r$, and thus,

$$\begin{aligned} -\int_{\mathbb{R}^3} \psi(x)|x|\Delta\psi\, dx &= -4\pi \int_0^\infty u''(r)ur\, dr = +4\pi \int_0^\infty \left(ru'(r)^2 + uu'\right)\, dr \\ &= +4\pi \int_0^\infty ru'(r)^2\, dr \geq 0, \end{aligned} \tag{5.16}$$

where the second equality follows by integration by parts and the fact that $u(0) = 0$ and $u(r) \to 0$ as $r \to \infty$. The last equality follows by direct integration using the boundary behavior of u. Combining (5.14), (5.15) and (5.16) and using the fact that $\int_{\mathbb{R}^3} |x|\rho^{5/3}\, dx \geq 0$ we finally conclude the upper bound

$$N \leq 2Z, \tag{5.17}$$

for the maximum number of electrons that can be bound in the TFW model. This bound on the maximum number of electrons for an atom in the TFW model was derived by Benguria (see [87]) and extended by Lieb to the molecular case.

In the derivation of the bound $N \leq 2Z$ above, we did not use the specific nature of the $\psi^{7/3}$, which takes into account the fact that the electrons obey Fermi statistics. A more sophisticated analysis of the TFW equation, taking into account the Fermi statistics and the use of a localization procedure yields the bound

$$N_c(Z) - Z \leq 270.74 \left(\frac{A}{\gamma}\right)^{3/2} K \tag{5.18}$$

(see [18], Theorem 1) for the maximum number of electrons a molecule of K nuclei with total charge Z can bind. The constant 270.74 was later improved by Solovej [143, 144]. For the actual numerical values of A and γ, the factor $270.74\,(A/\gamma)^{3/2}$ is less than 1.

6. The Stability of Matter for a Relativistic Toy Model

In this section we will consider the stability of a relativistic model. The model we will consider here is rather a toy model. There are two reasons for considering the stability of matter for this simple relativistic toy model. The first one is that it is simple but at the same time one uses, in studying its stability, most of the known techniques needed in order to prove stability for relativistic Coulomb systems. The second is that recently we used the analog two dimensional version of this model as a basic tool to get a Lieb–Oxford type bound for the exchange energy in two dimensions (see [15]).

The zero mass limit of the relativistic Thomas–Fermi–Weizsäcker (henceforth ultrarelativistic TFW) energy functional for nuclei of charges $z_i > 0$ (which need not be integral) located at R_i, $i = 1, \ldots, K$ is defined by [50, 51]

$$\xi(\rho) = a^2 \int_{\mathbb{R}^3} (\nabla \rho^{1/3})^2 \, dx + b^2 \int_{\mathbb{R}^2} \rho^{4/3} \, dx - \int_{\mathbb{R}^3} V(x)\rho(x) \, dx + D(\rho, \rho) + U, \tag{6.1}$$

where the potential V is given by,

$$V(x) = \sum_{i=1}^{K} \frac{z_i}{|x - R_i|}, \tag{6.2}$$

which is the Coulomb potential generated by K point particles (nuclei) of charge $z_i > 0$, located at $R_i \in \mathbb{R}^3$ (with $i = 1, \ldots, K$). Here, the function $\rho(x) \geq 0$ is the electronic density of a system of N electrons, and

$$D(\rho, \rho) = \frac{1}{2} \int_{\mathbb{R}^3 \times \mathbb{R}^3} \rho(x) \frac{1}{|x-y|} \rho(y)\, dx\, dy, \tag{6.3}$$

the electronic repulsion energy. Finally,

$$U = \sum_{1 \leq i < j \leq K} \frac{z^2}{|R_i - R_j|} \tag{6.4}$$

is the repulsion energy between the nuclei. The powers of the first two terms in (6.1), i.e.,

$$T(\rho) = a^2 \int_{\mathbb{R}^2} (\nabla \rho^{1/3})^2\, dx + b^2 \int_{\mathbb{R}^3} \rho^{4/3}\, dx, \tag{6.5}$$

are such that $T(\rho_\alpha) = \alpha T(\rho)$, where $\rho_\alpha(x) = \alpha^3 \rho(\alpha x)$ (with $\alpha > 0$) is such that $\int_{\mathbb{R}^3} \rho_\alpha\, dx = \int_{\mathbb{R}^3} \rho(x)\, dx$. In other words, the kinetic energy of the electrons scales like one over a length, i.e., in the same way as the potential energy. Then, as usual in this situation, the values of the coupling constant (i.e., the values of the nuclear charges) will be crucial to insure stability of the system. Our main result in this section is the following stability theorem [21]. In the sequel we will set all the nuclear charges equal to a given value z. By a standard convexity argument one can reduce the general case to this situation.

Theorem 6.1: *For any $a, b > 0$, and $R_i \in \mathbb{R}^3$, $i = 1, \ldots, K$, and for all $\rho \geq 0$ (with $\rho \in L^{4/3}(\mathbb{R}^3)$ and $\nabla \rho^{1/3} \in L^2(\mathbb{R}^3)$), we have that*

$$\xi(\rho) \equiv a^2 \int_{\mathbb{R}^3} (\nabla \rho^{1/3})^2\, dx + b^2 \int_{\mathbb{R}^3} \rho^{4/3}\, dx - \int_{\mathbb{R}^3} V(x) \rho(x)\, dx + D(\rho, \rho) + U \geq 0, \tag{6.6}$$

where V, D, and U are defined by (6.2), (6.3), and (6.4), respectively, provided,

$$0 \leq z \leq z_c(a, b) \equiv \frac{4\, a\, b}{3} \sqrt{1 - \sigma}. \tag{6.7}$$

Here $0 < \sigma < 1$ is the only positive root of the quartic equation

$$\frac{\sigma^3}{(1 - \sigma)} = \frac{117\, \pi}{8} \frac{a^2}{b^4} \tag{6.8}$$

on the interval $(0, 1)$.

In the rest of this section we will give the proof of this theorem (which is taken from Theorem 1.2 in [21]). Notice that the upper limit $z_c(a,b)$ on z to insure stability is not sharp; in other words, there could still be values of z above our z_c for which $\xi(\rho) \geq 0$. We start with an appropriate *Coulomb uncertainty principle.*

Theorem 6.2: *For any smooth function f on the closed ball B_R, of radius R, and for all a, $b \in \mathbb{R}$, we have*

$$a^2 \int_{B_R} |\nabla f(x)|^2\,dx + b^2 \int_{B_R} f(x)^4\,dx \geq ab \int_{B_R} \left(\frac{4}{3|x|} - \frac{2}{R}\right) f(x)^3\,dx.$$

The proof of this theorem is given in Theorem 2.1 in [21], which we reproduce here for completeness. We start with the following preliminary result which is of independent interest.

Lemma 6.3: *Let $u = u(|x|)$ be a smooth function on the interval $(0,R)$, such that $u(R) = 0$. Then the following uncertainty principle holds*

$$\left|\int_{B_R} [3u(|x|) + |x|u'(|x|)] f(x)^3\,dx\right| \leq 3 \left(\int_{B_R} |\nabla f(x)|^2\,dx\right)^{1/2} \left(\int_{B_R} u(|x|)^2 |x|^2 f(x)^4\,dx\right)^{1/2}. \quad (6.9)$$

In (6.9) there is equality if and only if

$$f(x) = \frac{1}{\lambda \int_0^{|x|} s u(s)\,ds + C}, \quad (6.10)$$

for some constants C and λ.

Proof: Set $g_j(x) = u(|x|)x_j$. Then we have,

$$\int_{B_R} [3u(|x|) + |x|u'(|x|)] f(x)^3\,dx = \sum_{j=1}^{3} \int_{B_R} [\partial_j g_j(x)] f(x)^2\,dx$$

$$= \sum_j \int_{B_R} f(x)\partial_j [g_j(x) f(x)^2]\,dx - 2 \sum_j \int_{B_R} f(x)^2 g_j(x) \partial_j f(x)\,dx$$

$$= -3 \int_{B_R} \langle \nabla f(x), x\rangle u(|x|) f(x)^2\,dx.$$

In the last equality we integrated by parts and made use of the fact that u vanishes on the boundary ∂B_R. Next, the Schwarz inequality implies,

$$\left|\int_{B_R} [3u(|x|) + |x|u'(|x|)] f(x)^3\,dx\right| \leq 3 \left(\int_{B_R} |\nabla f(x)|^2\,dx\right)^{1/2} \left(\int_{B_R} u(|x|)^2 |x|^2 f(x)^4\,dx\right)^{1/2}.$$

In the last expression, equality is obtained if and only if

$$\partial_j f(x) = -\lambda x_j u(|x|) f(x)^2,$$

which after an integration yields the function given by (6.10) above. □

Proof of Theorem 6.2: Choosing $u(r) = (r^{-1} - R^{-1})/2$ in (6.9), we conclude

$$\begin{aligned} ab \left| \int_{B_R} \left(\frac{4}{3|x|} - \frac{2}{R} \right) f(x)^3 dx \right| &\le 2ab \left(\int_{B_R} |\nabla f(x)|^2 dx \right)^{1/2} \left(\int_{B_R} f(x)^4 dx \right)^{1/2} \\ &\le a^2 \int_{B_R} |\nabla f(x)|^2 \, dx + b^2 \int_{B_R} f(x)^4 \, dx. \end{aligned}$$

□

To prove our main result of this section, i.e., Theorem 6.1, we will also need the following auxiliary lemma.

Lemma 6.4: *Let* $B_L(x_0) = \{x \in \mathbb{R}^3 \mid |x - x_0| < L\}$ *and* H *be a half space such that* $\mathrm{dist}(x_0, \partial H) = L$ *and* $x_0 \in H$. *Then*

$$\int_{H \setminus B_L(x_0)} \frac{1}{|x - x_0|^4} dx = \frac{3\pi}{L}.$$

Proof: Let us shift the origin of the coordinates to x_0, and the z axis perpendicular to the plane ∂H. Then in the respective spherical coordinates $(\varrho, \theta.\varphi)$,

$$\int_{H \setminus B_L(x_0)} \frac{dx}{|x - x_0|^4} = 2\pi \int_0^{\pi/2} \int_L^{\frac{L}{\cos\theta}} \frac{d\varrho}{\varrho^2} \sin\theta \, d\theta + 2\pi \int_{\pi/2}^{\pi} \int_L^{\infty} \frac{d\varrho}{\varrho^2} \sin\theta \, d\theta$$

from which the assertion of the lemma follows by a straightforward integration. □

Using the same notation and decomposition into Voronoi cells as in Section 4, and introducing the potential Φ as in (4.5) one has by Theorem 4.2 above that,

$$D(\rho, \rho) - \int_{\mathbb{R}^3} \Phi(x)\rho(x) \, dx + U \ge \frac{z^2}{8} \sum_{j=1}^{K} \frac{1}{D_j}. \tag{6.11}$$

With the help of the Coulomb uncertainty principle and the electrostatic inequality (6.11), we are ready to prove the following estimate.

Lemma 6.5: *For any* $\rho \in L^{4/3}(\mathbb{R}^3)$ *such that* $\nabla \rho^{1/3} \in L^2(\mathbb{R}^3)$*; for all* $b_1 > 0$*, and* $b_2 > 0$ *such that* $b_1^2 + b_2^2 = b^2$ *we have,*

$$\xi(\rho) \geq \sum_{j=1}^{K} \frac{1}{D_j}\left[\frac{z^2}{8} - \frac{27}{256\, b_1^6}\left(3\pi z^4 + \pi \frac{64}{3} a^4 b_2^4\right)\right]. \tag{6.12}$$

Proof: Setting $f(x)^3 = \rho(x)$, splitting $\mathbb{R}^3$ as the disjoint union of the Voronoi cells Γ_j, using Theorem 6.2 in each disk B_j, and discarding the kinetic energy terms (which are positive) in the complements $\Gamma_j \setminus B_j$ we conclude,

$$\begin{aligned}\xi(\rho) \geq\; & b_1^2 \int_{\mathbb{R}^3} \rho^{4/3} dx - \int_{\mathbb{R}^3} V \rho\, dx \\ & + a b_2 \sum_{j=1}^{K} \int_{B_j} \left(\frac{4}{3|x - R_j|} - \frac{2}{D_j}\right) \rho(x)\, dx + D(\rho, \rho) + U. \end{aligned}\tag{6.13}$$

It is convenient to define the piecewise function $W(x)$ as

$$W(x) = \begin{cases} \Phi(x) + \dfrac{z}{|x - R_j|} = V(x) & \text{if } x \in \Gamma_j \setminus B_j, \\ \Phi(x) + \dfrac{2\, a b_2}{D_j} & \text{if } x \in B_j. \end{cases}\tag{6.14}$$

Provided $z \leq 4\, a\, b_2/3$ (which we assume from here on), we can estimate from below the sum of the second and third integrals in (6.13) in terms of $W(x)$ as follows,

$$\begin{aligned} & a b_2 \sum_{j=1}^{K} \int_{B_j} \left(\frac{4}{3|x - R_j|} - \frac{2}{D_j}\right) \rho(x)\, dx - \int_{\mathbb{R}^3} V \rho\, dx \\ &= a b_2 \sum_{j=1}^{K} \int_{B_j} \left(\frac{4}{3|x - R_j|} - \frac{2}{D_j}\right) \rho(x)\, dx - z \sum_{i,j=1}^{K} \int_{\Gamma_j \setminus B_j} \frac{\rho(x)}{|x - R_i|}\, dx \\ &\quad - z \sum_{\substack{i,j=1 \\ i \neq j}}^{K} \int_{B_j} \frac{\rho(x)}{|x - R_i|}\, dx - z \sum_{j=1}^{K} \int_{B_j} \frac{\rho(x)}{|x - R_j|}\, dx \\ &= - \int_{\mathbb{R}^3} W(x) \rho(x) dx + \sum_{j=1}^{K} \int_{B_j} \left(\frac{4\, a b_2}{3} - z\right) \frac{\rho(x)}{|x - R_j|} dx \\ &\geq - \int_{\mathbb{R}^3} W(x) \rho(x) dx. \end{aligned}$$

Thus, we can write

$$\xi(\rho) \geq \xi_1(\rho) + \xi_2(\rho), \tag{6.15}$$

with

$$\xi_1(\rho) = b_1^2 \int_{\mathbb{R}^3} \rho^{4/3}\, dx - \int_{\mathbb{R}^3} (W - \Phi)(x)\rho(x)\, dx \qquad \text{and,}$$

$$\xi_2(\rho) = D(\rho,\rho) - \int_{\mathbb{R}^3} \Phi(x)\rho(x)\, dx + U.$$

From the definition of $\xi_1(\rho)$, it is clear that $\xi_1(\rho) \geq \xi_1(\hat{\rho})$, where $\hat{\rho}(x) = 27(W(x) - \Phi(x))_+^3/(64\, b_1^6)$, where as usual $u_+ = \max(u, 0)$. Hence,

$$\begin{aligned}\xi_1(\rho) &\geq -\frac{27}{256\, b_1^6} \int_{\mathbb{R}^2} (W - \Phi)_+^4\, dx \\ &= -\frac{27}{256\, b_1^6} \sum_{j=1}^{K} \left(\int_{\Gamma_j \setminus B_j} \frac{z^4}{|x - R_j|^4}\, dx + \int_{B_j} \left(\frac{2\, ab_2}{D_j}\right)^4 dx \right),\end{aligned}$$

where the last equality follows from the definition (6.14) of W. As any Γ_j is contained in a halfspace, we may estimate the first integral above with the help of Lemma 6.4. This way we get

$$\xi_1(\varrho) \geq -\frac{27}{256\, b_1^6} \left[3\pi z^4 + \pi \frac{64}{3} a^4 b_2^4 \right] \sum_{j=1}^{K} \frac{1}{D_j}. \tag{6.16}$$

The lower bound for $\xi_2(\rho)$ follows at once from (6.11), i.e.,

$$\xi_2(\varrho) \geq \frac{z^2}{8} \sum_{j=1}^{N} \frac{1}{D_j}. \tag{6.17}$$

Putting (6.15), (6.16), and (6.17) together the assertion of the lemma immediately follows. □

We end this section with the proof of Theorem 6.1.

Proof of Theorem 6.1: Let $M(z)$ stand for the right side of (6.12). Then $M(z) \geq 0$ if and only if

$$1 \geq \frac{27}{32 z^2 b_1^6} \left(3\pi z^4 + \frac{64}{3} \pi a^4 b_2^4 \right). \tag{6.18}$$

Since $(4\, ab_2/3) \geq z$, it follows from (6.18) that

$$1 \geq \frac{81}{128} (13\pi) \frac{z^2}{b_1^6}. \tag{6.19}$$

Let $\sigma = b_1^2/b^2$, hence $b_2 = b\sqrt{1-\sigma}$. In terms of σ, the conditions $z \le 4a\,b_2/3$ and (6.19), can be expressed, as $z \le (4\,a\,b/3)\sqrt{1-\sigma}$ and $z \le \sqrt{128}/(9\sqrt{13\pi})\sigma^{3/2}b^3$, respectively. That is,

$$z \le h(\sigma) = \min\left\{\frac{4}{3}ab\sqrt{1-\sigma},\, \frac{\sqrt{128}}{9\sqrt{13\pi}}\sigma^{3/2}b^3\right\}.$$

The maximum of $h(\sigma)$ in the interval $(0,1)$ is attained at the (unique) solution $\hat{\sigma}$ of

$$\frac{\sigma^3}{(1-\sigma)} = \frac{117\pi}{8}\frac{a^2}{b^4},$$

and $\xi(\rho) \ge 0$ for all $0 < z \le (4ab/3)\sqrt{1-\hat{\sigma}}$. □

6.1. *Bibliographical remarks*

i) The stability of matter of several relativistic models has been studied since the eighties. In particular, see the original articles of Lieb and Yau [105, 106]. For a review up to the recent literature on the subject see the review article of Michael Loss [108] and the recent monograph of Lieb and Seiringer [96]. See also [93], which contains many articles on different aspects of the stability of relativistic matter.
ii) In realistic Relativistic Models, one not only needs bounds on the nuclear charges to insure stability, but also one needs upper bounds on the values of the *fine structure constant* α. In the toy model that I have discussed in this lecture, there is no need of such a bound on the fine structure constant to ensure its stability. This is due to the absence of the exchange term in the present model.
iii) Recently [15], we have used an analog of the toy model considered in this lecture to obtain a lower bound on the exchange energy of a system of electrons confined to live in a two dimensional system.

7. A New Lieb–Oxford Bound with Gradient Corrections

In this section, we would like to review some recent results obtained by one of us (RB) in collaboration with P. Gallegos and M. Tušek [15] concerning estimates on the indirect Coulomb energy of a system of electrons in terms of their single particle density. Since the early days of Quantum Mechanics there has been a wide interest in estimating various energy terms of a system of electrons in terms of the single particle density $\rho_\psi(x)$. Given that the expectation value of the Coulomb attraction of the electrons by the nuclei can be expressed in closed form in terms of $\rho_\psi(x)$, the interest focuses on estimating the expectation value of the kinetic energy of the system of electrons and on the expectation value of the Coulomb repulsion between the electrons. Here, we will be concerned with the latest. The most natural approximation to the expectation value of the Coulomb repulsion between the electrons is given by

$$D(\rho, \rho) = \frac{1}{2} \int \rho(x) \frac{1}{|x-y|} \rho(y) \,\mathrm{d}x \,\mathrm{d}y, \tag{7.1}$$

which is usually called the *direct term*. The remainder, i.e., the difference between the expectation value of the electronic repulsion and $D(\rho, \rho)$, say E, is called the *indirect term*. In 1930, Dirac [41] gave the first approximation to the indirect Coulomb energy in terms of the single particle density. Using an argument with plane waves, he approximated E by

$$E \approx -c_D e^{2/3} \int \rho^{4/3} \, dx, \tag{7.2}$$

where $c_D = (3/4)(3/\pi)^{1/3} \approx 0.7386$ (see, e.g., [110], p. 299). Here e denotes the absolute value of the charge of the electron. The first rigorous lower bound for E was obtained by E. H. Lieb in 1979 [85], using the Hardy–Littlewood Maximal Function [148]. There he found that, $E \geq -8.52 e^{2/3} \int \rho^{4/3} \, dx$. The constant 8.52 was substantially improved by E.H. Lieb and S. Oxford in 1981 [95], who proved the bound

$$E \geq -C e^{2/3} \int \rho^{4/3} \, dx, \tag{7.3}$$

with $C = c_{LO} = 1.68$. The best value for C is unknown, but Lieb and Oxford [95] proved that it is larger or equal than 1.234. The Lieb–Oxford value was later improved to 1.636 by Chan and Handy, in 1999 [30]. It is this last constant, as far as we know, that is the smallest value for C that has been found to this day. During the last thirty years, after the work of Lieb and Oxford [95], there has been a special interest in quantum chemistry in constructing corrections to the Lieb–Oxford term involving the gradient of the single particle density. This interest arises with the expectation that states with a relatively small kinetic energy have a smaller indirect part (see, e.g., [83, 118, 155] and references therein). Recently, Benguria, Bley, and Loss obtained an alternative to (7.3), which has a lower constant (close to 1.45) to the expense of adding a gradient term (see Theorem 1.1 in [13]).

After the work of Lieb and Oxford [95] many people have considered bounds on the indirect Coulomb energy in lower dimensions (in particular see, e.g., [60] for the one dimensional case, [102, 113, 124, 125] for the two dimensional case, which is important for the study of quantum dots). In this manuscript we give an alternative to the Lieb–Solovej–Yngvason bound [102], with a constant much closer to the numerical values proposed in [125] (see also the references therein) to the expense of adding a gradient term. In some sense, the result proven here is the analog of the three dimensional result proven in [13] for two dimensional systems.

In [15], the following lower bound on the indirect Coulomb energy in two dimensions was obtained:

Theorem 7.1: (Estimate on the Indirect Coulomb Energy in Two Dimensions) *Let $\psi \in L^2(\mathbb{R}^{2N})$ be normalized to one and symmetric (or antisymmetric) in all its variables. Define*

$$\rho_\psi(x) = N \int_{\mathbb{R}^{2(N-1)}} |\psi|^2(x, x_2, \ldots, x_N)\, \mathrm{d}x_2 \ldots \mathrm{d}x_N.$$

Then, for all $\epsilon > 0$,

$$\begin{aligned} E(\psi) &\equiv \langle \psi, \textstyle\sum_{i<j}^{N} |x_i - x_j|^{-1} \psi \rangle - D(\rho_\psi, \rho_\psi) \\ &\geq -(1+\epsilon)\beta \textstyle\int_{\mathbb{R}^2} \rho_\psi^{3/2}\, \mathrm{d}x - \frac{4}{\beta\epsilon} \int_{\mathbb{R}^2} |\nabla \rho_\psi^{1/4}|^2\, \mathrm{d}x \end{aligned} \tag{7.4}$$

with

$$\beta = \left(\frac{4}{3}\right)^{3/2} \sqrt{5\pi - 1} \simeq 5.9045.$$

Remarks:

i) The constant $\beta \simeq 5.9045$ is substantially lower than the constant $C_{LSY} \simeq 481.27$ found in [102] (see equation (5.24) of lemma 5.3 in [102]), which is the best bound to date.

ii) The constant β is close to the numerical values (i.e., $\simeq 1.95$) of [124] (and references therein), but is not sharp.

The proof of this theorem relies on a stability result for an auxiliary molecular quantum system in two dimensions (which is similar to the result of Benguria, Loss and Siedentop [21], which proof is reproduced in the previous Section) and an observation of Lieb and Thirring [103] (see below). The stability of the auxiliary system is embodied in the following theorem (see the proof in [15]).

Theorem 7.2: (Benguria, Gallegos, and Tusek) *For any* $a, b > 0$, *and* $R_i \in \mathbb{R}^2$, $i = 1, \ldots, K$, *and for all* $\rho \geq 0$ *(with* $\rho \in L^{3/2}(\mathbb{R}^2)$ *and* $\nabla \rho^{1/4} \in L^2(\mathbb{R}^2)$*), we have that*

$$\xi(\rho) \equiv a^2 \int_{\mathbb{R}^2} (\nabla \rho^{1/4})^2 \, \mathrm{d}x + b^2 \int_{\mathbb{R}^2} \rho^{3/2} \, \mathrm{d}x - \int_{\mathbb{R}^2} V(x)\rho(x) \, \mathrm{d}x + D(\rho, \rho) + U \geq 0, \tag{7.5}$$

where V, D, *and* U *are given by the two dimensional analogues of (6.2), (6.3), and (6.4), respectively, provided,*

$$0 \leq z \leq z_c(a, b) \equiv \frac{a\, b}{2} \sqrt{1 - \sigma}. \tag{7.6}$$

Here $0 < \sigma < 1$ *is the only positive root of the quartic equation*

$$\frac{\sigma^2}{\sqrt{1 - \sigma}} = \frac{32(5\pi - 1)}{27} \frac{a}{b^3} \tag{7.7}$$

on the interval $(0, 1)$.

Using the previous theorem, one can prove (7.4) as follows: one uses an idea introduced by Lieb and Thirring in 1975 in their proof of the stability of matter [103] (see also the review article [84] and the recent monograph [96]).

Proof of Theorem 7.1: Consider the inequality (7.5), with $K = N$ (where N is the number of electrons in our original system), $z = 1$ (i.e., the charge of the electrons), and $R_i = x_i$ (for all $i = 1, \ldots, N$). With this choice, according to (7.6), the inequality (7.5) is valid as long as a and b satisfy the constraint,

$$2 \leq ab\sqrt{1 - \sigma}, \tag{7.8}$$

with $\sigma \in (0, 1)$ the solution of

$$\frac{\sigma^2}{\sqrt{1 - \sigma}} = \frac{\beta^2}{2} \frac{a}{b^3} \tag{7.9}$$

where

$$\beta = \left(\frac{4}{3}\right)^{3/2} \sqrt{5\pi - 1} \simeq 5.9045. \tag{7.10}$$

Then take any normalized wavefunction $\psi(x_1, x_2, \ldots, x_N)$, and multiply (7.5) by $|\psi(x_1, \ldots, x_N)|^2$ and integrate over all the electronic configurations, i.e., on $\mathbb{R}^{2N}$. Moreover, take $\rho = \rho_\psi(x)$. We get at once,

$$E(\psi) \equiv \langle \psi, \sum_{i<j}^N |x_i - x_j|^{-1}\psi\rangle - D(\rho_\psi, \rho_\psi)$$
$$\geq -b^2 \int_{\mathbb{R}^2} \rho_\psi^{3/2} \mathrm{d}x - a^2 \int_{\mathbb{R}^2} |\nabla \rho_\psi^{1/4}|^2 \mathrm{d}x, \quad (7.11)$$

provided a and b satisfy (7.8) and (7.9) above. Thinking of $\sigma \in (0,1)$ as a free parameter, and a, b satisfying (7.8) and (7.9), and writing $\varepsilon = (1-\sigma)/\sigma$ we get at once from (7.8) and (7.9) that

$$b^2 \geq (1+\epsilon)\beta,$$

for any $\epsilon > 0$. The theorem then follows by choosing the minimum value of b^2, i.e., $b^2 = (1+\epsilon)\beta$, hence $a^2 = 4/(\beta\epsilon)$. □

Remark 7.3: In general the two integral terms in (7.4) are not comparable. If one takes a very rugged ρ, normalized to N, the gradient term may be very large while the other term can remain small. However, if one takes a smooth ρ, the gradient term can be very small as we illustrate in the example below. Let us denote

$$L(\rho) = \int_{\mathbb{R}^2} \rho(x)^{3/2} \, \mathrm{d}x$$

and

$$G(\rho) = \int_{\mathbb{R}^2} (\nabla \rho(x)^{1/4})^2 \, \mathrm{d}x.$$

We will evaluate them for the normal distribution

$$\rho(|x|) = C\mathrm{e}^{-A|x|^2}$$

where C, $A > 0$. Some straightforward integration yields

$$L = C^{\frac{3}{2}} \frac{2\pi}{3A}, \quad G = C^{\frac{1}{2}}\pi.$$

With $C = NA/\pi$,

$$\int_{\mathbb{R}^2} \rho(|x|) \, \mathrm{d}x = N,$$

and we have

$$\frac{G}{L} = \frac{3\pi}{2N},$$

i.e., in the "large number of particles" limit, the G term becomes negligible.

Acknowledgments

One of us (RB) would like to thank the organizers of the workshop "Complex Quantum Systems" for their kind invitation, and the Institute of Mathematical Sciences of the National University of Singapore for their hospitality during the workshop and support. This work has been partially supported by Iniciativa Científica Milenio, ICM (CHILE), project P07–027-F. The work of RB has also been supported by FONDECYT (Chile) Project 1100679. We would like to thank the anonymous referee for many useful suggestions that helped us improve the manuscript.

Appendix: A Short History of the Atom

This Appendix is based on the article by R. D. Benguria [12] that appeared in the News Bulletin of the International Association of Mathematical Physics with the occasion of the one hundredth anniversary of the experiments that led Ernest Rutherford to the atomic model that bears his name.

On March 22, 2011, the New York Times published the article *"A Nucleated Century"* on its Editorial page, celebrating the hundredth anniversary of Rutherford's 1911 manuscript [131]. The quoted article read at the start: "*...If you asked someone to draw an atom, he or she would probably draw something like a cockeyed solar system (see figure 2). The sun – the nucleus – is at the center, and the planets – the electrons – orbit in several different planes. The critical discovery in this atomic model emerged a century ago in a talk before the Manchester Literary and Philosophical Society in March 1911 and a paper published soon after in the Philosophical Magazine. Both were by Ernest Rutherford, who had won the 1908 Nobel Prize in Chemistry in part for his discovery of the alpha particle, which he later proved was the nucleus of a helium atom...*". The title of New York Times editorial article emphasized the discovery of the nucleus of the atom, and the beginnings of Nuclear Physics, which is certainly a true fact. On the other hand, the 1911 article of Rutherford did much more than that, it gave rise to modern atomic physics and it contributed enormously to push the beginnings of Quantum Mechanics.

Although conceived more like a philosophical idea by Democritus of Abdera and others in Ancient Greece, the idea that matter is formed by

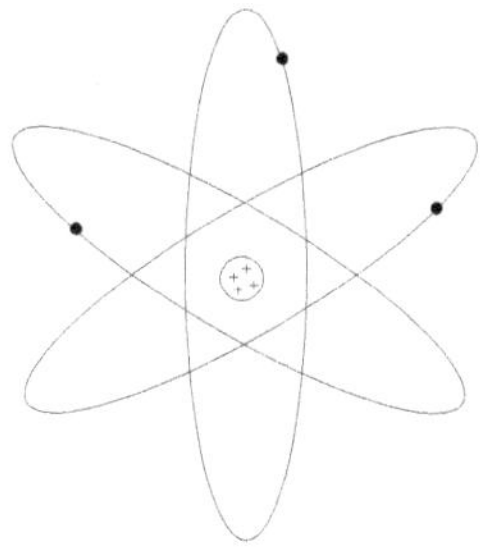

Fig. 2.

atoms, was first used in Physics by Daniel Bernoulli at the beginning of the 18th century to obtain from first principles in microscopic physics the law of ideal gases, giving birth to Kinetic Theory (later developed to full extent by Maxwell and Boltzmann in the 19th century). The discovery of photosynthesis by Jan Ingenhousz and Joseph Priestley in 1779 [73] prompted the discovery of oxygen, the introduction of the idea of chemical elements by Lavoisier, and the foundations of *modern chemistry* by John Dalton [37]. By that time, the concept of atoms was taking a proper place in physics and chemistry, and these atoms were much more than just small bodies whose restless motion would explain thermodynamic quantities like temperature and pressure. They had a structure, they could combine to form molecules, and the rules for combination were laid down by Dalton.

During the 19th century there were contributions from many people in different disciplines of physics that shed some light on the rich structure of the atom. In March of 1820 a crucial experiment of Hans Christian Oersted showed that Electricity and Magnetism were not independent phenomena [115], and that one could produce a magnetic field by driving a current through an electric circuit. Electromagnetism was born, and with it an extraordinary chain of pure and applied discoveries, including the first electric motors, which culminated in August of 1831, with the discovery of electromagnetic induction by Michael Faraday. A key figure of this period was Ampere who, among many contributions, introduced the idea of microscopic currents inside a metal to make the connection between Oersted's experiment and the properties of a magnet. The discovery of electrolysis by Humphry Davy, and the crucial observation by Michael Faraday that the amount of matter deposited on the cathodes of the electrolysis experiment is proportional to the total charge (current times time) applied to the electric terminals was the first experimental observation that electric charge in matter is quantized. In the meantime many key ideas were crystalizing in chemistry, among them, the introduction of the Avogadro number in 1811 [2] (i.e., when gaseous masses, at the same temperature and pressure, occupy equal volumes, they all contain the same number of molecules), the observation of Proust that the mass of any chemical element is an integer number times the mass of Hydrogen, and the skilled accumulated work by many people on chemical reactions that lead D. Mendeleev to establish the Periodic Table of Chemical Elements in 1869.

It was clear from all the previous observations that atoms and molecules ought to have an internal structure, yet to be discovered, that could explain these experimental facts. Still a whole new set of experiments and ideas would enter into the picture and help understanding this internal structure. These came from studying the interaction of light with matter. Using the idea of Newton that light could be decomposed into different colors letting it pass a prism, Wollaston and Fraunhofer introduced the field of spectroscopy, and determined the typical emission lines of incandescent gases. Moreover, in 1859, Gustav Kirchhoff posed the problem of determining the spectral decomposition of light emitted by a heated body. And, in 1887, Heinrich Hertz [64] discovered the photoelectric effect, i.e., the emission of an electric spark by a metal plate when illuminated by visible or ultraviolet light. The intensity of this spark was larger, the higher the frequency of the incident light.

During the last decade of the 19th century and the first one of the 20th century, the study of these three problems gave birth to the new Quantum Physics. In 1885, J. Balmer [4] classified the spectrum of Hydrogen in a simple phenomenological expression, which started to put some order in the huge experimental literature in atomic spectroscopy. By 1896, W. Wien, gave the first answer for the black body radiation [160], which reproduced appropriately the experimental data available at the time. However, better experimental results due to Kurlbaum, Pringsheim, and Rubens in 1900, showed small disagreements, at low frequencies, with Wien's theoretical results. In fact, for low frequencies, the experimental results were in agreement with, the then, recent results of Jeans and Rayleigh (based on the spectral asymptotics of the eigenfrequencies of electromagnetic cavities). By October of 1900, M. Planck [121] derived an interpolation between Wien's results (for high frequency) and the Rayleigh–Jeans results (for low frequency), which reproduced very precisely the experimental curves of Kurlbaum, Pringsheim, and Rubens. Planck's formula for the emission of a black body marked the beginning of Quantum Physics. In 1905, Albert Einstein [47], introduced the quanta of light (i.e., the present day *photons*) to explain the experimental results of H. Hertz on the photoelectric effect). The same year Einstein [48] gave a solution to the Brownian motion problem (observed independently by Robert Brown in 1827, and by Jan Ingenhousz in 1784), showed that the root mean square displacement squared of a Brownian particle is proportional to time, and the diffusion constant is inversely proportional to the Avogadro Number. This dependence of the diffusion constant allowed Perrin [119] to make the first accurate experimental determination of the Avogadro number (see also [120], and the review article of Duplantier [44]).

In the meantime, the discovery of radioactivity by H. Becquerel in 1896 [8] and the electron by J. J. Thomson in 1897 [152], prompted a renewed interest in trying to determine the internal structure of atoms. Thus, at the beginning of the 20th century several people (including J. J. Thomson [153], and H. Nagaoka [111]), introduced different models of atoms (basically neutral systems with positive and negative charge distributions interacting via a Coulomb potential). It is at this point in this history that Rutherford's contribution enter. By 1907, Ernest Rutherford had become a successor of Arthur Schuster (a leading spectroscopist of the time) as Professor of Physics at the University of Manchester. At Manchester, Rutherford continued his research on the properties of the radium emanation and of the alpha rays and, in conjunction with Hans Geiger, a method of detecting a single alpha particle and counting the number emitted from radium was devised (see the biography of Rutherford at the end of this manuscript). In order to try to determine the inner structure of the atom, Rutherford suggested to Geiger an experiment involving the scattering of alpha particles by a thin gold foil.

Collisions have always played a major role in physics. Already in 1668, The Royal Society of London established a competition in order to determine the laws of collisions in classical mechanics. The Royal Society received the memoirs of John Wallis (November 26, 1668), Christopher Wren (December 17, 1668), and Christiaan Huygens (January 4, 1668) who solved different aspects of the problem (see, e.g. [43], Chapter V). And the consideration of elastic collisions in special relativity yields the classical formula

$p = mv/\sqrt{1-(v/c)^2}$, for the momentum of a relativistic particle. Even today, smashing elementary particles at very high energy is the method to discover the physics at very small scales.

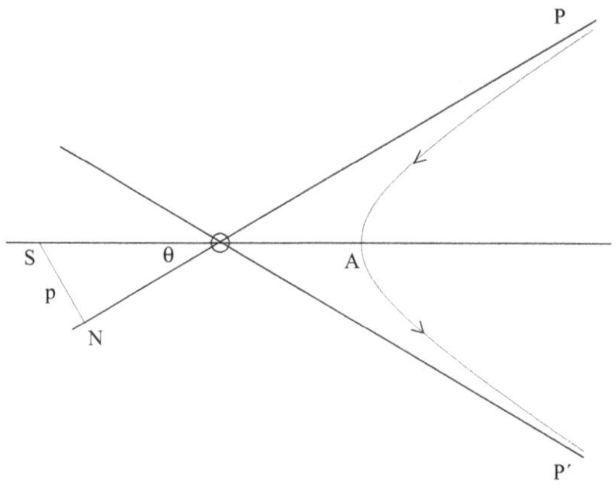

Fig. 3.

The gold foil experiment was conducted under the supervision of Rutherford at the University of Manchester in 1909 by Hans Geiger and the undergraduate student Ernest Marsden. In this experiment, most of the alpha particles passed straight through the foil. However, Geiger and Marsden [57] found that some alpha rays were scattered directly backwards, even from a thin film of gold. It was, a surprised Rutherford stated, "*as if one had fired a large naval shell at a piece of tissue paper and it had bounced back*". In his famous 1911 paper, Rutherford [131] starts comparing the results of the gold foil experiment with the theoretical predictions based on the Thomson model (which is usually referred to as the "plum–pudding model"), ruling it out. He then proceeds introducing the now classical picture (*the Rutherford atom*) in these words: "*...Consider an atom which contains a charge $\pm Ne$ at its centre surrounded by a sphere of electrification containing a charge $\mp Ne$ supposed uniformly distributed throughout a sphere of radius R. Here, e is the fundamental unit of charge, which in this paper is taken as 4.65×10^{10} E.S. unit. We shall suppose that for distances less than 10^{-12} cm, the central charge and also the charge on the alpha particle may be supposed to be concentrated at a point. It will be shown that the main deductions from the theory are independent of whether the* *central charge is supposed to be positive or negative. For convenience, the sign will be assumed to be positive.* ***The question of the stability of the atom*** *proposed need not be considered at this stage, for this will obviously depend upon the minute structure of the atom, and on the motion of the constituent charged parts...*" (see [131] for details). He then calculated the scattering cross section of a charged particle by a fixed target made of a charged point (finding a dependence like $\csc(\theta/2)^4$, where θ is the deflection angle in figure 3). Of course, Rutherford used Classical Mechanics for his computation of the scattering cross section, and he found an excellent agreement with the experimental data of the paper of Geiger and Marsden. It is a mayor coincidence that for the Coulomb potential, the results derived using Classical Mechanics (see, e.g., [79], p. 53, for the classical derivation), and the results using Quantum Mechanics (see, e.g., [56], Problem 110, pp. 290 ff) for the derivation using Quantum Mechanics) are the same. This is connected with the hidden symmetry ($SO(4)$ symmetry) of the motion of a particle moving in the presence of the Coulomb field. If this were not the case, it would have been an extra (may be impossible) puzzle to interpret the experimental results of Geiger and Marsden.

In 1911, Niels Bohr had obtained his Ph.D. at the University of Copenhagen, and joined the group of Ernest Rutherford in Manchester in 1912, attracted by the 1911 paper of Rutherford. As pointed out by Rutherford himself, his model has obvious stability problems, since in classical mechanics the accelerated electrons around the nucleus must radiate energy and fall to it in a very short time. It was in part to solve these stability problems that Niels Bohr introduced his model of the atom [24, 25, 26, 27], giving birth to the *Old Quantum Mechanics*. The *Bohr Atom* not only was an attempt to solve the stability problems of *Rutherford's Atom*, moreover, it was able to explain the Balmer series. The fact that the *semiclassical analysis* of Bohr could explain precisely the Balmer Series, and thus the spectrum of the Hydrogen Atom is again a happy coincidence due to the $SO(4)$ symmetry alluded to above.

In the century that has passed since the introduction of Rutherford's Atom [131, 132], there has been a fruitful interaction between mathematics and physics. Even in the days of the *Old Quantum Mechanics*, there were several mathematical developments carried by A. Sommerfeld and others, and with the introduction of the Schrödinger equation in 1926, the developments in Functional Analysis were growing hand by hand with Physics to help understanding the spectral properties of Atoms, Molecules and Solids. Also, in the last half a century, there has been a vast mathematical physics literature around stability problems in Atomic Physics. In summary, the introduction of the Rutherford Atom, not only marked the beginning of Nuclear Physics as stated in the New York Times Editorial. It also played a crucial role in Atomic Physics, and established a fertile ground for many problems in Mathematical Physics.

References

1. M. A. Aizenman, and E. H. Lieb, *On semiclassical bounds for eigenvalues of Schrödinger operators*, Phys. Letts. A **66** (1978), 427–429.
2. A. Avogadro, *Essai de déterminer les masses relatives des molécules élémentaires des corps, et les proportions dans lesquelles elles entrent dans ces combinaisons*, Journal de Physique **73** (1811), 58–76.
3. V. Bach, R. Lewis, E. H. Lieb, and H. Siedentop, *On the number of bound states of a bosonic Coulomb system*, Mathematische Zeitschrift **214** (1993), 441–459.
4. J. J. Balmer, *Hinweis auf die Spektrallinien des Wasserstoffs*, (i.e., Note on the Spectral Lines of Hydrogen), Annalen der Physik und Chemie **25** (1885), 80–85.
5. B. Baumgartner, *On the degree of ionization in the TFW theory*, Letters in Mathematical Physics **7** (1983), 439–441.
6. B. Baumgartner, *On Thomas–Fermi–von Weizsäcker and Hartree energies as functions of the degree of ionization*, J. Phys. A: Math. Gen. **17** (1984) 1593–1602.
7. R. J. Baxter, *Inequalities for potentials of particle systems*, Illinois Jurnal of Mathematics **24** (1980), 645–652.
8. H. Becquerel, *Sur les radiations émises par phosphorescence*, Comptes Rendus **122** (1896) 420–421.
9. R. Benguria, *The von Weizsäcker and exchange corrections in the Thomas Fermi theory*, Ph. D. Thesis, Princeton University, Princeton, NJ, 1979.
10. R. D. Benguria, *Stability of Matter*, Lecture Notes, in the Proceedings of the conference *Spectral Days*, held in Santiago de Chile, September 20–24, 2010, (R. Benguria, E. Friedmann, M. Mantoiu, Eds.), Birkhäuser Verlag, Basel (2012), pp. 25–54.
11. R. D. Benguria, *Isoperimetric inequalities for eigenvalues of the laplacian*, in **Entropy and the Quantum II**, (Robert Sims, and Daniel Ueltschi, Eds.), Contemporary Mathematics **552** (2011), 21–60.
12. R. D. Benguria, *A centennial of Rutherford's Atom*, International Association of Mathematical Physics News Bulletin, January 2012, pp. 4–10.
13. R. D. Benguria, G. A. Bley, and M. Loss, *A new estimate on the indirect Coulomb energy*, International Journal of Quantum Chemistry **112** (2012), 1579–1584.

14. R. Benguria, H. Brezis, and E. H. Lieb, *The Thomas–Fermi–von Weizsäcker theory of atoms and molecules*, Commun. Math. Phys. **79** (1981), 167–180.
15. R. D. Benguria, P. Gallegos, and M. Tušek, *A new estimate on the two-dimensional indirect Coulomb energy*, Annales Henri Poincaré **13** (2012), 1733–1744.
16. R. D. Benguria, S. Hoops, and H. Siedentop, *Bounds on the excess charge and the ionization energy for the Hellmann–Weizsčker model*, Annales de lInstitut Henri Poincaré (A) Physique Théorique **57** (1992), 47–65.
17. R. Benguria, and E. H. Lieb, *Proof of stability of highly negative ions in the absence of the Pauli principle*, Physical Review Letters **50** (1983), 1771–1774.
18. R. Benguria, and E. H. Lieb, *The most negative ion in the Thomas–Fermi–von Weizsäcker theory of atoms and molecules*, J. Phys. B **18** (1985), 1045–1059.
19. R. D. Benguria and M. Loss, *A simple proof of a theorem of Laptev and Weidl*, Mathematical Research Letters **7** (2000), 195-203.
20. R. D. Benguria and M. Loss, *Connection between the Lieb–Thirring conjecture for Schrödinger operators and an isoperimetric problem for ovals on the plane*, in **Partial Differential Equations and Inverse Problems**, C. Conca, R. Manásevich, G. Uhlmann, and M.S. Vogelius (Eds.), Comtemporary Mathematics **362**, Amer. Math. Soc., Providence , R.I., pp. 53–61 (2004).
21. R. D. Benguria, M. Loss, and H. Siedentop, *Stability of atoms and molecules in an ultrarelativistic Thomas–Fermi–Weizsäcker model*, J. Math. Phys. **49** (2008), article 012302.
22. R. D. Benguria and S. Pérez-Oyarzún, *The ultrarelativistic Thomas–Fermi von Weizsäcker model*, Journal of Physics A: Math. & Gen. **35** (2002), 3409–3414.
23. H. Bethe, *Berechnung der Elektronenaffinität des Wasserstoffs*, Zeitschrift für Physik **57** (1929), 815–821.
24. N. Bohr, *On the constitution of atoms and molecules*, Philosophical Magazine and Journal of Science, **26** (1913), 1–25.
25. N. Bohr, *On the constitution of atoms and molecules. Part II. Systems Containing only a Single Nucleus*, Philosophical Magazine and Journal of Science, **26** (1913), 476–501.
26. N. Bohr, *On the constitution of atoms and molecules. Part III. Systems Containing Several Nuclei*, Philosophical Magazine and Journal of Science, **26** (1913), 857–875.
27. N. Bohr, *On the constitution of atoms and molecules*, Papers of 1913 reprinted from the Philosophical Magazine with an Introduction by L. Rosenfeld; Munksgaard Ltd. Copenhagen, W.A. Benjamin Inc., NY, 1963.
28. Ph. Briet, P. Duclos, and H. Hogreve, *Monotonicity of ionization energies for three electron lithium–type systems*, Letters in Mathematical Physics **13** (1987), 137-140.
29. V. S. Buslaev, and L. D. Faddeev, *Formulas for traces for a singular Sturm–*

Liouville differential operator, Dokl. Akad. Nauk SSSR **132** (1960), 13–16 (Russian); English translation in Soviet Math. Dokl. **1** (1960), 451–454.
30. G. K.-L. Chan and N. C. Handy, *Optimized Lieb–Oxford bound for the exchange–correlation energy*, Phys. Rev. **A 59** (1999), 3075–3077.
31. S. Chandrasekhar, *The Negative Ions of Hydrogen and Oxygen in Stellar Atmospheres*, Rev. Mod. Phys. **16** (1944), 301–306.
32. J. Conlon, *A new proof of the Cwikel–Lieb–Rosenbljum bound*, Rocky Mountain J. Math., **15** (1985), 117–122.
33. M. M. Crum, *Associated Sturm–Liouville systems*, Quart. J. Math. Oxford Ser. 2, **6** (1955), 121–127.
34. M. Cwikel, *Weak type estimates for singular values and the number of bound states of Schrödinger operators*, Annals of Math., **106** (1977), 93-100.
35. H. L. Cycon, R. G. Froese, W. Kirsch, and B. Simon *Schrödinger Operators with Application to Quantum Mechanics and Global Geometry*, Springer–Verlag, Berlin, 1987.
36. A. Dall'Acqua and J. P. Solovej, *Excess charge for pseudo-relativistic atoms in Hartree-Fock theory*, Documenta Mathematica **15** (2010) 285–345.
37. J. Dalton, *A new system of Chemical Philosophy*, R. Bickerstaff, Strand, London, 1808.
38. G. Darboux, *Sur une proposition relative aux équations linéaires*, C. R. Acad. Sci. (Paris), **94** (1882), 1456–1459.
39. L. de Broglie, *Recherches sur la théorie des quanta*, Thesis (Paris), 1924 (see also, Ann. Phys. (Paris) **3**, (1925), 22).
40. P. A. Deift, *Applications of a commutation formula*, Duke Math. J. **45**, (1978), 267–310.
41. P. A. M. Dirac, *Note on Exchange Phenomena in the Thomas Atom*, Mathematical Proceedings of the Cambridge Philosophical Society, **26** (1930), 376–385.
42. J. Dolbeault, A. Laptev, and M. Loss, *Lieb–Thirring inequalities with improved constants*, J. Eur. Math. Soc., **10** (2008), 1121–1126.
43. R. Dugas, *Histoire de la Mècanique*, Jacques Gabay, Paris, 1998.
44. B. Duplantier, *Brownian Motion, "Diverse and Undulating"*, in **Einstein, 1905-2005, Poincare Seminar 1 (2005)**, T. Damour, O. Darrigol, B. Duplantier & V. Rivasseau, Eds., Progress in Mathematical Physics, **47**, 201–293 Birkhäuser Verlag, Basel, 2006.
45. F. Dyson and A. Lenard, *Stability of Matter, I*, J. Math. Phys. **8** (1967), 423–434.
46. A. Eden, and C. Foias, *A simple proof of the generalized Lieb–Thirring inequalities in one-space dimension*, J. Math. Anal. Appl. **162** (1991), 250–254.
47. A. Einstein, *Über einen die Erzeugung und Verwandlung des Lichtes betreffenden heuristischen Gesichtspunkt*, Annalen der Physik **17** (1905), 132–148.
48. A. Einstein, *Über die von der molekularkinetischen Theorie der Wärme geforderte Bewegung von in ruhenden Flüssigkeiten suspendierten Teilchen*, Annalen der Physik **17** (1905), 549–560.
49. E. Engel, *Zur relativischen Verallgemeinerung des TFDW modells*, Ph.D.

Thesis Johann Wolfgang Goethe Universität zu Frankfurt am Main, 1987.
50. E. Engel and R. M. Dreizler, *Field-theoretical approach to a relativistic Thomas–Fermi–Weizsäcker model*, Phys. Rev. A **35** (1987), 3607–3618.
51. E. Engel and R. M. Dreizler, *Solution of the relativistic Thomas–Fermi–Dirac–Weizsäcker model for the case of neutral atoms and positive ions*, Phys. Rev. A **38** (1988), 3909–3917.
52. G. Faber, *Beweis, dass unter allen homogenen Membranen von gleicher Fläche und gleicher Spannung die kreisförmige den tiefsten Grundton gibt*, Sitzungberichte der mathematisch-physikalischen Klasse der Bayerischen Akademie der Wissenschaften zu München Jahrgang, pp. 169–172 (1923).
53. L. D. Faddeev, and V. E. Zakharov, *The Korteweg de Vries equation is a completely integrable Hamiltonian system*, Funktsional. Anal. i Prilozhen. **5** (1971), 18–27 (Russian); English translation in Functional Anal. Appl. **5** (1971), 280–287.
54. C. L. Fefferman and L. A. Seco, *Asymptotic Neutrality of Large Ions*, Commun. Math. Phys. **128**, (1990) 109–130.
55. E. Fermi, *Un metodo statistico per la determinazione di alcune prioretà dell átome*, Rend. Acad. Naz. Lincei **6** (1927), 602–607.
56. S. Flügge, *Practical Quantum Mechanics*, Classical in Mathematics, Springer–Verlag, Berlin, 2002. [Reprint of the 1994 edition; The first English Edition of this book has been published in two volumes as: Die Grundlehren der mathematischen Wissenschaften Band 177 and 178].
57. H. Geiger, and E. Marsden, *On a Diffuse Reflection of the α-Particles*, Proceedings of the Royal Society, Series A **82** (1909), 495–500.
58. F. Gesztesy, *A Complete Spectral Characterization of the Double Commutator Method*, J. Funct. Anal. **117**, (1993), 401–446.
59. G. Green, **An essay on the application of mathematical analysis to the theories of electricity and magnetism**, Printed for the author, by T. Wheelhouse, Nottingham, England, 1828.
60. C. Hainzl and R. Seiringer, *Bounds on One-dimensional Exchange Energies with Applications to Lowest Landau Band Quantum Mechanics*, Letters in Mathematical Physics **55** (2001), 133–142.
61. G. H. Hardy, *Note on a Theorem of Hilbert*, Math. Z. **6** (1920), 314–317.
62. J. B. Hearnshaw, **The analysis of starlight: one hundred and fifty years of astronomical spectroscopy**, Cambridge University Press, Cambridge, UK, 1986.
63. P. Hertel, E. H. Lieb, and W. Thirring, *Lower bound to the energy of complex atoms*, J. Chem. Phys. **62** (1975), 3355–3356.
64. H. Hertz, *Über einen Einfluss des ultravioletten Lichtes auf die electrische Entladung*, Annalen der Physik und Chemie **267** (1887), 983–1000.
65. R. N. Hill, *Proof that the H^- Ion Has Only One Bound State*, Phys. Rev. Lett. **38** (1977), 643–646.
66. R. N. Hill, *Proof that the H^- ion has only one bound state. Details and extension to finite nuclear mass*, J. Math. Phys. **18** (1977), 2316–2330.
67. H. Hochstadt, **The functions of Mathematical Physics**, Pure and Applied Mathematics **23**, Wiley–Interscience, NY, 1971.

68. H. Hogreve, *On the maximal electronic charge bound by atomic nuclei*, J. Phys. B: At. Mol. Opt. Phys. **31** (1998), L439–L446.
69. W. Hughes, **An atomic Energy Lower Bound that proves Scott's Correction**, Ph. D. Thesis, Department of Mathematics, Princeton University, Princeton, NJ, 1986.
70. D. Hundertmark, E. H. Lieb, and L. E. Thomas, *A sharp bound for an eigenvalue moment of the one–dimensional Schrödinger operator*, Adv. Theor. Math. Phys. **2** (1998), 719–731.
71. E. A. Hylleraas, *The negative hydrogen ion in quantum mechanics and astrophysics*, Astrophysica Norvegica **IX** (1964), 345–349.
72. T. Ichinose, *Note on the Kinetic Energy Inequality Leading to Lieb's Negative Ionization Upper Bound*, Letters in Mathematical Physics **28** (1993), 219–230.
73. J. Ingenhousz, *Experiments upon vegetables, discovering their great power of purifying the common air in the sun-shine, and of injuring it in the shade and at night. To which is joined, a new method of examining the accurate degree of salubrity of the atmosphere*, Printed for P. Elmsly and H. Payne, London, 1779.
74. C. G. J. Jacobi, *Zur Theorie der Variationsrechnung und der Differentialgleichungen*, J. Reine Angew. Math. **17** (1837), 68–82.
75. J. H. Jeans, **The Mathematical Theory of Electricity and Magnetism**, Cambridge University Press, Cambridge, UK, 1915.
76. E. Krahn, *Über eine von Rayleigh formulierte Minimaleigenschaft des Kreises*, Math. Ann. **94** (1925), 97–100.
77. E. Krahn, *Über Minimaleigenschaften der Kugel in drei und mehr Dimensionen*, Acta Comm. Univ. Tartu (Dorpat) **A9**, 1–44 (1926). [English translation: *Minimal properties of the sphere in three and more dimensions*, **Edgar Krahn 1894–1961: A Centenary Volume**, Ü. Lumiste and J. Peetre, editors, IOS Press, Amsterdam, The Netherlands, pp. 139–174 (1994).]
78. J. B. Keller, *Lower bounds and isoperimetric inequalities for eigenvalues of the Schrödinger equation*, J. Math. Phys. **2** (1961), 262–266.
79. L. D. Landau, and E. M. Lifshitz , *Mechanics, Vol. 1*, Third Edition, Butterworth–Heinemann, Oxford, 1976.
80. A. Laptev, and T. Weidl, *Sharp Lieb–Thirring inequalities in high dimensions*, Acta Mathematica **184** (2000) 87–111.
81. A. Lenard and F. Dyson, *Stability of Matter, II*, J. Math. Phys. **9** (1968), 698–711.
82. W. Lenz, *Über die Anwendbarkeit der statistischen Methode auf lonengitter*, Z. Phys. **77** (1932), 713–721.
83. M. Levy and J. P. Perdew, *Tight bound and convexity constraint on the exchange–correlation–energy functional in the low–density limit, and other formal tests of generalized–gradient approximations*, Physical Review B **48** (1993), 11638–11645.
84. E. H. Lieb, *The stability of matter*, Reviews in Modern Physics **48** (1976), 553–569.
85. E. H. Lieb, *A lower bound for Coulomb energies*, Physics Letters **70 A**

(1979), 444–446.
86. E. H. Lieb, *The number of bound states of one-body Schrödinger operators and the Weyl problem*, Proc. Am. Math. Soc. Symposia Pure Math, **36** (1980), 241–252.
87. E. H. Lieb, *Thomas–Fermi and related theories of atoms and molecules*, Reviews in Modern Physics **53** (1981), 603–641.
88. E. H. Lieb, *Bound on the maximum negative ionization of atoms and molecules*, Phys. Rev. A **29** (1984), 3018–3028.
89. E. H. Lieb, *Atomic and molecular negative ions*, Physical Review Letters **52** (1984), 315–317.
90. E. H. Lieb, *Kinetic energy bounds and their applications to the stability of matter*, in Springer Lecture Notes in Physics **345** (1989), 371–382 (H. Holden, and A. Jensen, Eds.).
91. E. H. Lieb, *The stability of matter: from atoms to stars*, Bulletin Amer. Math. Soc. **22**, 1–49 (1990).
92. E. H. Lieb, *Lieb–Thirring inequalities*, in **Encyclopaedia of Mathematics**, (M. Hazewinkel, Ed.), Supplement vol. 2, pp. 311–313, Kluwer Academic Pub. (2000).
93. E. H. Lieb, *The stability of matter: From atoms to stars*, Selecta of Elliott H. Lieb, Edited by W. Thirring, 4th. Edition, Springer–Verlag, Berlin, 2005.
94. E. H. Lieb and M. Loss, *Analysis, Second Edition*, Graduate Studies in Mathematics, **14**, American Mathematical Society, Providence, 2001.
95. E. H. Lieb and S. Oxford, *Improved lower bound on the indirect Coulomb energy*, International Journal of Quantum Chemistry **19** (1981), 427–439.
96. E. H. Lieb and R. Seiringer, **The Stability of Matter in Quantum Mechanics**, Cambridge University Press, Cambridge, UK, 2009.
97. E. H. Lieb, I. M. Sigal, B. Simon, and W. Thirring, *Asymptotic neutrality of large-Z ions*, Physical Review Letters **52** (1984), 994–996.
98. E. H. Lieb, I. M. Sigal, B. Simon, and W. Thirring, *Asymptotic neutrality of large-Z ions*, Commun. Math. Phys. **116** (1988), 635-644.
99. E. H. Lieb and B. Simon, *Thomas–Fermi theory revisited*, Phys. Rev. Lett. **31** (1973), 681.
100. E. H. Lieb and B. Simon, *The Thomas–Fermi theory of atoms, molecules and solids*, Advances in Math. **23** (1977), 22-116
101. E. H. Lieb, J. P. Solovej, and J. Yngvason, *Quantum dots*, in **Differential equations and mathematical physics (Birmingham, AL, 1994)**, I. Knowles (Ed.), pp. 157–172, Int. Press, Boston, MA, 1995.
102. E. H. Lieb, J. P. Solovej, and J. Yngvason, *Ground states of large quantum dots in magnetic fields*, Physical Review B **51** (1995), 10646–10666.
103. E. H. Lieb and W. Thirring, *Bound for the kinetic energy of fermions which proves the stability of matter*, Phys. Rev. Lett. **35** (1975), 687-689; Errata **35** (1975), 1116.
104. E. H. Lieb and W. Thirring, *Inequalities for the moments of the eigenvalues of the Schrödinger Hamiltonian and their relation to Sobolev inequalities*, in **Studies in Mathematical Physics** (E. H. Lieb, B. Simon, and A. Wightman, Eds.), Princeton University Press, 1976, pp. 269-303.

105. E. H. Lieb and H. Yau, *Many-body stability implies a bound on the fine-structure constant*, Phys. Rev. Lett. **61** (1988), 1695–1697.
106. E. H. Lieb and H. Yau, *The stability and instability of relativistic matter*, Commun. Math. Phys. **118** (1988), 177–213.
107. H. Linde, *A lower bound for the ground state energy of a Schrödinger operator on a loop*, Proc. Amer. Math. Soc. bf 134 (2006), 3629–3635.
108. M. Loss, *Stability of Matter*, Lecture Notes, 2005 [Web site: www.mathematik.uni–muenchen.de/~lerdos/WS08/QM/lossstabmath.pdf]
109. M. Melgaard and T. Johnson, *On the maximal ionization for the atomic Pauli operator*, Proc. R. Soc. A **461** (2005), 3355–3364.
110. J. D. Morgan III, *Thomas–Fermi and other density functional theories*, in **Springer handbook of atomic, molecular, and optical physics, vol. 1**, pp. 295–306, edited by G.W.F. Drake, Springer–Verlag, NY, 2006.
111. H. Nagaoka, *Kinetics of a system of particles illustrating the line and the band spectrum and the phenomena of radioactivity*, Philosophical Magazine **7** (1904), 445–455.
112. P. T. Nam, *New bounds on the maximum ionization of atoms*, Commun. Math. Phys. **312** (2012), 427–445.
113. P.–T. Nam, F. Portmann, and J. P. Solovej, *Asymptotics for two dimensional Atoms*, Annales Henri Poincaré **13** (2012), 333–362.
114. I. Newton, **Philosophiae Naturalis Principia Mathematica**, London, 1687.
115. H. C. Oersted, *Experimenta Circa Effectum Conflictus Electrici in Acum Magneticam*, Neuere Electromagnetische Versuche, Nurnberg, Scrag (1820).
116. L. Onsager, *Electrostatic interaction of molecules*, J. Phys. Chem **43** (1939), 189–196.
117. L. Onsager, P. C. Hemmer, H. Holden, and S. K. Ratkie, *The collected works of Lars Onsager: with commentary*, World Scientific series in 20th century physics, World Scientific, Singapore, 1996.
118. J. P. Perdew, K. Burke, and M. Ernzerhof, *Generalized gradient approximation made simple*, Phys. Rev. Lett. **77**, 3865–3868 (1996).
119. J. B. Perrin, *Mouvement brownien et réalité moléculaire*, Ann. Chim. Phys. **18** (1909), 1–114.
120. J. B. Perrin, *Les Atomes*, Flammarion, Paris, 1991. [The first edition was published in 1913].
121. M. Planck, *On an improvement of Wien's Equation of the spectrum*, Verhandl d. Dtsch. Physikal. Gesellsch. **2** (1900), 237.
122. H. Poincaré, *Sur les rapports de lanalyse pure et de la physique mathématique*, Acta Mathematica **21**(1897), 331–342.
123. H. Poincaré, **Théorie du potentiel Newtonien - Leçons professées à la Sorbonne pendant le premier semestre 1894–1895**, Gauthiers-Villars, Paris, 1899. [Reprinted by Jacques Gabay, Paris, 1990].
124. E. Räsänen, S. Pittalis, K. Capelle, and C. R. Proetto, *Lower bounds on the exchange–correlation energy in reduced dimensions*, Phys. Rev. Lett. **102**, article 206406 (2009).
125. E. Räsänen, M. Seidl, and P. Gori–Giorgi, *Strictly correlated uniform electron droplets*, Phys. Rev. B **83**, article 195111 (2011).

126. A. R. P. Rau, *The negative ion of hydrogen*, J. Astrophys. Astr. **17** (1996), 113–145.
127. G. V. Rosenbljum, Distribution of the discrete spectrum of singular differential operator, Dokl. Aka. Nauk SSSR, **202** (1972), 1012–1015. The details are given in Distribution of the discrete spectrum of singular differential operators, Izv. Vyss. Ucebn. Zaved. Matematika **164** (1976), 75–86. (English trans. Sov. Math. (Iz. VUZ) **20** (1976), 63–71.)
128. M. B. Ruskai, *Absence of discrete spectrum of highly negative ions*, Commun. Math. Phys. **82** (1981), 457–469.
129. M. B. Ruskai, *Absence of discrete spectrum in highly negative ions, II. Extension to fermions*, Commun. Math. Phys. **85** (1982), 325–327.
130. M. B. Ruskai, *Improved estimate on the number of bound states of negative charged bosonic atoms*, Annales de l'I.H.P. A **61** (1994) 153–162.
131. E. Rutherford, *The scattering of alpha and beta particles by matter and the structure of the atom*, Philosophical Magazine **21** (1911), 669–688; (a brief account of this paper was communicated to the Manchester Literature and Philosophical Society in February of 1911).
132. E. Rutherford, *The structure of the atom*, Philosophical Magazine **27** (1914), 488–498.
133. U.-W. Schmincke, *On Schrödinger's factorization method for Sturm–Liouville operators*, Proc. Roy. Soc. Edinburgh Sect. A **80** (1978), 67–84.
134. E. Schrödinger, *Quantization as an eigenvalue problem*, Annalen der Physik **79** (1926), 361–376.
135. L. A. Seco, I. M. Sigal, and J. P. Solovej, *Bound on the ionization energy of large atoms*, Commun. Math. Phys. **131** (1990), 307–315.
136. R. Seiringer, *On the maximal ionization of atoms in strong magnetic fields*, J. Phys. A: Math. Gen. **34** (2001), 1943–1948.
137. R. Seiringer, *Inequalities for Schrödinger Operators and Applications to the Stability of Matter Problem*, in **Entropy and the Quantum, Arizona School of Analysis with Applications (Tucson, AZ, 2009)**, (R. Sims, D. Ueltschi, Eds.), Contemporary Mathematics **529** (2010), 53–72.
138. C. L. Siegel and J. K. Moser, **Lectures in Celestial Mechanics**, *Classics in Mathematics*, Springer–Verlag, Berlin, 1995.
139. I. M. Sigal, *Geometric methods in the quantum many–body problem. nonexistence of very negative ions*, Commun. Math. Phys. **85** (1982), 309–324.
140. I. M. Sigal, *How many electrons can a nucleus bind?*, Ann. Phys. **157** (1984), 307–320.
141. B. Simon, *Schrödinger operators in the twenty–first century*, Mathematical Physics 2000, 283–288, Imp. Coll. Press, London, 2000.
142. S.L. Sobolev, *On a theorem of functional analysis*, Mat. Sb. **4** (1938), 471–479. (The english translation appears in: AMS Trans. Series (2) **34** (1963), 39–68).
143. J. P. Solovej, *Universality in the Thomas–Fermi–von Weizsäcker model of atoms and molecules*, Ph. D. Thesis, Princeton University, 1989.
144. J. P. Solovej, *Asymptotics for bosonic atoms*, Lett. Math. Phys. **20** (1990), 165–172.

145. J. P. Solovej, *Proof of the ionization conjecture in a reduced Hartree–Fock model*, Invent. Math. **104** (1991), 291–311.
146. J. P. Solovej, *The ionization conjecture in Hartree-Fock theory*, Annals of Math. **158** (2003), 509–576.
147. J. P. Solovej, *Stability of Matter*, in Encyclopedia of Mathematical Physics, Eds. J.-P. Francoise, G.L. Naber and S.T. Tsou, **5**, 8–14, Elsevier (2006)
148. E. M. Stein and G. Weiss, *Introduction to Fourier Analysis on Euclidean Spaces*, Princeton University Press, Princeton, NJ (1971).
149. G. Talenti, *Elliptic equations and rearrangements*, Ann. Scuola Norm. Sup. Pisa (4) **3**, 697–718 (1976).
150. E. Teller, *On the Stability of Molecules in Thomas–Fermi Theory*, Reviews of Modern Physics **34** (1962), 627–631.
151. L. H. Thomas, *The calculation of atomic fields*, Proc. Cambridge Phil. Soc. **23** (1927), 542–548.
152. J. J. Thomson, *Cathode Rays*, Proceedings of the Royal Institution April 30, 1897, 1–14.
153. J. J. Thomson, *On the Structure of the Atom: An Investigation of the Stability and Periods of Oscillation of a number of Corpuscles arranged at equal intervals around the Circumference of a Circle; with Application of the Results to the Theory of Atomic Structure*, Philosophical Magazine **7** (1904), 237–265.
154. Y. Tomishima and K. Yonei, *Solution of the Thomas–Fermi–Dirac equation with a modified Weizsäcker correction*, J. Phys. Soc. Japan. **21** (1966), 142–153.
155. A. Vela, V. Medel, and S. B. Trickey, *Variable Lieb–Oxford bound satisfaction in a generalized gradient exchange-correlation functional*, The Journal of Chemical Physics **130** (2009), 244103.
156. G. Voronoi, *Nouvelles applications des paramètres continus à la théorie des formes quadratiques*, Journal für die Reine und Angewandte Mathematik **133** (1907), 97–178.
157. G. Voronoi, *Recherches sur les paralléloèdres Primitives*, J. Reine Angew. Math. **134** (1908), 198–287.
158. T. Weidl, *On the Lieb–Thirring constants $L_{\gamma,1}$ for $\gamma \geq 1/2$*, Commun. Math. Phys. **178** (1996), 135–146.
159. C. F. von Weizsäcker, Zur Theorie de Kernmassen, Z. Physik **96** (1935), 431–458.
160. W. Wien, *Über die Energievertheilung im Emissionsspectrum eines schwarzen Körpers*, Annalen der Physik **58** (1896), 662–669.
161. R. Wildt, *Negative Ions of Hydrogen and the Opacity of Stellar Atmospheres*, Astrophysical Journal **90** (1939), 611–620.
162. G. M. Zhislin, *Discussion of the spectrum of Schrödinger operator for system of many particles*, Trudy. Mosk. Mat. Obs. **9**, (1960), 81–128.
163. G. M. Zhislin, *On the Finiteness of the Discrete Spectrum of the Energy Operator of Negative Atomic and Molecular Ions*, Teor. Mat. Fiz. **7** (1971), 332–34, [English Translation, Theor. Math. Phys. **7** (1971), 571–578].

MATHEMATICAL DENSITY AND DENSITY MATRIX FUNCTIONAL THEORY (DFT AND DMFT)

Volker Bach

Institute for Analysis and Algebra
Technical University Braunschweig
Pockelsstr. 14, 38106 Braunschweig, Germany
v.bach@tu-bs.de

These notes give a survey on mathematical results concerning approximations to the ground state and the ground state energy of large atoms and molecules in terms of density functional theory of density matrix functional theory. In particular, stability of matter is proved, as well as, the asymptotic validity of the Hartree-Fock approximation in the limit of large nuclear charges. The latter requires the use of correlation inequalities which are derived here and related to some notions from the theory of N-representability, such as the G, P, and Q conditions and also Erdahl's T_1 and T_2 conditions. The latter have recently drawn a lot of attention in quantum chemistry.

Contents

1 Introduction 61
2 Exchange Correlation and LDA 64
3 Kinetic Energy and Lieb-Thirring Inequality 68
4 Thomas-Fermi Theory and Stability of Matter 76
5 Hartree-Fock Theory 79
6 Correlation Estimate Improving the Lieb-Oxford Inequality 89
7 Accuracy of the Hartree-Fock Approximation for Large Neutral Atoms 100
8 N-Representability 110
References 118

1. Introduction

The law of motion of quantum mechanics, the Schrödinger equation, is known for more than eighty years, and also many of its basic mathematical properties like existence and uniqueness of its solution have been established meanwhile. Yet, even for an atom or molecule with more than one electron, its solution is inaccessible to explicit computation. With the number of electrons the degree of difficulty of determining the solution of the Schrödinger equation increases very rapidly. For practical computations, one has to resort to approximative theories such as *density matrix functional theory*, *density functional theory* or even coarser approximations, as the number of particles in the system becomes very large.

In these notes we derive the Hartree-Fock (HF) approximation and then the Thomas-Fermi (TF) approximation, which define prominent density matrix functional and density functional theories, respectively, from the full quantum mechanical description of atoms and molecules. We use TF theory to prove stability of matter in Theorem 4.1, a basic result of quantum theory, and following [13] we show that the TF energy agrees with the (full) quantum mechanical ground state energy of a large neutral atom to leading order in Corollary 7.5. For the necessary facts about TF theory, as a variational problem for the density, we also refer mostly to [13] and other papers of E. Lieb. Our derivation of the HF approximation is mathematical in the sense that we prove in Corollary 7.8 its accuracy for large neutral atoms to be sufficiently good to resolve the exchange correction to the ground state energy.

Corollary 7.8 is based on the correlation inequality for fermions derived in Theorem 6.7. In the final section of these notes we discuss this correlation inequality in view of the notion of N-representability of states and new dual methods that emerge from the so-called G, P, Q, T_1, and T_2 representability conditions. In fact in a recent paper [5] it is shown that the G, P, and Q conditions imply Theorem 6.7. It is an interesting open problem to derive improved correlation inequalities, as compared to the one of Theorem 6.7, from the T_1, and T_2 conditions.

The reduction of the quantum mechanical model of fully interacting electrons in atoms and molecules to the corresponding Hartree-Fock approximation and then further to the Thomas-Fermi approximation is carried out in basically two steps:

- The first step consist in replacing the Coulomb repulsion of the electrons among each other by the electrostatic energy of the corresponding electron density minus the exchange correction expressed in terms of the one-particle density matrix,

$$\left\langle \Psi \middle| \sum_{1 \le m < n \le N} \frac{1}{|x_m - x_n|} \Psi \right\rangle \tag{1.1}$$
$$\approx \frac{1}{2} \int \rho_\Psi(x)\, \rho_\Psi(y)\, \frac{d^3x\, d^3y}{|x-y|} - \frac{1}{2} \int \big|\gamma_\Psi(x,y)\big|^2 \frac{d^3x\, d^3y}{|x-y|}.$$

With this replacement one obtains the Hartree-Fock (HF) approximation which is a density matrix functional theory. We show in Corollary 7.8, below, that the HF approximation is very accurate for large neutral atoms. The key step of the derivation of that estimate is the correlation inequality proved in Theorem 6.7.

Already Dirac proposed that, for the ground state wave function Ψ of the Coulomb system, the exchange term in (1.1) is well-approximated by the *local density approximation (LDA)*, more specifically by

$$\frac{1}{2} \int \big|\gamma_\Psi(x,y)\big|^2 \frac{d^3x\, d^3y}{|x-y|} \approx C_D \int \rho_\Psi^{4/3}(x)\, d^3x, \tag{1.2}$$

where γ_Ψ and ρ_Ψ are the one-particle density matrix and the one-particle density corresponding to Ψ, respectively, and $C_D := [81/(4\pi)]^{1/3}$. Lieb and Oxford established in [14] this approximation by proving that

$$\left\langle \Psi \middle| \sum_{1 \le m < n \le N} \frac{1}{|x_m - x_n|} \Psi \right\rangle \tag{1.3}$$
$$\ge \frac{1}{2} \int \rho_\Psi(x)\, \rho_\Psi(y)\, \frac{d^3x\, d^3y}{|x-y|} - C_{L-O} \int \rho_\Psi^{4/3}(x)\, d^3x,$$

where C_{L-O} is about twice as large as C_D. Even though C_{L-O} is too large, (1.3) is an important estimate that bounds the magnitude of the exchange correction and is sufficiently precise to show that the ground state energy and the Thomas-Fermi energy of a neutral atom agree to leading order. We also give a proof of (1.3) (although with a worse constant) in these notes in Theorem 2.1.

- The second step is the approximation of the kinetic energy of the ground state wave function Ψ by the kinetic energy in Thomas-Fermi theory, i.e.,

$$\left\langle \Psi \middle| \sum_{n=1}^{N} -\Delta_n \Psi \right\rangle \approx C_{sc} \int \rho_\Psi^{5/3}(x)\, d^3x, \tag{1.4}$$

where, as before, ρ_Ψ is the one-particle density matrix and $C_{sc} := (4\pi^3/5)(\pi/6)^{-5/3}$. Lieb and Thirring established (1.4) in terms of a lower bound

$$\left\langle \Psi \middle| \sum_{n=1}^{N} -\Delta_n \, \Psi \right\rangle \ \geq \ C_{LT} \int \rho_\Psi^{5/3}(x) \, d^3x, \tag{1.5}$$

where the constant C_{LT} is about five times smaller than C_{sc}. The Pauli principle which implies the antisymmetry of the wave function Ψ under exchanging coordinates is of crucial importance for the validity of this inequality.

From (1.5) and (1.3) it is easy to see that the Thomas-Fermi functional evaluated on the one-particle density of the ground state is a lower bound to the ground state energy. This fact is in turn the basic input for proving stability of matter, which is also carried out in these notes in Theorem 4.1.

Even though (1.5) is not directly used to approximate the quantum mechanical energy by the Hartree-Fock functional, it is also of great technical importance for the proof of showing that these two quantities are actually close to each other, as the particle number gets large.

The material in these notes is organized as follows. In Sect. 2 we carry out the first step described above and derive the Lieb-Oxford bound (1.3), while in Sect. 3 we give a proof of the Lieb-Thirring inequality (1.5). In Sect. 4 we combine the results of Sects. 2 and 3 with Thomas-Fermi theory to obtain a proof a stability of (nonrelativistic) matter. In Sect. 5 we introduce the Hartree-Fock approximation whose accuracy is established in Sect. 7, using the correlation inequality derived in Sect. 6. Sect. 8 is the last part of these notes in which we discuss N-representability and present the fairly well-known G, P, and Q conditions, but also the T_1 and T_2 conditions that sparked recent research interest in dual methods for the approximation of the ground state energy.

2. Exchange Correlation and LDA

In the 1930's Dirac was searching for a simple, yet quantitatively accurate, representation of exchange correlations in electron gases. He considered a free Fermi gas confined to a box $\Lambda = [-L, L]^3 \subseteq \mathbb{R}^3$ at chemical potential $\mu > 0$ [8]. Assuming for simplicity that the fermions under consideration are spinless, the one-particle density matrix (henceforth abbreviated by *1-pdm*) of this Fermi gas is given, to leading order in $L \gg 1$, by

$$\gamma(x,y) \;=\; \int e^{-ip(x-y)}\, \mathbb{1}\left[p^2 \le \mu\right] \frac{d^3p}{(2\pi)^3}, \tag{2.1}$$

and, accordingly, its one-particle density is

$$\begin{aligned} \rho_\gamma(x) = \gamma(x,x) \;&=\; \int \mathbb{1}\left[p^2 \le \mu\right] \frac{d^3p}{(2\pi)^3} \\ &= \frac{4\pi}{(2\pi)^3} \int_0^{\sqrt{\mu}} p^2\, dp \;=\; \frac{4\pi}{3(2\pi)^3}\, \mu^{3/2}. \end{aligned} \tag{2.2}$$

On the other hand, the exchange energy of this Fermi gas is to leading order in $L \gg 1$ given by

$$\begin{aligned} \int_{\Lambda\times\Lambda} \frac{|\gamma(x,y)|^2}{|x-y|}\, d^3x\, d^3y &= \int_{p^2,\hat{p}^2\le\mu} \left(\int_{\Lambda\times\Lambda} e^{-i(p-\hat{p})(x-y)}\, \frac{d^3x\, d^3y}{|x-y|} \right) \frac{d^3p\, d^3\hat{p}}{(2\pi)^6} \\ &\approx \int_\Lambda \left(\int_{p^2,\hat{p}^2\le\mu} \mathcal{F}\Big(\frac{1}{|.|}\Big)[p-\hat{p}]\, \frac{d^3p\, d^3\hat{p}}{(2\pi)^{9/2}} \right) d^3x \\ &\approx \int_\Lambda \left(\int_{p^2,\hat{p}^2\le\mu} \frac{1}{4\pi\,(p-\hat{p})^2}\, \frac{d^3p\, d^3\hat{p}}{16\pi^5} \right) d^3x \\ &= \int_\Lambda \mu^2 \left(\int_{p^2,\hat{p}^2\le 1} \frac{1}{(p-\hat{p})^2}\, \frac{d^3p\, d^3\hat{p}}{16\pi^5} \right) d^3x \\ &= C_{\mathrm{Dirac}} \int_\Lambda \rho_\gamma^{4/3}(x)\, d^3x, \end{aligned} \tag{2.3}$$

where C_{Dirac} is the numerical constant

$$C_{\mathrm{Dirac}} \;=\; \frac{3^{4/3}}{2^{8/3}\,\pi^{7/3}} \int_{|p|,|\hat{p}|\le 1} \frac{d^3p\, d^3\hat{p}}{(p-\hat{p})^2} \;=\; \left(\frac{81}{4\,\pi}\right)^{1/3} \;\approx\; 1,86. \tag{2.4}$$

Inspired by this computation, the exchange correlation energy

$$E_{\mathrm{corr}}(\Psi) \;:=\; \left\langle \Psi \middle| \sum_{1\le m<n\le N} \frac{1}{|x_m - x_n|}\, \Psi \right\rangle \;-\; \frac{1}{2} \int \rho_\Psi(x)\, \rho_\Psi(y)\, \frac{d^3x\, d^3y}{|x-y|} \tag{2.5}$$

of an N-fermion state Ψ with one-particle density ρ_Ψ is often approximated by the *local density approximation (LDA)*,

$$E_{\mathrm{corr}}(\Psi) \;\approx\; -C_{\mathrm{Dirac}} \int_\Lambda \rho_\gamma^{4/3}(x)\, d^3x. \tag{2.6}$$

Lieb [12] and Lieb and Oxford [14] showed that (a sufficiently large multiple of) the LDA provides a rigorous lower bound to the exchange correlation energy.

Theorem 2.1: *Let $\Psi \in \bigotimes^N L^2(\mathbb{R}^3)$ be a normalized vector, such that $\rho_\Psi \in L^1 \cap L^{5/3}(\mathbb{R}^3)$, where*

$$\rho_\Psi(x) := \tag{2.7}$$

$$\sum_{n=1}^{N} \int |\Psi(x_1, \ldots, x_{n-1}, x, x_{n+1}, \ldots, x_N)|^2 \, d^3x_1 \ldots d^3x_{n-1} \, d^3x_{n+1} \ldots dx_N.$$

Then

$$E_{\mathrm{corr}}(\Psi) \;\geq\; -(1.68) \int \rho_\gamma^{4/3}(x) \, d^3x. \tag{2.8}$$

Proof: For the proof, we abbreviate $\rho := \rho_\Psi$, $\underline{x} := (x_1, \ldots, x_N) \in (\mathbb{R}^3)^N$, and $d^{3N}\underline{x} := d^3x_1 \ldots d^3x_N$. We introduce

$$\Gamma_{r,z}(\underline{x}) \;:=\; \sum_{n=1}^{N} \mathbb{1}\big(|x_n - z| < r\big) \tag{2.9}$$

for all $r > 0$ and $z \in \mathbb{R}^3$ and observe that $\Gamma_{r,z}(\underline{x}) \in \{0, 1, \ldots, N\}$ is integer-valued. Furthermore, we use the Fefferman-de la Llave Identity [10]

$$\frac{1}{|x-y|} \;=\; \int_0^\infty \int_{\mathbb{R}^3} \mathbb{1}\big(|x-z| < r\big) \, \mathbb{1}\big(|y-z| < r\big) \, d^3z \, \frac{dr}{\pi \, r^5}, \tag{2.10}$$

for all $x, y \in \mathbb{R}^3$ with $x \neq y$. We then obtain

$$\begin{aligned} W(\Psi) \;&:=\; \Big\langle \Psi \Big| \sum_{1 \leq i < j \leq N} \frac{1}{|x_i - x_j|} \, \Psi \Big\rangle \\ &= \int d^3z \int_0^\infty \frac{dr}{\pi \, r^5} \int d^{3N}\underline{x} \sum_{1 \leq i < j \leq N} \mathbb{1}\big(|x_i - z| < r\big) \, \mathbb{1}\big(|x_j - z| < r\big) \, |\Psi(\underline{x})|^2 \\ &= \frac{1}{2} \int d^3z \int_0^\infty \frac{dr}{\pi \, r^5} \int d^{3N}\underline{x} \Big\{ \Gamma_{r,z}^2(\underline{x}) - \Gamma_{r,z}(\underline{x}) \Big\} \, |\Psi(\underline{x})|^2 \\ &\geq \frac{1}{2} \int d^3z \int_{R(z)}^\infty \frac{dr}{\pi \, r^5} \int d^{3N}\underline{x} \Big\{ \Gamma_{r,z}^2(\underline{x}) - \Gamma_{r,z}(\underline{x}) \Big\} \, |\Psi(\underline{x})|^2, \end{aligned} \tag{2.11}$$

for any measurable choice of $R : \mathbb{R}^3 \to \mathbb{R}_0^+$, using the positivity of the integrand due to $\Gamma_{r,z}^2 - \Gamma_{r,z} \geq 0$. The Cauchy-Schwarz inequality yields

$$\int \Gamma_{r,z}^2(\underline{x}) \, |\Psi(\underline{x})|^2 \, d^{3N}\underline{x} \;\geq\; \Big(\int \Gamma_{r,z}(\underline{x}) \, |\Psi(\underline{x})|^2 \, d^{3N}\underline{x} \Big)^2, \tag{2.12}$$

and moreover we have that

$$\int \Gamma_{r,z}(\underline{x}) \, |\Psi(\underline{x})|^2 \, d^{3N}\underline{x} \;=\; \sum_{i=1}^{N} \int \mathbb{1}\big(|x_i - z| < r\big) \, |\Psi(\underline{x})|^2 \, d^{3N}\underline{x} \;=\; N_{r,z}(\rho_\Psi), \tag{2.13}$$

where

$$N_{r,z}(\rho) \;:=\; \int_{|x-z|<r} \rho(x) \, d^3x. \tag{2.14}$$

Inserting (2.12)-(2.14) into (2.11), we obtain

$$\begin{aligned} W(\Psi) &\geq \frac{1}{2} \int d^3z \int_{R(z)}^{\infty} \frac{dr}{\pi \, r^5} \Big\{ N_{r,z}^2(\rho_\Psi) - N_{r,z}(\rho_\Psi) \Big\} \qquad (2.15) \\ &= \frac{1}{2} \int d^3z \int_0^{\infty} \frac{dr}{\pi \, r^5} N_{r,z}^2(\rho_\Psi) \\ &\quad - \frac{1}{2} \int d^3z \bigg\{ \int_0^{R(z)} \frac{N_{r,z}^2(\rho_\Psi) \, dr}{\pi \, r^5} + \int_{R(z)}^{\infty} \frac{N_{r,z}(\rho_\Psi) \, dr}{\pi \, r^5} \bigg\}. \end{aligned}$$

Using again (2.10) and (2.14), we have that

$$\int d^3z \int_0^{\infty} \frac{dr}{\pi \, r^5} N_{r,z}^2(\rho_\Psi) \;=\; \int \rho_\Psi(x) \, \rho_\Psi(y) \, \frac{d^3x \, d^3y}{|x-y|}, \tag{2.16}$$

and comparing (2.15) and (2.16) to (2.5), we arrive at

$$E_{\mathrm{corr}}(\Psi) \;\geq\; -\frac{1}{2} \int d^3z \bigg\{ \int_0^{R(z)} \frac{N_{r,z}^2(\rho_\Psi) \, dr}{\pi \, r^5} + \int_{R(z)}^{\infty} \frac{N_{r,z}(\rho_\Psi) \, dr}{\pi \, r^5} \bigg\}. \tag{2.17}$$

At this point we introduce the Hardy-Littlewood maximal function $M_\rho : \mathbb{R}^3 \to \mathbb{R}_0^+$ corresponding to $\rho \in L^1(\mathbb{R}^3; \mathbb{R}_0^+)$ by

$$M_\rho(z) \;:=\; \sup_{r>0} \bigg\{ \frac{N_{r,z}(\rho)}{N_{r,z}(1)} \bigg\} \;=\; \sup_{r>0} \bigg\{ \Big(\frac{4\pi r^3}{3} \Big)^{-1} \int_{|x-z|<r} \rho(x) \, d^3x \bigg\}. \tag{2.18}$$

Since

$$\rho(z) \;=\; \lim_{r \to 0} \bigg\{ \Big(\frac{4\pi r^3}{3} \Big)^{-1} \int_{|x-z|<r} \rho(x) \, d^3x \bigg\}, \tag{2.19}$$

for almost all $z \in \mathbb{R}^3$, we have that $M_\rho \geq \rho$, a.e. in $\mathbb{R}^3$. The important fact from harmonic analysis we use is that, though dominating ρ, M_ρ is comparable to ρ in L^p, provided $p > 1$. More precisely, if $\rho \in L^p(\mathbb{R}^3; \mathbb{R}_0^+)$ then $M_\rho \in L^p(\mathbb{R}^3; \mathbb{R}_0^+)$, too, and for all $p \in (1, \infty)$, there exists a universal constant $C_p < \infty$, such that

$$\forall \, \rho \in L^p(\mathbb{R}^3; \mathbb{R}_0^+) : \quad \int M_\rho^p(x) \, d^3x \;\leq\; C_p \int \rho^p(x) d^3x, \tag{2.20}$$

see, e.g., [18]. Abbreviating $M := M_{\rho_\Psi}$, we hence obtain

$$\int_0^{R(z)} \frac{N_{r,z}^2(\rho_\Psi)\,dr}{\pi\,r^5} \le M^2(z) \int_0^{R(z)} \frac{(4\pi)^2\,r\,dr}{3^2\pi} = \frac{8\pi}{9} M^2(z)\,R^2(z), \quad (2.21)$$

$$\int_{R(z)}^{\infty} \frac{N_{r,z}(\rho_\Psi)\,dr}{\pi\,r^5} \le M(z) \int_{R(z)}^{\infty} \frac{4\pi\,dr}{3\pi\,r^2} = \frac{4}{3}\frac{M(z)}{R(z)}. \quad (2.22)$$

Choosing

$$R(z) := \left(\frac{3}{4\pi\,M(z)}\right)^{1/3}, \quad (2.23)$$

we derive from (2.21)-(2.22) that

$$\int_0^{R(z)} \frac{N_{r,z}^2(\rho_\Psi)\,dr}{\pi\,r^5} + \int_{R(z)}^{\infty} \frac{N_{r,z}(\rho_\Psi)\,dr}{\pi\,r^5} \le 2^{5/3}\left(\frac{\pi}{3}\right)^{1/3} M^{4/3}(z). \quad (2.24)$$

Inserting this estimate into (2.17) and using (2.20), we arrive at the asserted estimate (2.8), but with a larger value of the constant than 1.68, namely,

$$E_{\mathrm{corr}}(\Psi) \ge -\left(\frac{4\pi}{3}\right)^{1/3} \int M^{4/3}(z)\,d^3z \ge -\left(\frac{4\pi}{3}\right)^{1/3} C_{4/3} \int \rho^{4/3}(x)\,d^3x. \quad (2.25)$$

□

3. Kinetic Energy and Lieb-Thirring Inequality

An accurate approximation or even estimation of the kinetic energy of an atom or molecule necessarily requires the careful observation of the Pauli principle because the kinetic energy of boson systems is by orders of magnitude smaller compared to fermion systems. As in the case of the approximation of the exchange correlations by the local density approximation, a heuristic consideration leads to the correct value for the kinetic energy as a functional of the one-particle density. We consider the Dirichlet Laplacian $-\Delta$ on the box $\Lambda_L := [-\frac{L}{2}, \frac{L}{2}]^3$. Its spectrum is purely discrete and in the sense of the spectral theorem, we may write

$$-\Delta = \sum_{k\in\mathbb{N}^3} \frac{\pi^2\,k^2}{L^2} |e_k\rangle\langle e_k|, \quad (3.1)$$

where $e_k(x) := cL^{-3/2}\prod_{\nu=1}^3 \sin\left[\frac{\pi}{2L}(2k_\nu - 1)x_\nu\right]$. On the Hilbert space $\bigotimes^N L^2(\Lambda_L)$, the kinetic energy operator

$$H_0^{(N)} := \sum_{n=1}^N -\Delta_n, \quad (3.2)$$

has the lowest eigenvalue

$$N\,\frac{3\pi^2}{L^2} \;=\; \rho\, L^3\,\frac{3\pi^2}{L^2} \;\approx\; \frac{3\pi^2}{L^2}\int_{\Lambda_L}\rho(x)\,d^3x. \tag{3.3}$$

If we consider $H_0^{(N)}$ on $\bigwedge^N L^2(\Lambda_L)$, instead, then its lowest eigenvalue is given by

$$\sum_{|k|\le\mu_N}\frac{\pi^2\,k^2}{L^2}, \tag{3.4}$$

where the chemical potential $\mu_N > 0$ is determined up to $\mathcal{O}(N^{-1/3})$ by the particle number,

$$\sum_{|k|\le\mu_N} 1 \;=\; N. \tag{3.5}$$

Since

$$\sum_{k\in\mathbb{N}^3,|k|\le\mu_N} 1 \;\approx\; \frac{1}{8}\sum_{k\in\mathbb{Z}^3,|k|\le\mu_N} 1 \;\approx\; \frac{1}{8}\int_{k\in\mathbb{Z}^3,|k|\le\mu_N} d^3k \;=\; \frac{\pi}{6}\mu_N^3, \tag{3.6}$$

the chemical potential is given by

$$\mu_N \;\approx\; (6/\pi)^{1/3}\,N^{1/3}, \tag{3.7}$$

and this in turn yields

$$\begin{aligned}\sum_{|k|\le\mu_N}\frac{\pi^2\,k^2}{L^2} &\approx \frac{\pi^2}{L^2}\int_{|k|\le\mu_N} k^2\,d^3k \;=\; \frac{4\pi^3}{5}\,\frac{\mu_N^5}{L^2}\\ &\approx \frac{4\pi^3}{5\,(\pi/6)^{5/3}}\,\frac{N^{5/3}}{L^2} \;=\; \frac{4\pi^3}{5\,(\pi/6)^{5/3}}\,L^3\left(\frac{N}{L^3}\right)^{5/3}\\ &\approx \frac{4\pi^3}{5\,(\pi/6)^{5/3}}\int_{\Lambda}\rho^{5/3}(x)\,d^3x. \end{aligned} \tag{3.8}$$

From (3.8) we conclude that the kinetic energy of a free fermion gas is an extensive quantity which possesses a nontrivial thermodynamic limit,

$$\frac{T_{\text{fermion}}}{|\Lambda|} \;=\; \frac{1}{|\Lambda_L|}\sum_{|k|\le\mu_N}\frac{\pi^2\,k^2}{L^2} \;\approx\; \frac{4\pi^3}{5\,(\pi/6)^{5/3}}\,\rho^{5/3} \;>\; 0, \tag{3.9}$$

while for boson systems,

$$\frac{T_{\text{boson}}}{|\Lambda|} \;=\; \rho\,\frac{3\pi^2}{L^2} \;\to\; 0, \tag{3.10}$$

even if $\rho > 0$. The heuristic consideration (3.1)-(3.9) was proved with mathematical rigor by Lieb and Thirring in 1975 [16, 15] to be correct, up to a multiplicative constant.

Theorem 3.1: [Lieb+Thirring] *There exists a constant $C_{LT} > 0$, such that*

$$T_\Psi := \left\langle \Psi \middle| \left(\sum_{n=1}^{N} - \Delta_n \right) \Psi \right\rangle \geq C_{LT} \int \rho_\Psi^{5/3}(x) d^3x, \tag{3.11}$$

for all normalized $\Psi \in \bigwedge_{n=1}^{N} L^2(\mathbb{R}^3) \cap \bigotimes_{n=1}^{N} H^1(\mathbb{R}^3)$, where $H^1(\mathbb{R}^3) := \left\{ \varphi \in L^2(\mathbb{R}^3) \mid \|\nabla\varphi\|^2 < \infty \right\}$ denotes the Sobolev space.

Eq. (3.11) has an equivalent formulation, as is asserted in the following theorem.

Theorem 3.2: *The following two statements* (3.12) *and* (3.13) *are equivalent:*

$$\Big\{ \forall N \in \mathbb{N},\ \Psi \in \bigwedge_{n=1}^{N} L^2(\mathbb{R}^3) \cap \bigotimes_{n=1}^{N} H^1(\mathbb{R}^3),\ \|\Psi\| = 1 :$$

$$T_\Psi \ \geq \ C_{LT} \int \rho_\Psi^{5/3}(x)\, d^3x \Big\} \tag{3.12}$$

$$\Longleftrightarrow \Big\{ \forall V \in L^{5/2}(\mathbb{R}^3) : \tag{3.13}$$

$$\mathrm{Tr}_{L^2(\mathbb{R}^3)} \left\{ [-\Delta - V]_- \right\} \geq -\frac{2}{5} \left(\frac{3}{5C_{LT}} \right)^{3/2} \int [V(x)]_+^{5/2}\, d^3x \Big\}.$$

Here, $[\lambda]_+ := \max(\lambda, 0)$ *and* $[\lambda]_- := \min(\lambda, 0)$.

We prepare for the proof of Theorem 3.2 by the following lemma.

Lemma 3.3: *Let A be a self-adjoint operator on a Hilbert space $\mathfrak{H}$. If $[A]_- \in \mathscr{L}^1(\mathfrak{H})$ is trace-class then*

$$\begin{aligned} \mathrm{Tr}\left\{[A]_-\right\} &= \inf\left\{\mathrm{Tr}\,(A\gamma) \mid 0 \leq \gamma \leq 1\right\} \\ &= \inf\left\{\mathrm{Tr}\,(A\gamma) \mid \gamma = \gamma^* = \gamma^2, \mathrm{rk}(\gamma) < \infty\right\}, \end{aligned} \tag{3.14}$$

otherwise both sides in (3.14) *equal* $-\infty$.

Proof: Without loss of generality we may assume that $\sigma(A) \cap \mathbb{R}^- \subseteq \sigma_{\text{disc}}(A)$, for otherwise we have $-\infty$ on both sides of (3.14). So, let $-\infty < -e_1 \le -e_2 \le \cdots \le -e_L < 0$ be the negative eigenvalues of A with corresponding eigenvectors φ_ℓ, with $\langle \varphi_k \mid \varphi_\ell \rangle = \delta_{k,\ell}$. We then have

$$A = \sum_{\ell=1}^{L} (-e_\ell) \mid \varphi_\ell \rangle\langle \varphi_\ell \mid + [A]_+ , \tag{3.15}$$

where $L \in \mathbb{N}$ or $L = \infty$. For $0 \le \gamma \le 1$, we thus have

$$\operatorname{Tr}\{A\gamma\} = \sum_{\ell=1}^{L} (-e_\ell) \underbrace{\langle \varphi_\ell \mid \gamma\varphi_\ell \rangle}_{\in[0,1]} + \underbrace{\operatorname{Tr}\{[A]_+ \gamma\}}_{\ge 0} \ge \sum_{\ell=1}^{L} (-e_\ell) = \operatorname{Tr}\{[A]_-\}. \tag{3.16}$$

Consequently,

$$\begin{aligned} \operatorname{Tr}\{[A]_-\} &\le \inf\{\operatorname{Tr}(A\gamma) \mid 0 \le \gamma \le 1\} \\ &\le \inf\{\operatorname{Tr}(A\gamma) \mid \gamma = \gamma^* = \gamma^2, \operatorname{rk}(\gamma) < \infty\}. \end{aligned} \tag{3.17}$$

For the proof of the converse inequality, we distinguish two cases:

(a) $L \in \mathbb{N}$: We set $\gamma_L := \sum_{\ell=1}^{L} |\varphi_\ell\rangle\langle\varphi_\ell| = \gamma_L^* = \gamma_L^2$ and observe that $\operatorname{rk}(\gamma_L) = L < \infty$ and that

$$\operatorname{Tr}\{A\gamma_L\} = \sum_{l=1}^{L} (-e_\ell) = \operatorname{Tr}\{[A]_-\}. \tag{3.18}$$

(b) $L = \infty$: For $n \in \mathbb{N}$ we set $\gamma_n := \sum_{\ell=1}^{n} |\varphi_\ell\rangle\langle\varphi_\ell|$. Then $\gamma_n = \gamma_n^* = \gamma_n^2$, $\operatorname{rk}(\gamma_n) = n < \infty$ and hence

$$\begin{aligned} &\inf\{\operatorname{Tr}(A\gamma) \mid \gamma = \gamma^* = \gamma^2, \operatorname{rank}(\gamma) < \infty\} \\ &\le \lim_{n\to\infty} \{\operatorname{Tr}(A\gamma_n)\} = \lim_{n\to\infty} \left\{ \sum_{\ell=1}^{n} (-e_\ell) \right\} \\ &= \operatorname{Tr}\{[A]_-\}. \end{aligned} \tag{3.19}$$

□

Proof: We proceed to proving Theorem 3.2.
(3.12) $\Longrightarrow$ (3.13): First we remark that $-\Delta - V$ is self-adjoint and semibounded, for $V \in L^{5/2}(\mathbb{R}^3)$. Next, by Lemma 3.3 we have

$$\begin{aligned} &\mathrm{Tr}\left\{[-\Delta - V]_-\right\} \\ &= \inf\left\{\mathrm{Tr}\,(-\Delta\gamma) - \mathrm{Tr}\,(V\gamma) \mid \gamma = \gamma^2 = \gamma^*, \mathrm{rk}\,(\gamma) < \infty\right\}. \end{aligned} \tag{3.20}$$

Let $\gamma = \gamma^* = \gamma^2$ with $\mathrm{rk}\,(\gamma) =: N < \infty$, $\mathrm{Tr}\,(-\Delta\gamma) < \infty$, and $\rho_\gamma(x) = \gamma(x,x)$. If $\{\varphi_n\}_{n=1}^N \subseteq \mathrm{Ran}\,(\gamma)$ is an orthonormal basis then $\{\varphi_n\}_{n=1}^N \subseteq H^1(\mathbb{R}^3)$ and $\gamma = \sum_{n=1}^N \mid \varphi_n \rangle\langle \varphi_n \mid$. We now set

$$\begin{aligned} \Phi &:= \varphi_1 \wedge \cdots \wedge \varphi_N \\ &:= \left(\frac{1}{N!}\right)^{1/2} \sum_{\pi \in \mathcal{P}_N} (-1)^\pi \varphi_{\pi(1)} \otimes \cdots \otimes \varphi_{\pi(N)} \in \bigwedge^N \mathfrak{H} \cap \bigotimes_{n=1}^N H^1\left(\mathbb{R}^3\right), \end{aligned} \tag{3.21}$$

where $\mathfrak{H} = L^2\left(\mathbb{R}^3\right)$. Then

$$\|\Phi\| = 1, \;\; \rho_\Phi(x) = \rho_\gamma(x) = \sum_{n=1}^N |\varphi_n(x)|^2 \tag{3.22}$$

and, moreover,

$$\mathrm{Tr}\,(-\Delta\gamma) = \left\langle \Phi \middle| \left(\sum_{n=1}^N - \Delta_n\right) \Phi \right\rangle \geq C_{LT} \int \rho_\Phi^{5/3}(x) d^3x, \tag{3.23}$$

due to (3.12). With this we obtain

$$\begin{aligned} &\mathrm{Tr}\left\{-\Delta\gamma\right\} - \mathrm{Tr}\left\{V\gamma\right\} \\ &\geq C_{LT} \int \rho_\gamma^{5/3}(x) d^3x - \int [V(x)]_+ \rho_\gamma(x) d^3x \\ &\geq C_{LT} \left(\int \rho_\gamma^{5/3}(x) d^3x\right) - \left(\int [V(x)]_+^{5/2}\, d^3x\right)^{2/5} \left(\int \rho_\gamma^{5/3}(x) d^3x\right)^{3/5}, \end{aligned} \tag{3.24}$$

after an application of Hölder's inequality. Therefore,

$$\begin{aligned} \mathrm{Tr}\left\{[-\Delta - V(x)]_-\right\} &\geq \inf_{r \geq 0} \left\{C_{LT} r^{5/3} - \left\|[V]_+\right\|_{5/2} r\right\} \\ &= -\frac{2}{5}\left(\frac{3}{5C_{LT}}\right)^{3/2} \left\|[V]_+\right\|_{5/2}^{5/2}. \end{aligned} \tag{3.25}$$

(3.12) $\Longleftarrow$ (3.13): Let $\Psi \in \bigwedge^N \mathfrak{H}$ be normalized and such that $\rho_\Psi \in L^{5/3}\left(\mathbb{R}^3; \mathbb{R}_0^+\right)$. Then $\|\rho_\Psi\|_{L^1(\mathbb{R}^3)} = N < \infty$, so $\rho_\Psi \in L^{5/3} \cap L^1\left(\mathbb{R}^3; \mathbb{R}_0^+\right)$. Moreover, $\rho_\Psi^{2/3} \in L^{(5/3)(3/2)}\left(\mathbb{R}^3\right) = L^{5/2}\left(\mathbb{R}^3\right)$, and for all $\kappa > 0$, we have that

$$\left\langle \Psi \middle| \left(\sum_{n=1}^{N} -\Delta_n - \kappa \rho_\Psi^{2/3}(x_n) \right) \Psi \right\rangle \geq \operatorname{Tr}\left\{ \left[-\Delta - \kappa \rho_\Psi^{2/3}(x) \right]_- \right\}$$
$$\geq -\frac{2}{5} \left(\frac{3}{5C_{LT}} \right)^{3/2} \kappa^{5/2} \int \rho_\Psi^{5/3}(x) d^3x, \tag{3.26}$$

due to (3.13). Since

$$\left\langle \Psi \middle| \left(\sum_{n=1}^{N} \rho_\Psi^{2/3}(x_n) \right) \Psi \right\rangle = \int \rho_\Psi^{2/3}(x) \rho_\Psi(x) d^3x, \tag{3.27}$$

this yields

$$\left\langle \Psi \middle| \sum_{n=1}^{N} (-\Delta_n) \Psi \right\rangle \geq \left(\kappa - \frac{2}{5} \left(\frac{3}{5C_{LT}} \right)^{3/2} \kappa^{5/2} \right) \int \rho_\Psi^{5/3}(x) d^3x. \tag{3.28}$$

Further setting $\kappa_{\text{opt}} := \frac{5C_{LT}}{3}$, we have

$$\max_{\kappa > 0} \left\{ \kappa - \frac{2}{5} \left(\frac{3}{5C_{LT}} \right)^{3/2} \kappa^{5/2} \right\} = C_{LT}. \tag{3.29}$$

□

Theorem 3.4: *There exists a constant $C < \infty$, such that, for all $V \in L^{5/2}\left(\mathbb{R}^3\right)$,*

$$\operatorname{Tr}\left\{ [-\Delta - V]_- \right\} \geq -C \left(\int [V(x)]_+^{5/2} d^3x \right). \tag{3.30}$$

Proof: Our proof is essentially a reproduction of [11]. As $V \in L^{5/2}\left(\mathbb{R}^3\right)$, the multiplication operator V is a relatively compact perturbation of $-\Delta$ and the negative spectrum of $-\Delta - V$ is purely discrete. Moreover, we can assume w.l.o.g. that $V \geq 0$. Now, if $-e < 0$ is an eigenvalue of $-\Delta - V$ with corresponding eigenfunction φ then

$$(-\Delta - V)\varphi = -e\varphi \quad \Longleftrightarrow \quad (-\Delta + e)\varphi = V\varphi, \tag{3.31}$$

and, setting $\psi := V^{1/2}\varphi$, it follows that

$$B_e\psi := \left(V^{1/2}\,(-\Delta+e)^{-1}\,V^{1/2}\right)\psi$$
$$= V^{1/2}\,(-\Delta+e)^{-1}\,V\varphi = V^{1/2}\varphi = \psi, \tag{3.32}$$

i.e.,

$$\Big\{-e \text{ is an eigenvalue of } -\Delta - V \text{ of multiplicity } n_e\Big\} \tag{3.33}$$
$$\iff \Big\{1 \text{ is an eigenvalue of } B_e \text{ of multiplicity } n_e\Big\}. \tag{3.34}$$

The equivalence (3.33) $\iff$ (3.34) is known as the **Birman-Schwinger Principle**, the operator B_e is called Birman-Schwinger operator. By the monotonicity and continuity of the map

$$\mathbb{R}^+ \ni e \mapsto B_e \in \mathscr{B}(\mathfrak{H}), \tag{3.35}$$

we have, for all $E > 0$, that

$$\begin{aligned} N_V(E) &= \dim \operatorname{Ran} \mathbb{1}\,[-\Delta - V(x) < -E] \\ &= \dim \operatorname{Ran} \mathbb{1}\,[B_E > 1] \\ &= \operatorname{Tr}\{\mathbb{1}\,[B_E > 1]\} \\ &\le \operatorname{Tr}\{B_E^2\} \\ &= \operatorname{Tr}\Big\{V\,(-\Delta+E)^{-1}\,V\,(-\Delta+E)^{-1}\Big\} \\ &= \int V(x)V(y)\left|(-\Delta+E)^{-1}\,(x-y)\right|^2 d^3x d^3y, \end{aligned} \tag{3.36}$$

where $\mathbb{1}\,[A < -E] := \mathbb{1}_{(-\infty,-E)}[A]$ and $\mathbb{1}\,[B > 1] := \mathbb{1}_{(1,\infty)}[B]$ denote the spectral projections of self-adjoint operators A and B onto the intervals $(-\infty, -E)$ and $(1, \infty)$, respectively, and the integral kernel of the resolvent of $-\Delta + E$,

$$(-\Delta+E)^{-1}\,(z) = \int \frac{e^{izp}}{p^2+E}\,\frac{d^3p}{(2\pi)^3} = \frac{1}{4\pi}\,\frac{e^{-\sqrt{E}|z|}}{|z|} \tag{3.37}$$

is the Yukawa potential. Denoting by $-e_1 \le -e_2 \le \cdots \le -e_L < 0$ the negative eigenvalues of $-\Delta - V(x)$ (counting multiplicities), monotone convergence yields

$$\sum_{n=1}^{L} e_n = \sum_{n=1}^{L} \int_0^{e_n} dE = \sum_{n=1}^{L} \int_0^{\infty} \mathbb{1}\left[e_n > E\right] dE$$
$$= \int_0^{\infty} \sum_{n=1}^{L} \mathbb{1}\left[e_n > E\right] dE = \int_0^{\infty} N_V(E) dE, \tag{3.38}$$

where $L \in \mathbb{N}$ or $L = \infty$. We finally observe that $V - E/2 \leq [V - E/2]_+$ and hence

$$\begin{aligned} N_V(E) &= \dim \operatorname{Ran} \mathbb{1}\left[-\Delta - V < -E\right] \\ &= \dim \operatorname{Ran} \mathbb{1}\left[-\Delta - (V - E/2) < -E/2\right] \\ &\leq \dim \operatorname{Ran} \mathbb{1}\left[-\Delta - [V - E/2]_+ < -E/2\right] \\ &= N_{[V-E/2]_+}(E/2). \end{aligned} \tag{3.39}$$

Putting these estimates together, we obtain

$$\begin{aligned} &\operatorname{Tr}\left\{[-\Delta - V]_-\right\} \\ &= \sum_{n=1}^{L} (-e_n) = -\int_0^{\infty} N_V(E) dE \geq -\int_0^{\infty} N_{[V-E/2]_+}(E/2) dE \\ &\geq -\left(\frac{1}{4\pi}\right)^2 \int_0^{\infty} dE \int d^3x d^3y \left[V(x) - \frac{E}{2}\right]_+ \left[V(y) - \frac{E}{2}\right]_+ \frac{e^{-\sqrt{2E}|x-y|}}{|x-y|^2}. \end{aligned} \tag{3.40}$$

Applying the Cauchy-Schwarz inequality and Plancherel's identity, we obtain

$$\begin{aligned} &\int \left[V(x) - \frac{E}{2}\right]_+ \left[V(y) - \frac{E}{2}\right]_+ \frac{e^{-\sqrt{2E}|x-y|}}{(4\pi)^2 |x-y|^2} d^3x \, d^3y \\ &\leq \left(\int \left[V(x) - \frac{E}{2}\right]_+^2 d^3x\right) \left(\int \left|\frac{e^{-\sqrt{E/2}|z|}}{4\pi |z|}\right|^2 d^3z\right) \\ &= \left(\int \left[V(x) - \frac{E}{2}\right]_+^2 d^3x\right) \left(\int \frac{d^3p}{(p^2 + E/2)^2}\right) \\ &= \left(\int \left[V(x) - \frac{E}{2}\right]_+^2 \frac{d^3x}{\sqrt{E/2}}\right) \left(\int \frac{d^3p}{(p^2 + 1)^2}\right), \end{aligned} \tag{3.41}$$

and hence

$$\operatorname{Tr}\{[-\Delta - V]_-\} \geq -2\left(\int \frac{d^3p}{(p^2+1)^2}\right)\int\left(\int_0^{V(x)} (V(x)-E)^2\,\frac{dE}{\sqrt{E}}\right) d^3x$$

$$= -2\left(\int \frac{d^3p}{(p^2+1)^2}\right)\left(\int_0^1 (1-E)^2\,\frac{dE}{\sqrt{E}}\right)\left(\int (V(x))^{5/2}\, d^3x\right). \tag{3.42}$$

□

4. Thomas-Fermi Theory and Stability of Matter

The ground state energy of an atom ($K = 1$) or molecule ($K \geq 2$) with $N \in \mathbb{N}$ (dynamic) electrons and $K \in \mathbb{N}$ (static) nuclei of charges $Z_1, Z_2, \ldots, Z_K > 0$ at positions $R_1, R_2, \ldots, R_K \in \mathbb{R}^3$ is given by

$$E_{\mathrm{gs}}(N, \underline{Z}, \underline{R}) := \inf\left\{ \langle \Psi \mid H_N(\underline{Z}, \underline{R})\Psi\rangle \;\middle|\; \Psi \in \bigwedge^N L^2\left(\mathbb{R}^3; \mathbb{C}^2\right), \|\Psi\| = 1 \right\}, \tag{4.1}$$

plus

$$U(\underline{Z}, \underline{R}) := \sum_{1 \leq k < l \leq K} \frac{Z_k Z_l}{|R_k - R_l|}, \tag{4.2}$$

where

$$H_N(\underline{Z}, \underline{R}) := \sum_{n=1}^{N}\left\{-\Delta_n - \sum_{k=1}^{K} \frac{Z_k}{|x_n - R_k|}\right\} + \sum_{1 \leq m < n \leq N} \frac{1}{|x_m - x_n|} \tag{4.3}$$

is the Hamiltonian of the system and $U(\underline{Z}, \underline{R})$ is the electrostatic energy of the configuration of nuclei. The main goal of this section is to show that stability of matter holds true, i.e.,

Theorem 4.1: *Let* $\varsigma := \max_{1 \leq k \leq K} Z_k$. *Then there exists a constant* $C_{SM} \equiv C_{SM}(\varsigma) < \infty$, *such that, for all* $N \in \mathbb{N}$ *and all* $\underline{R} = (R_1, \ldots, R_K) \in \mathbb{R}^{3K}$,

$$E_{\mathrm{gs}}(N, \underline{Z}, \underline{R}) + U(\underline{Z}, \underline{R}) \geq -C_{SM}(\varsigma)\,(N + K). \tag{4.4}$$

Remarks:

- The antisymmetry of the wave function is of crucial importance: If we assume instead that the wave function is symmetric or even possesses no symmetry properties then

$$E_{\rm gs}(N, \underline{Z}, \underline{R}) \leq -CN^{7/5}, \tag{4.5}$$

as was already noticed by Dyson [9].

- The inclusion of the repulsion of the nuclei in the energy is crucial, as well, as has been observed by Thirring [19]: If we set $Z_{tot} := \sum_{k=1}^{K} Z_k$ then

$$\begin{aligned} H_N(\underline{Z}, \underline{R}) &= \sum_{k=1}^{K} \frac{Z_k}{Z_{tot}} \left[\sum_{n=1}^{N} \left\{ -\Delta_n - \frac{Z_{tot}}{|x_n - R_k|} \right\} + \sum_{m<n} \frac{1}{|x_m - x_n|} \right] \\ &= \sum_{k=1}^{K} \left(\frac{Z_k}{Z_{tot}} \right) H_N(Z_{tot}, R_k). \end{aligned} \tag{4.6}$$

  Hence we have

$$\begin{aligned} E_{\rm gs}(N, \underline{Z}, \underline{R}) &= \inf_{\|\psi\|=1} \left\{ \sum_{k=1}^{K} \left(\frac{Z_k}{Z_{tot}} \right) \langle \Psi \mid H_N(Z_{tot}, R_k) \Psi \rangle \right\} \\ &\geq \sum_{k=1}^{K} \left(\frac{Z_k}{Z_{tot}} \right) \inf_{\|\psi\|=1} \{ \langle \Psi \mid H_N(Z_{tot}, R_k) \Psi \rangle \} \\ &= E_{\rm gs}(N, Z_{tot}, 0) = E_{\rm gs}(N, \underline{Z}, 0), \end{aligned} \tag{4.7}$$

  i.e., without nuclear repulsion the nuclear configuration of lowest energy is the one in which all nuclei are located at the origin (or any other fixed point in $\mathbb{R}^3$). We will see later that

$$E_{\rm gs}(N, Z_{tot}, 0) \leq -C\, Z_{tot}\, N^{4/3}, \tag{4.8}$$

  and, especially for $Z_1 = Z_2 = \cdots = Z_K$

$$E_{\rm gs}(N, Z_{tot}, 0) \leq -C\, Z_1\, K\, N^{4/3}, \tag{4.9}$$

  holds. Therefore, matter cannot be stable without nuclear repulsion.

Now we apply both the Lieb-Thirring inequality and the Lieb-Oxford inequality (2.8) and obtain

$$\begin{aligned} \langle \Psi \mid H_N(\underline{Z}, \underline{R}) \Psi \rangle \; &\geq C_{LT} \int \rho_\Psi^{5/3}(x) d^3x - \sum_{k=1}^{K} \int \frac{\rho_\Psi(x)\, Z_k}{|x - R_k|} d^3x \qquad (4.10) \\ &+ \frac{1}{2} \int \frac{\rho_\Psi(x) \rho_\Psi(y)}{|x - y|} d^3x d^3y - 2 \int \rho_\Psi^{4/3}(x) d^3x. \end{aligned}$$

A further application of Hölder's inequality yields

$$\begin{aligned}\int \rho_\Psi^{4/3}(x)d^3x &\le \left(\int \rho_\Psi^{5/3}(x)d^3x\right)^{1/2}\left(\int \rho_\Psi(x)d^3x\right)^{1/2} \\ &\le \frac{C_{LT}}{4}\left(\int \rho_\Psi^{5/3}(x)d^3x\right) + \frac{1}{C_{LT}}\left(\int \rho_\Psi(x)d^3x\right),\end{aligned} \tag{4.11}$$

so that

$$\langle \Psi \mid H_N(\underline{Z},\underline{R})\Psi\rangle \ge \frac{-2N}{C_{LT}} + \mathcal{E}_{\mathrm{TF}}^{(C_{LT}/2,\underline{Z},\underline{R})}(\rho_\Psi), \tag{4.12}$$

where $\mathcal{E}_{\mathrm{TF}}^{(\alpha,\underline{Z},\underline{R})} : \left(L^{5/3}\cap L^1\right)\left(\mathbb{R}^3\right) \to \mathbb{R}$ is the Thomas-Fermi (TF) functional defined by

$$\begin{aligned}\mathcal{E}_{\mathrm{TF}}^{(\alpha,\underline{Z},\underline{R})}(\rho) :=& \alpha\int \rho^{5/3}(x)d^3x - \sum_{k=1}^{K} Z_k \int \frac{\rho(x)}{|x-R_k|}d^3x \\ &+ \frac{1}{2}\int \frac{\rho(x)\,\rho(y)}{|x-y|}\,d^3x d^3y.\end{aligned} \tag{4.13}$$

We summarize our intermediate results in the following lemma.

Lemma 4.2:

$$\begin{aligned}&E_{\mathrm{gs}}(N,\underline{Z},\underline{R}) + \frac{2}{C_{LT}}N \\ &\ge \inf\left\{\mathcal{E}_{\mathrm{TF}}^{(C_{LT}/2,\underline{Z},\underline{R})}(\rho) \mid \rho \in \left(L^1\cap L^{5/3}\right)\left(\mathbb{R}^3;\mathbb{R}_0^+\right), \|\rho\|_{L^1} = N\right\} \\ &=: E_{\mathrm{TF}}^{(C_{LT}/2)}(N,\underline{Z},\underline{R}).\end{aligned} \tag{4.14}$$

As pointed out in the remark above, see (4.6)-(4.9), matter is unstable without the repulsive energy of the nuclei. In Thomas-Fermi theory, however, the opposite extreme holds true: If one includes the repulsion of the nuclei in the total energy, the atomic nuclei are driven apart and no molecules can form. This is made precise in Teller's lemma.

Lemma 4.3: **[Teller]** *For all* $\gamma > 0$, $N > 0$ *and* $\underline{Z} \in \left(\mathbb{R}^+\right)^K$,

$$\begin{aligned}&\inf\left\{E_{\mathrm{TF}}^{(\gamma)}(N,\underline{Z},\underline{R}) + U(\underline{Z},\underline{R}) \,\middle|\, \underline{R} \in \mathbb{R}^{3K}\right\} \\ &\qquad = \inf\left\{\sum_{k=1}^{K} E_{\mathrm{TF}}^{(\gamma)}(n_k,Z_k,0) \,\middle|\, n_k \ge 0, \sum_{k=1}^{K} n_k = N\right\}.\end{aligned} \tag{4.15}$$

It is furthermore known from Thomas-Fermi theory that the TF energy is minimal for neutral systems, i.e., for all $N > 0$,

$$E_{\mathrm{TF}}^{(\gamma)}\left(N, \underline{Z}, \underline{R}\right) = E_{\mathrm{TF}}^{(\gamma)}\left(Z_{tot}, \underline{Z}, \underline{R}\right), \tag{4.16}$$

and in particular

$$E_{\mathrm{TF}}^{(C_{LT}/2)}\left(N, Z_k, 0\right) \geq E_{\mathrm{TF}}^{(C_{LT}/2)}\left(Z_k, Z_k, 0\right) = -C_{\mathrm{TF}} \cdot Z_k^{7/3}, \tag{4.17}$$

where $C_{\mathrm{TF}} := E_{\mathrm{TF}}^{(C_{LT}/2)}\left(1, 1, 0\right) > 0$.

Proof: We proceed to proving Theorem 4.1. According to Lemmata 4.2 and 4.3, as well as, (4.16) and (4.17), we have

$$\begin{aligned} E_{\mathrm{gs}}\left(N, \underline{Z}, \underline{R}\right) + U\left(\underline{Z}, \underline{R}\right) + \frac{2N}{C_{LT}} & \\ &\geq E_{\mathrm{TF}}^{(C_{LT}/2)}\left(N, \underline{Z}, \underline{R}\right) + U\left(\underline{Z}, \underline{R}\right) \\ &\geq E_{\mathrm{TF}}^{(C_{LT}/2)}\left(Z_{tot}, \underline{Z}, \underline{R}\right) + U\left(\underline{Z}, \underline{R}\right) \\ &\geq \inf\left\{ \sum_{k=1}^{K} E_{\mathrm{TF}}^{(C_{LT}/2)}\left(n_k, Z_k, 0\right) \,\middle|\, n_k \geq 0, \sum_{k=1}^{K} n_k = Z_{tot} \right\} \\ &= \sum_{k=1}^{K} E_{\mathrm{TF}}^{(C_{LT}/2)}\left(Z_k, Z_k, 0\right) = -C_{\mathrm{TF}} \sum_{k=1}^{K} Z_k^{7/3} \\ &\geq -C_{\mathrm{TF}}\left(\max_k \{Z_k\}\right)^{7/3} K. \end{aligned} \tag{4.18}$$

□

5. Hartree-Fock Theory

We define the fermion Fock space $\mathcal{F}_f\left[\mathfrak{H}\right]$ corresponding to the one-particle Hilbert space $\mathfrak{H}$[a] by $\mathcal{F}_f^{(0)} := \mathbb{C}\Omega$, $\|\Omega\| = 1$ and

$$\mathcal{F}_f\left[\mathfrak{H}\right] = \bigoplus_{N=0}^{\infty} \mathcal{F}_f^{(N)}, \; \mathcal{F}_f^{(N)} := \bigwedge^{N} \mathfrak{H}. \tag{5.1}$$

Note that in the particular case of nonrelativistic electrons

$$\mathfrak{H} = L^2\left(\mathbb{R}^3 \times \mathbb{Z}_2\right). \tag{5.2}$$

[a] We will here and henceforth always assume the Hilbert spaces to be complex and separable.

We define fermion creation and annihilation operators $a^*(\varphi)$, $a(\varphi) := (a^*(\varphi))^*$ for every orbital $\varphi \in \mathfrak{H}$ by

$$\begin{aligned} &a^*(\varphi_1)\left[a^*(\varphi_2)\cdots a^*(\varphi_N)\Omega\right] \\ &:= a^*(\varphi_1)a^*(\varphi_2)\cdots a^*(\varphi_N)\Omega \\ &:= \varphi_1 \wedge \varphi_2 \wedge \cdots \wedge \varphi_N \\ &:= (N!)^{-1/2} \sum_{\pi \in S_N} (-1)^\pi \varphi_{\pi(1)} \otimes \varphi_{\pi(2)} \otimes \cdots \otimes \varphi_{\pi(N)}, \end{aligned} \tag{5.3}$$

where the wedge product is normalized such that $\|\varphi_1 \wedge \cdots \wedge \varphi_N\| = 1$, for orthonormal $\varphi_1, \ldots, \varphi_N$. Then $\{a^*(\varphi), a(\varphi)\}_{\varphi \in \mathfrak{H}} \subseteq \mathscr{B}(\mathcal{F}_f[\mathfrak{H}])$ fulfill the canonical anticommutation relations (CAR), i.e., for all $\varphi, \psi \in \mathfrak{H}$, we have that

$$\{a(\varphi), a^*(\psi)\} := a(\varphi)a^*(\psi) + a^*(\psi)a(\varphi) = \langle \varphi \mid \psi \rangle \, \mathbb{1}, \tag{5.4}$$

$$\{a(\varphi), a(\psi)\} = \{a^*(\varphi), a^*(\psi)\} = 0, \tag{5.5}$$

$$a(\varphi)\,\Omega = 0. \tag{5.6}$$

Now let $\{\varphi_k\}_{k=1}^\infty \subseteq \mathfrak{H}$ be a sufficiently regular orthonormal basis (ONB). (For electrons, we have $\mathfrak{H} = L^2(\mathbb{R}^3 \times \mathbb{Z}_2)$, and one could choose φ_k to be the Hermite functions, i.e., the eigenfunctions of the harmonic oscillator.) Then the N-particle Hamiltonian

$$H_N = \sum_{n=1}^N h_n + \sum_{1 \le m < n \le N} V_{m,n}, \tag{5.7}$$

with $h_n := -\Delta_{x_n} - \sum_{k=1}^K \frac{Z_k}{|x_n - R_k|}$ and $V_{m,n} := \frac{1}{|x_m - x_n|}$, can be viewed as the restriction $H\big|_{\mathcal{F}_f^{(N)}}$ to the N-particle subspace $\mathcal{F}_f^{(N)}$ of the second-quantized Hamiltonian

$$H = \sum_{k,\ell=1}^\infty h_{k,\ell}\, a_k^* a_\ell + \frac{1}{2} \sum_{k,\ell;m,n=1}^\infty V_{k,\ell;m,n}\, a_\ell^* a_k^* a_m a_n, \tag{5.8}$$

on $\mathcal{F}_f^{(N)}$, where

$$h_{k,\ell} = \left\langle \varphi_k \middle| \left(-\Delta - \sum_{k=1}^K \frac{Z_k}{|x - R_k|} \right) \varphi_\ell \right\rangle, \tag{5.9}$$

$$V_{k,\ell;m,n} = \left\langle \varphi_k \otimes \varphi_\ell \middle| \frac{1}{|x-y|} \varphi_m \otimes \varphi_n \right\rangle \tag{5.10}$$

$$= \sum_{\sigma_1 \ldots \sigma_4 = \pm} \int \frac{d^3x d^3y}{|x-y|} \overline{\varphi_k(x,\sigma_1)\,\varphi_\ell(y,\sigma_2)}\, \varphi_m(x,\sigma_3)\,\varphi_n(y,\sigma_4), \tag{5.11}$$

and we abbreviate

$$a_k^* := a^*(\varphi_k)\,,\ a_k := a(\varphi_k)\,. \tag{5.12}$$

Definition 5.1: Let $\mathfrak{H}$ be a Hilbert space.

(i) $\rho \in \mathscr{B}(\mathcal{F}_f[\mathfrak{H}])$ is called **density matrix** $:\Longleftrightarrow \rho > 0$, $\mathrm{Tr}_{\mathcal{F}}\{\rho\} = 1$.
(ii) Let $\rho \in \mathscr{B}(\mathcal{F}_f[\mathfrak{H}])$. Then $\gamma_\rho^{(1)} \in \mathscr{B}(\mathfrak{H})$ and $\gamma_\rho^{(2)} \in \mathscr{B}[\mathfrak{H}\otimes\mathfrak{H}]$ are called **one-particle density matrix (1-pdm)** and **two-particle density matrix (2-pdm)**, respectively, where

$$\left\langle f \middle| \gamma_\rho^{(1)} g \right\rangle := \langle a^*(g) a(f) \rangle_\rho, \tag{5.13}$$

$$\left\langle f \otimes \tilde{f} \mid \gamma_\rho^{(2)}(g \otimes \tilde{g}) \right\rangle := \langle a^*(\tilde{g}) a^*(g) a(f) a(\tilde{f}) \rangle_\rho, \tag{5.14}$$

for all $f, \tilde{f}, g, \tilde{g} \in \mathfrak{H}$, where $\langle A \rangle_\rho := \mathrm{Tr}\left\{\rho^{\frac{1}{2}} A \rho^{\frac{1}{2}}\right\}$.

Theorem 5.2: *Let $\mathfrak{H}$ be a Hilbert space, $\rho \in \mathscr{B}(\mathcal{F}_f[\mathfrak{H}])$ a density matrix and $\gamma_\rho^{(1)} \in \mathscr{B}(\mathfrak{H})$ and $\gamma_\rho^{(2)} \in \mathscr{B}[\mathfrak{H}\otimes\mathfrak{H}]$ the corresponding 1-pdm and 2-pdm. Let furthermore $\check{\mathrm{N}} := \bigoplus_{N=0}^{\infty} N \mathbb{1}_{\mathcal{F}_f^{(N)}}$ be the number operator on $\mathcal{F}_f[\mathfrak{H}]$.*

(i) If $\mathrm{Tr}\{\check{\mathrm{N}}\rho\} < \infty$ then $\gamma_\rho^{(1)} \in \mathscr{L}^1(\mathfrak{H})$, and

$$0 \le \gamma_\rho^{(1)} \le \mathbb{1},\ \mathrm{Tr}_{\mathfrak{H}}\left\{\gamma_\rho^{(1)}\right\} = \langle \check{\mathrm{N}} \rangle_\rho\,. \tag{5.15}$$

(ii) If $\mathrm{Tr}\{\check{\mathrm{N}}^2\rho\} < \infty$ then $\gamma_\rho^{(2)} \in \mathscr{L}^1(\mathfrak{H}\otimes\mathfrak{H})$, and

$$0 \le \gamma_\rho^{(2)} \le \langle \check{\mathrm{N}} \rangle_\rho\, \mathbb{1},\quad \mathrm{Tr}\left\{\gamma_\rho^{(2)}\right\} = \langle \check{\mathrm{N}}(\check{\mathrm{N}}-1) \rangle_\rho\,. \tag{5.16}$$

Proof: Assume that $\{\Phi_k\}_{k=1}^{\infty} \subseteq \mathcal{F}_f[\mathfrak{H}]$ is an orthonormal basis of eigenvectors of ρ with corresponding eigenvalues $\{\mu_k\}_{k=1}^{\infty} \subseteq \left(\mathbb{R}_0^+\right)^{\mathbb{N}}$, i.e., $\sum_{k=1}^{\infty} \mu_k = 1$ and $\rho = \sum_{k=1}^{\infty} \mu_k \mid \Phi_k \rangle\langle \Phi_k \mid$. Let $\{\varphi_\ell\}_{\ell=1}^{\infty} \subseteq \mathfrak{H}$ also be an orthonormal basis.

(i) For $f \in \mathfrak{H}$ with $\|f\| = 1$, we observe that

$$\left\langle f \middle| \gamma_\rho^{(1)} f \right\rangle = \mathrm{Tr}\left\{\rho a^*(f)\, a(f)\right\} = \sum_{k=1}^{\infty} \mu_k \left\langle \Phi_k \mid a^*(f)\, a(f)\, \Phi_k \right\rangle = \sum_{k=1}^{\infty} \mu_k \left\|a(f)\, \Phi_k\right\|^2 \geq 0, \tag{5.17}$$

while, with (5.4), we have

$$\left\langle f \middle| \gamma_\rho^{(1)} f \right\rangle \leq \sum_{k=1}^{\infty} \mu_k \left(\left\|a(f)\, \Phi_k\right\|^2 + \left\|a^*(f)\, \Phi_k\right\|^2\right) = \sum_{k=1}^{\infty} \mu_k \left\langle \Phi_k \middle| \underbrace{\left(a^*(f)\, a(f) + a(f)\, a^*(f)\right)}_{=\langle f | f \rangle \cdot \mathbb{1}} \Phi_k \right\rangle = \sum_{k=1}^{\infty} \mu_k \left\|\Phi_k\right\|^2 = 1. \tag{5.18}$$

Thus $0 \leq \gamma_\rho^{(1)} \leq 1$. Since $\gamma_\rho^{(1)} \geq 0$, we further have

$$\gamma_\rho^{(1)} \in \mathscr{L}^1(\mathfrak{H}) \quad \Longleftrightarrow \quad \sum_{l=1}^{\infty} \left\langle \varphi_l \middle| \gamma_\rho^{(1)} \varphi_l \right\rangle < \infty, \tag{5.19}$$

and the right side of (5.19) is indeed the trace of $\gamma_\rho^{(1)}$. Note, however, that by assumption

$$\mathrm{Tr}\left\{\gamma_\rho^{(1)}\right\} = \sum_{\ell=1}^{\infty} \left\langle \varphi_\ell \middle| \gamma_\rho^{(1)} \varphi_\ell \right\rangle = \sum_{\ell=1}^{\infty} \left\langle a^*(\varphi_\ell) a(\varphi_\ell) \right\rangle_\rho = \langle \check{\mathbb{N}} \rangle_\rho, \tag{5.20}$$

where we use that

$$\check{\mathbb{N}} = \sum_{\ell=1}^{\infty} a^*(f_\ell)\, a(f_\ell), \tag{5.21}$$

for *any* orthonormal basis $\{f_\ell\}_{\ell=1}^{\infty} \subseteq \mathfrak{H}$, as follows from the monotone convergence theorem.

(ii) Assume that $\psi \in \mathfrak{H} \otimes \mathfrak{H}$ with $\|\psi\| = 1$ and

$$\psi = \sum_{m,\ell=1}^{\infty} \alpha_{\ell,m} \, \varphi_\ell \otimes \varphi_m, \tag{5.22}$$

so $\sum_{m,\ell=1}^{\infty} |\alpha_{\ell,m}|^2 = 1$. Then

$$\begin{aligned}\left\langle \psi \middle| \gamma_\rho^{(2)} \psi \right\rangle &= \sum_{\ell_1,\ell_2,m_1,m_2=1}^{\infty} \overline{\alpha_{\ell_1,m_1}} \alpha_{\ell_2,m_2} \left\langle a^*(\varphi_{\ell_2}) a^*(\varphi_{m_2}) a(\varphi_{m_1}) a(\varphi_{\ell_1}) \right\rangle_\rho \\ &= \left\langle A^* A \right\rangle_\rho = \sum_{k=1}^{\infty} \mu_k \, \|A \Phi_k\|^2 \geq 0,\end{aligned} \tag{5.23}$$

where

$$A := \sum_{m,\ell=1}^{\infty} \alpha_{\ell,m} a\left(\varphi_\ell\right) a\left(\varphi_m\right). \tag{5.24}$$

Since $a\left(\varphi_\ell\right) a\left(\varphi_m\right) = -a\left(\varphi_m\right) a\left(\varphi_\ell\right)$, we can assume w.l.o.g. that $\alpha_{\ell,m} = -\alpha_{m,\ell}$, i.e., $\psi \in \mathfrak{H} \wedge \mathfrak{H}$. From $\sum_{\ell,m=1}^{\infty} |\alpha_{\ell,m}|^2 = 1$ it follows that $\alpha := \left(\alpha_{\ell,m}\right)_{\ell,m=1}^{\infty} \in \mathscr{L}^2\left[\ell^2(\mathbb{N})\right]$ is a Hilbert-Schmidt operator whose singular value decomposition yields two ONB $\{f^{(j)}\}_{j=1}^{\infty}$, $\{g^{(j)}\}_{j=1}^{\infty} \subseteq \ell^2(\mathbb{N})$ and a sequence $\{\lambda_j\}_{j=1}^{\infty} \subseteq \mathbb{R}_0^+$ of nonnegative numbers, with $\sum_{j=1}^{\infty} \lambda_j^2 = 1$, such that

$$\alpha = \sum_{j=1}^{\infty} \lambda_j |f^{(j)}\rangle\langle \overline{g^{(j)}}|, \tag{5.25}$$

i.e.,

$$\alpha_{\ell,m} = \sum_{j=1}^{\infty} \lambda_j \, f_\ell^{(j)} g_m^{(j)}. \tag{5.26}$$

We set

$$\eta_j := \sum_{\ell=1}^{\infty} f_\ell^{(j)} \varphi_\ell \quad , \quad \widetilde{\eta}_j := \sum_{m=1}^{\infty} g_m^{(j)} \varphi_m \tag{5.27}$$

and observe that $\{\eta_j\}_{j=1}^{\infty}$, $\{\widetilde{\eta}_j\}_{j=1}^{\infty} \subseteq \mathfrak{H}$ are ONB, as is easily checked by

$$\langle \eta_i | \eta_j \rangle = \sum_{\ell,m=1}^{\infty} \overline{f_\ell^{(i)}} f_m^{(j)} \langle \varphi_\ell | \varphi_m \rangle = \sum_{\ell=1}^{\infty} \overline{f_\ell^{(i)}} f_\ell^{(j)} = \delta_{i,j}, \tag{5.28}$$

$$\sum_{j=1}^{\infty} \overline{f_m^{(j)}} \eta_j = \sum_{j,\ell=1}^{\infty} \overline{f_m^{(j)}} f_\ell^{(j)} \varphi_\ell = \varphi_m, \tag{5.29}$$

and similarly for $\{\widetilde{\eta}_j\}_{j=1}^{\infty}$. Inserting (5.26) and (5.27) into (5.22), we further observe that

$$\psi = \sum_{j=1}^{\infty} \lambda_j \eta_j \otimes \widetilde{\eta}_j. \tag{5.30}$$

Using the Cauchy-Schwarz inequality, we further obtain

$$\begin{aligned}
\left\langle \psi \middle| \gamma_\rho^{(2)} \psi \right\rangle &= \sum_{i,j=1}^{\infty} \lambda_i \overline{\lambda_j} \left\langle a^*(\eta_i)\, a^*(\widetilde{\eta}_i)\, a(\widetilde{\eta}_j)\, a(\eta_j) \right\rangle_\rho \\
&\le \sum_{i,j=1}^{\infty} |\lambda_i|\, |\lambda_j| \left\langle a^*(\eta_i)\, a^*(\widetilde{\eta}_i)\, a(\widetilde{\eta}_i)\, a(\eta_i) \right\rangle_\rho^{1/2} \\
&\qquad \left\langle a^*(\eta_j)\, a^*(\widetilde{\eta}_j)\, a(\widetilde{\eta}_j)\, a(\eta_j) \right\rangle_\rho^{1/2} \\
&= \left(\sum_{j=1}^{\infty} |\lambda_j| \left\langle a^*(\eta_j)\, a^*(\widetilde{\eta}_j)\, a(\widetilde{\eta}_j)\, a(\eta_j) \right\rangle_\rho^{1/2} \right)^2 \\
&\le \left(\sum_{j=1}^{\infty} \lambda_j^2 \right) \left(\sum_{j=1}^{\infty} \left\langle a^*(\eta_j)\, \underbrace{a^*(\widetilde{\eta}_j)\, a(\widetilde{\eta}_j)}_{\le \langle \widetilde{\eta}_j | \widetilde{\eta}_j \rangle \mathbb{1} \le \mathbb{1}} a(\eta_j) \right\rangle_\rho \right) \\
&\le \sum_{j=1}^{\infty} \left\langle a^*(\eta_j)\, a(\eta_j) \right\rangle_\rho = \langle \check{\mathbb{N}} \rangle_\rho,
\end{aligned} \tag{5.31}$$

so

$$0 \le \gamma_\rho^{(2)} \le \langle \check{\mathbb{N}} \rangle_\rho \mathbb{1}. \tag{5.32}$$

Furthermore, we have

$$\begin{aligned}
\mathrm{Tr}\left\{ \gamma_\rho^{(2)} \right\} &= \sum_{\ell,m=1}^{\infty} \left\langle \varphi_\ell \otimes \varphi_m \middle| \gamma_\rho^{(2)} (\varphi_\ell \otimes \varphi_m) \right\rangle \\
&= \sum_{\ell,m=1}^{\infty} \left\langle a^*(\varphi_m)\, a^*(\varphi_\ell)\, a(\varphi_\ell)\, a(\varphi_m) \right\rangle_\rho \\
&= \sum_{\ell,m=1}^{\infty} \left\langle a^*(\varphi_\ell)\, a(\varphi_\ell) \left[a^*(\varphi_m)\, a(\varphi_m) - \delta_{\ell,m} \right] \right\rangle_\rho \\
&= \left\langle \check{\mathbb{N}} (\check{\mathbb{N}} - \mathbb{1}) \right\rangle_\rho.
\end{aligned} \tag{5.33}$$

□

Remarks:

- It can be shown that (5.32) is optimal, in general.
- Inserting (5.13) and (5.14) into (5.8), we observe that the energy expectation value of a density matrix ρ can be written as a trace over its 1-pdm and 2-pdm:

$$\begin{aligned}\left\langle H\right\rangle_\rho &= \sum_{k,\ell=1}^{\infty} h_{k,\ell}\mathrm{Tr}\left\{\rho a_k^* a_\ell\right\} + \frac{1}{2}\sum_{k,\ell;m,n=1}^{\infty} V_{k,\ell;m,n}\mathrm{Tr}\left\{\rho a_\ell^* a_k^* a_m a_n\right\} \\ &= \sum_{k,\ell=1}^{\infty} h_{k,\ell}\left(\gamma_\rho^{(1)}\right)_{\ell,k} + \frac{1}{2}\sum_{k,\ell;m,n=1}^{\infty} V_{k,\ell;m,n}\left(\gamma_\rho^{(2)}\right)_{m,n;k,\ell} \\ &= \mathrm{Tr}_{\mathfrak{H}}\left\{h\gamma_\rho^{(1)}\right\} + \frac{1}{2}\mathrm{Tr}_{\mathfrak{H}\otimes\mathfrak{H}}\left\{V\gamma_\rho^{(2)}\right\}. \qquad (5.34)\end{aligned}$$

Lemma 5.3: *Let* $\Phi \in \mathcal{F}_f[\mathfrak{H}]$ *and* $\rho = |\Phi\rangle\langle\Phi|$

(i)

$$\left\{\exists\varphi_1,\ldots,\varphi_N \in \mathfrak{H}, \langle\varphi_i \mid \varphi_j\rangle = \delta_{i,j} : \Phi = a^*(\varphi_1)\ldots a^*(\varphi_N)\Omega\right\} \quad (5.35)$$

$$\Longleftrightarrow \left\{\gamma_\rho^{(1)} = \left(\gamma_\rho^{(1)}\right)^2, \mathrm{Tr}\left\{\gamma_\rho^{(1)}\right\} = N\right\}. \quad (5.36)$$

(ii)

$$(5.36) \Longrightarrow \left\{\gamma_\rho^{(2)} = \frac{1}{2}\left(\gamma_\rho^{(2)}\right)^2 = (1-\mathrm{Ex})\left(\gamma_\rho^{(1)}\otimes\gamma_\rho^{(1)}\right)\right\}, \quad (5.37)$$

where $\mathrm{Ex} \in \mathscr{B}(\mathfrak{H}\otimes\mathfrak{H})$ *is defined by extension by linearity and continuity of*

$$\mathrm{Ex}(\varphi\otimes\psi) := \psi\otimes\varphi. \quad (5.38)$$

Proof: If $\langle\varphi_i \mid \varphi_j\rangle = \delta_{i,j}$ and $\Phi = a^*(\varphi_1)\ldots a^*(\varphi_N)\Omega$ then we add to $\{\varphi_1,\ldots,\varphi_N\}$ suitable orthonormal vectors (e.g., by the Gram-Schmidt procedure) so that $\{\varphi_n\}_{n=1}^\infty \subseteq \mathfrak{H}$ is an ONB and, hence,

$$\begin{aligned}\left\langle\varphi_k \middle| \gamma_\rho^{(1)}\varphi_\ell\right\rangle &= \langle\Phi\,|a^*(\varphi_\ell)\,a(\varphi_k)\,\Phi\rangle = \langle a(\varphi_\ell)\,\Phi \mid a(\varphi_k)\,\Phi\rangle \\ &= \sum_{i,j=1}^{N}(-1)^{i+j}\,\delta_{\ell,i}\delta_{k,j}\underbrace{\left\langle\prod_{\substack{\alpha=1,\\ \alpha\neq i}}^{N} a^*(\varphi_\alpha)\,\Omega\middle|\prod_{\substack{\beta=1,\\ \beta\neq j}}^{N} a^*(\varphi_\beta)\,\Omega\right\rangle}_{=\delta_{i,j}} \\ &= \delta_{\ell,k}\mathbb{1}\left[\ell \le N\right]. \qquad (5.39)\end{aligned}$$

Conversely:

(i) If γ is a rank-N orthogonal projection then there is an ONB $\{\varphi_n\}_{n=1}^{\infty} \subseteq \mathfrak{H}$ of eigenvectors of γ, so $\gamma = \sum_{j=1}^{N} |\varphi_j\rangle\langle\varphi_j|$. Setting $\Phi := a^*(\varphi_1)\cdots a^*(\varphi_N)\Omega$, we have $\gamma = \gamma^{(1)}_{|\Phi\rangle\langle\Phi|}$ in this case.

(ii) If $\Phi = a^*(\varphi_1)\cdots a^*(\varphi_N)\Omega$, then, for all $1 \le k < \ell \le N$, we have

$$\begin{aligned} a_k a_\ell \Phi &= a_k a_\ell a_1^* a_2^* \cdots a_N^* \Omega \\ &= (-1)^{\ell+1} a_k a_1^* a_2^* \cdots a_{\ell-1}^* a_{\ell+1}^* \cdots a_N^* \Omega \\ &= (-1)^{k+\ell} a_1^* \cdots a_{k-1}^* a_{k+1}^* \cdots a_{\ell-1}^* a_{\ell+1}^* \cdots a_N^* \Omega. \end{aligned} \tag{5.40}$$

Thus, for all $1 \le m < n \le N$ and $1 \le k < \ell \le N$,

$$\left\langle \varphi_k \otimes \varphi_\ell \middle| \gamma_\rho^{(2)} (\varphi_m \otimes \varphi_n) \right\rangle = \langle a_m a_n \Phi \mid a_k a_\ell \Phi\rangle = (-1)^{k+\ell+m+n} \delta_{k,m}\delta_{\ell,n}. \tag{5.41}$$

By anticommutation we turn this into

$$\begin{aligned} &\left\langle \varphi_k \wedge \varphi_\ell \middle| \gamma_\rho^{(2)} (\varphi_m \wedge \varphi_n) \right\rangle \\ &= \frac{1}{2} \left\langle (\varphi_k \otimes \varphi_\ell - \varphi_\ell \otimes \varphi_k) \middle| \gamma_\rho^{(2)} (\varphi_m \otimes \varphi_n - \varphi_n \otimes \varphi_m) \right\rangle \\ &= 2\delta_{k,m}\delta_{\ell,n}, \end{aligned} \tag{5.42}$$

for all $k < \ell$ and $m < n$, that is

$$\begin{aligned} \gamma_\rho^{(2)} &= \sum_{1 \le k < \ell \le N} 2|\varphi_k \wedge \varphi_\ell\rangle\langle\varphi_k \wedge \varphi_\ell| \\ &= \sum_{k,\ell=1}^{N} |\varphi_k \wedge \varphi_\ell\rangle\langle\varphi_k \wedge \varphi_\ell|. \end{aligned} \tag{5.43}$$

We finish the proof by comparing (5.43) to

$$\begin{aligned} (1 - \mathrm{Ex})\left(\gamma_\rho^{(1)} \otimes \gamma_\rho^{(1)}\right) &= \sum_{k,\ell=1}^{N} (1 - \mathrm{Ex})\, |\varphi_k \otimes \varphi_\ell\rangle\langle\varphi_k \otimes \varphi_\ell| \\ &= \sum_{k,\ell=1}^{N} |\varphi_k \otimes \varphi_\ell - \varphi_\ell \otimes \varphi_k\rangle\langle\varphi_k \otimes \varphi_\ell| \qquad (5.44) \\ &= \sum_{k,\ell=1}^{N} \frac{1}{2} |\varphi_k \otimes \varphi_\ell - \varphi_\ell \otimes \varphi_k\rangle\langle\varphi_k \otimes \varphi_\ell - \varphi_\ell \otimes \varphi_k| \\ &= \sum_{k,\ell=1}^{N} |\varphi_k \wedge \varphi_\ell\rangle\langle\varphi_k \wedge \varphi_\ell| = \gamma_\rho^{(2)}. \end{aligned}$$

□

Definition 5.4: Let $\underline{Z} \in (\mathbb{R}^+)^K$, $\underline{R} \in (\mathbb{R}^3)^K$, $\mathfrak{H} := L^2(\mathbb{R}^3 \times \mathbb{Z}_2)$, $h := -\Delta_x - \sum_{k=1}^{K} Z_k |x - R_k|^{-1}$, $V(x-y) := |x-y|^{-1}$, and $N \in \mathbb{N}$.
Recalling from (5.7) the definition of H_N, the Hartree-Fock (HF) energy of an atom or molecule is defined as

$$\begin{aligned} &E_{\mathrm{hf}}(N, \underline{Z}, \underline{R}) \\ &:= \inf\Big\{ \langle \Phi | H_N(\underline{Z}, \underline{R}) \Phi \rangle \Big| \ \Phi = a^*(\varphi_1) \cdots a^*(\varphi_N)\, \Omega, \langle \varphi_i \mid \varphi_j \rangle = \delta_{i,j} \Big\}. \end{aligned} \tag{5.45}$$

Remarks:

- For $\Phi = a^*(\varphi_1) \cdots a^*(\varphi_N)\, \Omega$ and $\rho = | \Phi \rangle \langle \Phi |$, we have according to Lemma 5.3 that

$$\begin{aligned} \gamma_\rho^{(1)} &= \sum_{k=1}^{N} |\varphi_k\rangle\langle\varphi_k|, \\ \gamma_\rho^{(2)} &= \sum_{k,\ell=1}^{N} |\varphi_k \wedge \varphi_\ell\rangle\langle\varphi_k \wedge \varphi_\ell| = (1 - \mathrm{Ex})\left(\gamma_\rho^{(1)} \otimes \gamma_\rho^{(1)}\right). \end{aligned} \tag{5.46}$$

  Thus, by (5.34),

$$\begin{aligned} \langle \Phi \mid H_N(\underline{Z}, \underline{R})\, \Phi \rangle &= \mathrm{Tr}\left\{h\gamma_\rho^{(1)}\right\} + \frac{1}{2}\mathrm{Tr}\left\{V\gamma_\rho^{(2)}\right\} \\ &= \mathrm{Tr}\left\{h\gamma_\rho^{(1)}\right\} + \frac{1}{2}\mathrm{Tr}\left\{V(1 - \mathrm{Ex})\left(\gamma_\rho^{(1)} \otimes \gamma_\rho^{(1)}\right)\right\}. \end{aligned} \tag{5.47}$$

  It follows, in particular, that

$$E_{\mathrm{hf}}(N, \underline{Z}, \underline{R}) = \inf\left\{ \mathcal{E}_{\mathrm{hf}}^{(\underline{Z},\underline{R})}(\gamma) \Big| \gamma = \gamma^2 = \gamma^*, \mathrm{Tr}\{\gamma\} = N \right\}, \tag{5.48}$$

  where the HF functional is defined by

$$\mathcal{E}_{\mathrm{hf}}^{(\underline{Z},\underline{R})}(\gamma) := \mathrm{Tr}\{h\gamma\} + \frac{1}{2}\mathrm{Tr}\{V(1 - \mathrm{Ex})(\gamma \otimes \gamma)\}. \tag{5.49}$$

Lemma 5.5: [Lieb's Variational Principle]

$$E_{\mathrm{hf}}(N, \underline{Z}, \underline{R}) = \inf\left\{ \mathcal{E}_{\mathrm{hf}}^{(\underline{Z},\underline{R})}(\gamma) \Big| \ 0 \le \gamma \le 1, \mathrm{Tr}\{\gamma\} = N \right\}. \tag{5.50}$$

Proof: Let $0 \leq \gamma \leq \mathbb{1}$ with $\mathrm{Tr}\{\gamma\} = N$ and $\gamma \neq \gamma^2$. By invoking a limiting argument, we may assume that γ is of finite rank, $N < \mathrm{rk}(\gamma) =: M < \infty$. Then there exist an ONB $\{\varphi_n\}_{n=1}^{\infty} \subseteq \mathfrak{H}$ and positive numbers $\lambda_1, \lambda_2, \ldots, \lambda_M \in (0,1]$ summing up to $\sum_{m=1}^{M} \lambda_m = N$ such that $\gamma = \sum_{m=1}^{M} \lambda_m |\varphi_m\rangle\langle\varphi_m|$. After rearranging the order, if necessary, we may assume that

$$\forall 1 \leq n \leq M-1: \quad P_n \leq P_{n+1}. \tag{5.51}$$

where

$$\begin{aligned} P_n &:= h_n + \sum_{m=1}^{M} \lambda_m V_{m,n}, \\ h_n &:= \langle \varphi_n \mid h\varphi_n \rangle, \\ V_{m,n} &:= \langle \varphi_m \wedge \varphi_n \mid V \varphi_m \wedge \varphi_n \rangle. \end{aligned} \tag{5.52}$$

Hence, the evaluation of the HF functional on γ equals

$$\mathcal{E}_{\mathrm{hf}}^{(\underline{Z},\underline{R})}(\gamma) = \sum_{n=1}^{M} \lambda_n h_n + \frac{1}{2} \sum_{m,n=1}^{M} \lambda_m \lambda_n V_{m,n}. \tag{5.53}$$

Since $M > N$, there exist $1 \leq k < \ell \leq M$ such that

$$0 < \lambda_k < 1, \; 0 < \lambda_\ell < 1. \tag{5.54}$$

For sufficiently small $\delta > 0$ we then have

$$0 \leq \gamma_\delta := \gamma + \delta\left(|\,\varphi_k\,\rangle\langle\,\varphi_k\,| - |\,\varphi_\ell\,\rangle\langle\,\varphi_\ell\,|\right) \leq \mathbb{1}. \tag{5.55}$$

Furthermore, we have

$$\begin{aligned} &\mathcal{E}_{\mathrm{hf}}^{(\underline{Z},\underline{R})}(\gamma_\delta) - \mathcal{E}_{\mathrm{hf}}^{(\underline{Z},\underline{R})}(\gamma) \\ &= \delta\,(h_k - h_\ell) \\ &\qquad + \frac{1}{2} \sum_{m,n=1}^{M} V_{m,n} \Big\{ [\lambda_m + \delta\,(\delta_{mk} - \delta_{m\ell})]\,[\lambda_n + \delta\,(\delta_{nk} - \delta_{n\ell})] - \lambda_m \lambda_n \Big\} \\ &= \underbrace{\delta(P_k - P_\ell)}_{\leq 0} - \delta^2 \underbrace{V_{k,\ell}}_{>0} < 0, \end{aligned} \tag{5.56}$$

as $V_{k,k} = V_{\ell,\ell} = 0$. Choosing $\delta := \min\{1 - \lambda_k, \lambda_\ell\} > 0$, the rank of γ_δ is strictly lower than the rank M of γ,

$$\mathrm{rk}(\gamma_\delta) \leq \mathrm{rk}(\gamma) - 1. \tag{5.57}$$

Iterating this procedure, we arrive after (at most) $M - N$ such steps at a rank-N projection $\widetilde{\gamma}$ with

$$\mathcal{E}_{\mathrm{hf}}^{(\underline{Z},\underline{R})}(\widetilde{\gamma}) < \mathcal{E}_{\mathrm{hf}}^{(\underline{Z},\underline{R})}(\gamma). \tag{5.58}$$

$\square$

6. Correlation Estimate Improving the Lieb-Oxford Inequality

The goal of this section is the derivation of a correlation estimate which improves the Lieb-Oxford inequality. The precise formulation of this estimate is the content of Theorem 6.7 below. To this end we choose a fixed density matrix $\rho \in \mathcal{L}_+^1(\mathcal{F}_f[\mathfrak{H}])$ with $[\rho, \check{\mathbb{N}}] = 0$ (ensuring that pairing terms of the form $\mathrm{Tr}\{\rho a^*(\varphi)a^*(\psi)\} = 0$ vanish) and recall that the expectation value of an observable $A \in \mathcal{B}(\mathcal{F}_f[\mathfrak{H}])$ is given by

$$\langle A \rangle_\rho := \mathrm{Tr}\left\{\rho^{\frac{1}{2}} A \rho^{\frac{1}{2}}\right\}. \tag{6.1}$$

We further assume that $\langle \check{\mathbb{N}} \rangle_\rho < \infty$, hence $\gamma_\rho^{(1)} \in \mathcal{L}_+^1(\mathfrak{H})$. Consequently, there exists an ONB $\{\varphi_n\}_{n=1}^\infty \subseteq \mathfrak{H}$ of eigenvectors of $\gamma_\rho^{(1)}$ with corresponding eigenvalues $1 \geq \lambda_1 \geq \lambda_2 \geq \cdots \geq 0$, so that

$$\gamma_\rho^{(1)} = \sum_{n=1}^\infty \lambda_n |\varphi_n\rangle\langle\varphi_n|, \tag{6.2}$$

or, equivalently

$$\langle a_k^* \, a_\ell \rangle_\rho = \lambda_k \delta_{k,\ell}, \tag{6.3}$$

where $a_k^* := a^*(\varphi_k)$, $a_\ell := a(\varphi_\ell)$. Finally, we denote by $\gamma_{\rho,T}^{(2)}$ the truncated 2-pdm,

$$\gamma_{\rho,T}^{(2)} := \gamma_\rho^{(2)} - (1 - \mathrm{Ex})\left(\gamma_\rho^{(1)} \otimes \gamma_\rho^{(1)}\right) \tag{6.4}$$

$$\text{and } M_{k\ell} := \langle \varphi_k \mid M \varphi_\ell \rangle, \tag{6.5}$$

for any $M \in \mathcal{B}(\mathfrak{H})$.

Lemma 6.1: *Let $X = X^* = X^2 \in \mathcal{B}(\mathfrak{H})$ be an orthogonal projection. Then*

$$\mathrm{Tr}\left\{(X \otimes X)\, \gamma_{\rho,T}^{(2)}\right\} \geq -\mathrm{Tr}\left\{X \gamma_\rho^{(1)}\right\}. \tag{6.6}$$

Proof: A computation gives

$$\begin{aligned}
&\mathrm{Tr}\left\{(X\otimes X)\,\gamma^{(2)}_{\rho,T}\right\}\\
&=\sum_{k,\ell;m,n} X_{km}X_{\ell n}\big(\langle a^*_\ell a^*_k a_m a_n\rangle_\rho - \langle a^*_k a_m\rangle_\rho\,\langle a^*_\ell a_n\rangle_\rho + \langle a^*_k a_n\rangle_\rho\,\langle a^*_\ell a_m\rangle_\rho\big)\\
&=\left\langle\left(\sum_{km} X_{km}\left\{a^*_k a_m - \langle a^*_k a_m\rangle_\rho\right\}\right)^*\left(\sum_{km} X_{km}\left\{a^*_k a_m - \langle a^*_k a_m\rangle_\rho\right\}\right)\right\rangle_\rho\\
&\qquad-\sum_{k,\ell;m,n} X_{km}X_{\ell n}\delta_{\ell,m}\underbrace{\langle a^*_k a_n\rangle_\rho}_{=\delta_{k,n}\cdot\lambda_k}+\sum_{k,\ell;m,n} X_{km}X_{\ell n}\underbrace{\langle a^*_k a_n\rangle_\rho}_{=\delta_{k,n}\cdot\lambda_k}\underbrace{\langle a^*_\ell a_m\rangle_\rho}_{=\delta_{\ell,m}\cdot\lambda_\ell}\\
&\geq\sum_{k,\ell} -\left|X_{k\ell}\right|^2\underbrace{(1-\lambda_\ell)}_{\leq 1}\lambda_k \geq \sum_{k,\ell} -\left|X_{k\ell}\right|^2\lambda_k\\
&=-\sum_k \left(X^2\right)_{kk}\lambda_k = -\mathrm{Tr}\left\{X^2\gamma^{(1)}_\rho\right\} = -\mathrm{Tr}\{X\gamma^{(1)}_\rho\}. \qquad (6.7)
\end{aligned}$$

□

We introduce some further notation, namely,

$$\sum_{k<\frac12} r_k := \sum_{k:\lambda_k\leq\frac12} r_k, \qquad \sum_{k>\frac12} r_k := \sum_{k:\lambda_k>\frac12} r_k \qquad (6.8)$$

and

$$\begin{aligned}
c_p &:= \sum_{k<\frac12} X_{pk}a_k, \qquad d^*_p := \sum_{k>\frac12} X_{pk}a_k,\\
A_p &:= c_p + d^*_p = \sum_k X_{pk}a_k. \qquad (6.9)
\end{aligned}$$

Moreover, we henceforth write

$$\gamma = \gamma^{(1)}_\rho, \qquad \gamma_T := \gamma - \gamma^2. \qquad (6.10)$$

Since $X = X^2$ we then have

$$\begin{aligned}
\mathrm{Tr}\left\{(X\otimes X)\,\gamma^{(2)}_\rho\right\} &= \sum_{k,\ell;m,n} X_{km}X_{\ell n}\,\langle a^*_\ell a^*_k a_m a_n\rangle_\rho\\
&=\sum_{k,\ell;m,n}\sum_{p,q} X_{kq}X_{qm}X_{\ell p}X_{pn}\,\langle a^*_\ell a^*_k a_m a_n\rangle_\rho\\
&=\sum_{p,q}\left\langle A^*_p A^*_q A_q A_p\right\rangle_\rho\\
&=\sum_{p,q}\left\langle\left(c^*_p + d_p\right)A^*_q\left(c_q + d^*_q\right)A_p\right\rangle_\rho. \qquad (6.11)
\end{aligned}$$

We divide the sum into its main part MP and a remainder R as

$$\begin{aligned}
\mathrm{MP} &:= \sum_{p,q} \left\langle d_p A_q^* d_q^* A_p \right\rangle_\rho, \\
\mathrm{R} &= \sum_{p,q} \{ \left\langle c_p^* A_q^* c_q A_p \right\rangle_\rho + \left\langle c_p^* A_q^* d_q^* A_p \right\rangle_\rho + \left\langle d_p A_q^* c_q A_p \right\rangle_\rho \} \\
&= \sum_{p,q} \{ \left\langle c_p^* A_q^* c_q A_p \right\rangle_\rho + 2\mathrm{Re} \left\langle d_p A_q^* c_q A_p \right\rangle_\rho \}, \qquad (6.12)
\end{aligned}$$

so that

$$\mathrm{Tr}\left\{ (X \otimes X)\, \gamma_\rho^{(2)} \right\} = \mathrm{MP} + \mathrm{R}. \qquad (6.13)$$

Lemma 6.2:

$$\mathrm{MP} - \mathrm{Tr}\left\{ (X \otimes X)(1 - \mathrm{Ex})(\gamma \otimes \gamma) \right\} \geq -6\mathrm{Tr}\{X\gamma\}\,\mathrm{Tr}\{X\gamma_T\}. \qquad (6.14)$$

Proof: We omit the subscript ρ and write $\langle A \rangle_\rho =: \langle A \rangle$ throughout the proof.

$$\begin{aligned}
&\mathrm{Tr}\left\{ (X \otimes X)(1 - \mathrm{Ex})(\gamma \otimes \gamma) \right\} \\
&= \sum_{k,\ell;m,n} X_{km} X_{\ell n} \left\{ \langle a_k^* a_m \rangle \langle a_\ell^* a_n \rangle - \langle a_k^* a_n \rangle \langle a_\ell^* a_m \rangle \right\} \\
&= \sum_{k,\ell;m,n} X_{km} X_{\ell n} \lambda_\ell \lambda_k \left(\delta_{km} \delta_{\ell,n} - \delta_{k,n} \delta_{\ell,m} \right) \\
&= \sum_{k,\ell} \left(X_{kk} X_{\ell\ell} - |X_{k\ell}|^2 \right) \lambda_k \lambda_\ell. \qquad (6.15)
\end{aligned}$$

Furthermore, using $\sum_p |X_{kp}|^2 = (X^2)_{kk} = X_{kk}$ we infer that

$$\begin{aligned}
\left\langle \sum_p d_p A_p \right\rangle &= \sum_p \left\{ \underbrace{\langle d_p c_p \rangle}_{=0} + \langle d_p d_p^* \rangle \right\} \\
&= \sum_{k,k' > \frac{1}{2}} \sum_p X_{kp} X_{pk'} \underbrace{\langle a_k^* a_{k'} \rangle}_{=\lambda_k \delta_{k,k'}} \\
&= \sum_{k > \frac{1}{2}} \sum_p |X_{kp}|^2 \lambda_k = \sum_{k > \frac{1}{2}} X_{kk} \lambda_k. \qquad (6.16)
\end{aligned}$$

Thus we obtain

$$
\begin{aligned}
\mathrm{MP} &= -\sum_{p,q} \left\{ d_p, d_q^* \right\} \left\langle A_p^* A_q \right\rangle + \left\langle \left(\sum_p d_p A_p \right)^* \left(\sum_p d_p A_p \right) \right\rangle \\
&\geq \left| \left\langle \sum_p d_p A_p \right\rangle \right|^2 - \sum_{p,q} \sum_{k,k'>\frac{1}{2}} \sum_{\ell,\ell'} X_{kp} X_{qk'} \delta_{k,k'} X_{p\ell} X_{\ell' q} \lambda_\ell \delta_{\ell\ell'} \\
&= \left(\sum_{k>\frac{1}{2}} X_{kk} \lambda_k \right)^2 - \sum_{k>\frac{1}{2}} \sum_{\ell} \left| X_{k\ell} \right|^2 \lambda_\ell, \qquad (6.17)
\end{aligned}
$$

and we finally arrive at

$$
\begin{aligned}
&\mathrm{MP} - \mathrm{Tr}\left\{ (X \otimes X)(1 - \mathrm{Ex})(\gamma \otimes \gamma) \right\} \\
&\geq - \left\{ \left(\sum_k X_{kk} \lambda_k \right)^2 - \left(\sum_{k>\frac{1}{2}} X_{kk} \lambda_k \right)^2 \right\} - \sum_{k>\frac{1}{2},\ell} \left| X_{k\ell} \right|^2 \lambda_\ell (1 - \lambda_k) \\
&\geq -2 \left(\sum_{k<\frac{1}{2}} X_{kk} \lambda_k \right) \left(\sum_k X_{kk} \lambda_k \right) - \sum_{k>\frac{1}{2},\ell} X_{kk} X_{\ell\ell} \lambda_\ell (1 - \lambda_k) \\
&\geq -4 \left(\sum_k X_{kk} \left(\lambda_k - \lambda_k^2 \right) \right) \left(\sum_k X_{kk} \lambda_k \right) \\
&\qquad - 2 \left(\sum_k X_{kk} \left(\lambda_k - \lambda_k^2 \right) \right) \left(\sum_\ell X_{\ell\ell} \lambda_\ell \right), \qquad (6.18)
\end{aligned}
$$

where we use $\lambda_k \leq 2\left(\lambda_k - \lambda_k^2\right)$, if $\lambda_k \leq \frac{1}{2}$, and $1 - \lambda_\ell \leq 2\left(\lambda_\ell - \lambda_\ell^2\right)$, for $\lambda_\ell \geq \frac{1}{2}$.

□

Next we write

$$
\begin{aligned}
\sum_{p,q} \left\langle c_p^* A_q^* c_q A_p \right\rangle &= \sum_{p,q} \left\langle c_p^* \left(c_q^* + d_q \right) c_q \left(c_p + d_p^* \right) \right\rangle \\
&= \sum_{p,q} \left\{ \left\langle c_p^* c_q^* c_q c_p \right\rangle + \left\langle c_p^* c_q^* c_q d_p^* \right\rangle + \left\langle c_p^* d_q c_q c_p \right\rangle + \left\langle c_p^* d_q c_q d_p^* \right\rangle \right\} \\
&= \sum_{p,q} \left\{ \left\langle c_p^* c_q^* c_q c_p \right\rangle + 2\mathrm{Re} \left\langle c_p^* d_q c_q c_p \right\rangle + \left\langle c_p^* d_q c_q d_p^* \right\rangle \right\} \qquad (6.19)
\end{aligned}
$$

and

$$\sum_{p,q} \langle d_p A_q^* c_q A_p \rangle = \sum_{p,q} \langle d_p \left(c_q^* + d_q\right) c_q \left(c_p + d_p^*\right) \rangle$$
$$= \sum_{p,q} \left\{ \langle d_p c_q^* c_q c_p \rangle + \langle d_p d_q c_q c_p \rangle + \langle d_p c_q^* c_q d_p^* \rangle + \langle d_p d_q c_q d_p^* \rangle \right\}. \tag{6.20}$$

With this we obtain

$$\mathrm{R} = \sum_{p,q} \Big\{ \langle c_p^* c_q^* c_q c_p \rangle + 4\mathrm{Re} \langle c_p^* d_q c_q c_p \rangle + \langle c_p^* d_q c_q d_p^* \rangle$$
$$+ 2\mathrm{Re} \langle d_p d_q c_q c_p \rangle + 2 \langle d_p c_q^* c_q d_p^* \rangle + 2\mathrm{Re} \langle d_p d_q c_q d_p^* \rangle \Big\}. \tag{6.21}$$

Lemma 6.3:

$$\mathrm{R} \geq -20\mathrm{Tr}\{X\gamma\}\,\mathrm{Tr}\{X\gamma_T\} + 2\mathrm{Re}\left(\sum_{p,q} \langle d_p d_q c_q c_p \rangle\right). \tag{6.22}$$

Proof: First, for all $\varepsilon > 0$, we have

$$\left|\sum_{p,q} 4\mathrm{Re} \langle c_p^* d_q c_q c_p \rangle\right| \leq 4 \left(\sum_{p,q} \langle c_p^* d_q d_q^* c_p \rangle\right)^{1/2} \left(\sum_{p,q} \langle c_p^* c_q^* c_q c_p \rangle\right)^{1/2}$$
$$\leq \frac{2}{\varepsilon} \sum_{p,q} \langle c_p^* d_q d_q^* c_p \rangle + 2\varepsilon \sum_{p,q} \langle c_p^* c_q^* c_q c_p \rangle, \tag{6.23}$$

and

$$\left|\sum_{p,q} \langle c_p^* d_q c_q d_p^* \rangle\right| = \left|\sum_{p,q} \langle c_p^* d_q d_p^* c_q \rangle\right| \leq \sum_{p,q} \langle c_p^* d_q d_q^* c_p \rangle. \tag{6.24}$$

Moreover,

$$\left|2\mathrm{Re}\left(\sum_{p,q} \langle d_p d_q c_q d_p^* \rangle\right)\right|$$
$$= \left|2\mathrm{Re}\left\{\sum_{p,q} \left(-\{d_q, d_p^*\} \underbrace{\langle d_p c_q \rangle}_{=0} + \{d_p, d_p^*\} \underbrace{\langle d_q c_q \rangle}_{=0}\right) + \langle d_p^* d_q d_p c_q \rangle \right\}\right|$$
$$= \left|2\mathrm{Re} \sum_{p,q} \langle d_p^* d_q d_p c_q \rangle\right| \leq \sum_{p,q} \langle d_p^* d_q d_q^* d_p \rangle + \sum_{p,q} \langle c_q^* d_p^* d_p c_q \rangle. \tag{6.25}$$

Inserting (6.23)-(6.25) into (6.21), we conclude

$$\begin{aligned}
&\mathrm{R} - \sum_{p,q} 2\mathrm{Re}\,\langle d_q d_p c_p c_q \rangle \\
&= \sum_{p,q} \Big\{ \langle c_p^* c_q^* c_q c_p \rangle + 4\mathrm{Re}\,\langle c_p^* d_q c_q c_p \rangle + \langle c_p^* d_q c_q d_p^* \rangle \\
&\qquad + 2\,\langle d_p c_q^* c_q d_p^* \rangle + 2\mathrm{Re}\,\langle d_p d_q c_q d_p^* \rangle \Big\} \\
&\geq (-4 - 1 + 2 - 1) \sum_{p,q} \langle c_p^* d_q d_q^* c_p \rangle - \sum_{p,q} \langle d_p^* d_q d_q^* d_p \rangle \\
&= -4 \sum_{p,q} \langle c_p^* d_q d_q^* c_p \rangle - \sum_{p,q} \langle d_p^* d_q d_q^* d_p \rangle \\
&\geq -4 \sum_{k > \frac{1}{2}} \sum_{\ell < \frac{1}{2}} X_{kk} X_{\ell\ell} \lambda_\ell - \sum_{k,\ell > \frac{1}{2}} X_{kk} X_{\ell\ell} (1 - \lambda_\ell) \\
&\geq -20 \left(\sum_k X_{kk} \lambda_k \right) \left(\sum_\ell X_{\ell\ell} (\lambda_\ell - \lambda_\ell^2) \right) \\
&= -20 \mathrm{Tr}\,\{X\gamma\}\, \mathrm{Tr}\,\{X\gamma_T\}\,.
\end{aligned} \tag{6.26}$$

Again using $1 \leq 2\lambda_k$, for $\lambda_k \geq \frac{1}{2}$, and $\lambda_\ell \leq 2(\lambda_\ell - \lambda_\ell^2)$, for $\lambda_\ell \leq \frac{1}{2}$.

□

Lemma 6.4:

$$\left| 2\mathrm{Re} \sum_{p,q} \langle d_q d_p c_p c_q \rangle \right| \leq 6\, \mathrm{Tr}\,\{X\gamma\}\, (1 + 4\mathrm{Tr}\,\{X\gamma_T\})^{1/2} (\mathrm{Tr}\,\{X\gamma_T\})^{1/2}. \tag{6.27}$$

Proof: We have

$$\begin{aligned}
\left| 2\mathrm{Re} \sum_{p,q} \langle d_q d_p c_p c_q \rangle \right| &= \left| 2\mathrm{Re} \left\langle \left(\sum_p d_p c_p \right)^2 \right\rangle \right| \\
&\leq 2 \left\langle \sum_{p,q} c_p^* d_p^* d_q c_q \right\rangle^{1/2} \left\langle \sum_{p,q} d_p c_p c_q^* d_q^* \right\rangle^{1/2},
\end{aligned} \tag{6.28}$$

and

$$\begin{aligned}\sum_{p,q}\left\langle c_p^* d_p^* d_q c_q\right\rangle &= \sum_{p,q}\left\{d_p^*, d_q\right\}\left\langle c_p^* c_q\right\rangle - \sum_{p,q}\left\langle c_p^* d_q d_p^* c_q\right\rangle \\ &\leq \sum_{k>\frac{1}{2}}\sum_{\ell<\frac{1}{2}}\left|X_{k\ell}\right|^2 \lambda_\ell + \sum_{p,q}\left\langle c_p^* d_q d_p^* c_q\right\rangle \\ &\leq 8\left(\sum_k X_{kk}\lambda_k\right)\left(\sum_\ell X_{\ell\ell}\left(\lambda_\ell - \lambda_\ell^2\right)\right) \\ &= 8\mathrm{Tr}\left\{X\gamma\right\}\mathrm{Tr}\left\{X\gamma_T\right\}, \end{aligned} \tag{6.29}$$

as well as,

$$\begin{aligned}\sum_{p,q}\left\langle d_p c_p c_q^* d_q^*\right\rangle &= \sum_{p,q}\left\langle c_p d_p d_q^* c_q^*\right\rangle \\ &= \sum_{p,q}\left\{d_p, d_q^*\right\}\left\langle c_p c_q^*\right\rangle + \sum_{p,q}\left\langle d_p^* c_p d_p c_q^*\right\rangle \\ &= \sum_{k>\frac{1}{2}}\sum_{\ell<\frac{1}{2}}\left|X_{k\ell}\right|^2(1-\lambda_\ell) - \sum_{p,q}\left\{c_p, c_q^*\right\}\left\langle d_q^* d_p\right\rangle \\ &\quad + \sum_{p,q}\left\langle c_q^* d_q^* d_p c_p\right\rangle \\ &= \sum_{k>\frac{1}{2}}\sum_{l<\frac{1}{2}}\left|X_{k\ell}\right|^2\left\{(1-\lambda_l)-(1-\lambda_k)\right\} \\ &\quad + \sum_{k>\frac{1}{2}}\sum_{\ell<\frac{1}{2}}\left|X_{k\ell}\right|^2\lambda_\ell - \sum_{p,q}\left\langle c_q^* d_p d_q^* c_p\right\rangle \\ &\leq \sum_{k>\frac{1}{2}}\sum_{\ell<\frac{1}{2}}\left|X_{k\ell}\right|^2\lambda_k + 4\mathrm{Tr}\left\{X\gamma\right\}\mathrm{Tr}\left\{X\gamma_T\right\} \\ &= \mathrm{Tr}\left\{X\gamma\right\}\left(1+4\mathrm{Tr}\left\{X\gamma_T\right\}\right). \end{aligned} \tag{6.30}$$

□

We summarize Lemmata 6.1-6.4 in the following theorem.

Theorem 6.5: *If* $\rho \in \mathscr{B}(\mathcal{F}_f[\mathfrak{H}])$ *is a particle number preserving density matrix,* $[\rho, \check{\mathbb{N}}] = 0$, *and* $X = X^* = X^2 \in \mathscr{B}(\mathfrak{H})$ *an orthogonal projection* *then*

$$\mathrm{Tr}\left\{(X\otimes X)\,\gamma^{(2)}_{\rho,T}\right\}$$
$$\geq -\mathrm{Tr}\left\{X\gamma^{(1)}_{\rho}\right\}\min\left\{1,\ 12\left(\mathrm{Tr}\left\{X\left(\gamma^{(1)}_{\rho}-\left(\gamma^{(1)}_{\rho}\right)^{2}\right)\right\}\right)^{1/2}\right\}. \quad (6.31)$$

Proof: According to Lemmata 6.2-6.4, we have

$$\mathrm{Tr}\left\{(X\otimes X)\,\gamma^{(2)}_{\rho,T}\right\} \geq -\,\mathrm{Tr}\left\{X\gamma\right\}\left(\mathrm{Tr}\left\{X\gamma_T\right\}\right)^{1/2}$$
$$\left[(6+20)\left(\mathrm{Tr}\left\{X\gamma_T\right\}\right)^{1/2}+6\left(1+4\mathrm{Tr}\left\{X\gamma_T\right\}\right)^{1/2}\right] \quad (6.32)$$

and furthermore, by Lemma 6.1,

$$\mathrm{Tr}\left\{(X\otimes X)\,\gamma^{(2)}_{\rho}\right\} \geq -\mathrm{Tr}\left\{X\gamma\right\}. \quad (6.33)$$

Setting $r := \left(\mathrm{Tr}\left\{X\gamma_T\right\}\right)^{1/2} \geq 0$, we may assume

$$26r^2 \leq r\left[26r+6\left(1+4r^2\right)^{1/2}\right] \leq 1, \quad (6.34)$$

thanks to (6.33), which in particular implies that $r \leq \frac{1}{\sqrt{26}}$ and thus

$$26r+6\left(1+4r^2\right)^{1/2} \leq \sqrt{26}+6\left(1+\frac{4}{26}\right)^{1/2} = \frac{26+6\sqrt{30}}{\sqrt{26}} \leq 12. \quad (6.35)$$

We finally obtain

$$\mathrm{Tr}\left\{(X\otimes X)\,\gamma^{(2)}_{\rho,T}\right\} \geq -\mathrm{Tr}\left\{X\gamma\right\}\min\left\{1,\ 12\left(\mathrm{Tr}\left\{X\gamma_T\right\}\right)^{1/2}\right\}. \quad (6.36)$$

□

We combine Theorem 6.5 with the Fefferman-de la Llave identity

$$\frac{1}{|x-y|} = \int_0^\infty \frac{dr}{\pi r^5}\int d^3z\,\mathbb{1}_{B(z,r)}(x)\mathbb{1}_{B(z,r)}(y), \quad (6.37)$$

by choosing

$$X = X_{r,z} := \mathbb{1}_{B(z,r)} \quad (6.38)$$

in Theorem 6.5 and superimposing the resulting estimates. This yields the following estimate.

Theorem 6.6: *For any particle number preserving density matrix* $\rho \in \mathscr{B}(\mathcal{F}_f[\mathfrak{H}])$, $[\rho, \check{\mathbb{N}}] = 0$,

$$\frac{1}{2}\mathrm{Tr}\left\{\frac{1}{|x-y|}\gamma^{(2)}_{\rho,T}\right\} \geq -C\left(\int f^{5/3}(x)\,d^3x\right)^{1/2}\left(\int f(x)\,d^3x\right) \cdot \left(\mathrm{Tr}\{\gamma^{(1)}_\rho\}\right)^{-\frac{1}{4}} \cdot \left(\mathrm{Tr}\{\gamma^{(1)}_\rho - (\gamma^{(1)}_\rho)^2\}\right)^{\frac{1}{4}} \tag{6.39}$$

where $f(x) := \gamma^{(1)}_\rho(x,x)$ *is the one-particle density of* ρ.

Proof: Using (6.37) with (6.38) and superimposing the estimate of Theorem 6.5, we obtain

$$\frac{1}{2}\mathrm{Tr}\left\{\frac{1}{|x-y|}\gamma^{(2)}_{\rho,T}\right\} = \frac{1}{2\pi}\int d^3z \int_0^\infty \frac{dr}{r^5}\mathrm{Tr}\left\{(X_{r,z}\otimes X_{r,z})\,\gamma^{(2)}_{\rho,T}\right\} \tag{6.40}$$

$$\geq -\frac{1}{2\pi}\int d^3z \int_0^\infty \frac{dr}{r^5}\left(\mathrm{Tr}\left\{X_{r,z}\gamma^{(1)}_\rho\right\}\right)\min\left\{1,\ 12\left(\mathrm{Tr}\left\{X_{r,z}\gamma^{(1)}_{\rho,T}\right\}\right)^{1/2}\right\}$$

$$\geq -\frac{1}{2\pi}\int d^3z \left(12\int_0^{R(z)} \frac{dr}{r^5}\left(\mathrm{Tr}\left\{X_{r,z}\gamma^{(1)}_\rho\right\}\right)\left(\mathrm{Tr}\left\{X_{r,z}\gamma^{(1)}_{\rho,T}\right\}\right)^{1/2} + \int_{R(z)}^\infty \frac{dr}{r^5}\left(\mathrm{Tr}\left\{X_{r,z}\gamma^{(1)}_\rho\right\}\right)\right).$$

Now, we introduce

$$\begin{aligned}
f(x) &:= \gamma^{(1)}_\rho(x,x),\\
f_T(x) &:= \gamma^{(1)}_{\rho,T}(x,x) = \gamma^{(1)}_\rho(x,x) - \left(\gamma^{(1)}_\rho\right)^2(x,x),\\
M(z) &:= \sup_{r>0}\left\{\frac{3}{4\pi r^3}\int_{B(z,r)} f(x)\,d^3x\right\},\\
M_T(z) &:= \sup_{r>0}\left\{\frac{3}{4\pi r^3}\int_{B(z,r)} f_T(x)\,d^3x\right\},
\end{aligned} \tag{6.41}$$

and infer from (6.40) that

$$\begin{aligned}
&\frac{1}{2}\mathrm{Tr}\left\{\frac{1}{|x-y|}\gamma^{(2)}_{\rho,T}\right\} \\
&\geq -\frac{1}{2\pi}\int d^3z\left[12\int_0^{R(z)}\frac{dr}{r^5}\left(\int_{B(z,r)} f(x)\,d^3x\right)\left(\int_{B(z,r)} f_T(x)\,d^3x\right)^{1/2}\right. \\
&\qquad\qquad \left. + \int_{R(z)}^{\infty}\frac{dr}{r^5}\left(\int_{B(z,r)} f(x)\,d^3x\right)\right] \\
&\geq -\frac{1}{2\pi}\int d^3z\left[12\left(\frac{4\pi}{3}\right)^{3/2} M(z)\,M_T^{1/2}(z)\left(\int_0^{R(z)}\frac{dr}{\sqrt{r}}\right)\right. \\
&\qquad\qquad \left. + \left(\frac{4\pi}{3}\right) M(z)\left(\int_{R(z)}^{\infty}\frac{dr}{r^2}\right)\right] \\
&= -\frac{1}{2\pi}\frac{4\pi}{3}\int\left(M(z)\left[12\left(\frac{4\pi}{3}\right)^{1/2} M_T^{1/2}(z)\cdot 2R(z)^{1/2} + R(z)^{-1}\right]\right) d^3z.
\end{aligned} \tag{6.42}$$

Choosing $R(z)$ as

$$\begin{aligned}
12\left(\frac{4\pi}{3}M_T(z)\right)^{1/2} R(z)^{1/2} &= R(z)^{-1} \\
\iff R(z) &= (12)^{-\frac{2}{3}}\left(\frac{4\pi}{3}M_T(z)\right)^{-\frac{1}{3}},
\end{aligned} \tag{6.43}$$

we hence obtain

$$\frac{1}{2}\mathrm{Tr}\left\{\frac{1}{|x-y|}\gamma^{(2)}_{\rho,T}\right\} \geq -2\left(\frac{4\pi}{3}\right)^{\frac{1}{3}}(12)^{\frac{2}{3}}\int M(z)\,M_T^{1/3}(z)\,d^3z. \tag{6.44}$$

An application of Hölder's inequality yields

$$\int M(z)\,M_T^{1/3}(z)\,d^3z \leq \left(\int M^p(z)\,d^3z\right)^{1/p}\left(\int M_T^{q/3}(z)\,d^3z\right)^{1/q}, \tag{6.45}$$

where $\frac{1}{p} + \frac{1}{q} = 1$ and $3 < q \leq 4$. Next we apply the Hardy-Littlewood maximal inequality and arrive at

$$\int M(z)\, M_T^{1/3}(z)\, d^3z \leq C_q \left(\int f^p(x)\, d^3x \right)^{1/p} \left(\int f_T^{q/3}(x)\, d^3x \right)^{1/q}, \tag{6.46}$$

where $C_q < \infty$ is a constant. Furthermore, for any $\alpha, \beta > 0$ obeying $\frac{1}{\alpha} + \frac{1}{\beta} = 1$ and $0 < t < 1$, we have

$$\begin{aligned} \int f_T^{q/3}(x)\, d^3x &= \int f_T^{qt/3}(x)\, f_T^{q(1-t)/3}(x)\, d^3x \\ &\leq \left(\int f_T^{q\alpha t/3}(x)\, d^3x \right)^{1/\alpha} \left(\int f_T^{q\beta(1-t)/3}(x)\, d^3x \right)^{1/\beta}, \end{aligned} \tag{6.47}$$

and, analogously,

$$\int f^p(x)\, d^3x \leq \left(\int f^{\eta ps}(x)\, d^3x \right)^{1/\eta} \left(\int f^{\kappa p(1-s)}(x)\, d^3x \right)^{1/\kappa}, \tag{6.48}$$

with $\frac{1}{\eta} + \frac{1}{\kappa} = 1$ and $0 < s < 1$. The optimal choice of $\alpha, \beta, \eta, \kappa, s, t$ under the condition that

$$\frac{\alpha qt}{3} = \eta ps = \frac{5}{3} \quad \text{and} \quad \frac{\beta q(1-t)}{3} = \kappa p(1-s) = 1, \tag{6.49}$$

now yields the assertion. □

We refrain from actually computing the optimum and rather simply choose

$$q := 4, \quad p = \frac{4}{3}, \quad \alpha = \beta = \eta = \kappa = 2, \quad s = t = \frac{5}{8}, \tag{6.50}$$

which yields

$$\begin{aligned} \frac{1}{c_4} \int M(z)\, M_T^{1/3}(z)\, d^3z & \\ \leq \left(\int f^{5/3}(x)\, d^3x \right)^{3/8} & \left(\int f(x)\, d^3x \right)^{3/8} \\ & \cdot \left(\int f_T^{5/3}(x)\, d^3 \right)^{1/8} \left(\int f_T(x)\, d^3x \right)^{1/8} \\ \leq \left(\int f^{5/3}(x)\, d^3x \right)^{1/2} & \left(\int f(x)\, d^3x \right)^{1/2} \left[\frac{\int f_T(x)\, d^3}{\int f(x)\, d^3x} \right]^{1/4} \end{aligned} \tag{6.51}$$

additionally observing that $f_T \leq f$. As a special case we obtain the following result:

Theorem 6.7: *Let* $\Psi \in \bigwedge^N \mathfrak{H}$ *with* $\|\Psi\| = 1$ *and* $\int \rho_\Psi^{5/3} < \infty$. *Then there is a universal constant* $C < \infty$, *such that*

$$\left\langle \Psi \middle| \sum_{1 \le m < n \le N} \frac{1}{|x_m - x_n|} \Psi \right\rangle \ge \frac{1}{2} \int \left(\rho_\psi(x)\, \rho_\psi(y) - \left| \gamma_\Psi^{(1)}(x,y) \right|^2 \right) \frac{d^3x d^3y}{|x-y|}$$
$$- C \left(\int \rho_\psi^{5/3}(x)\, d^3x \right)^{\frac{1}{2}} N^{1/2} \left(\frac{1}{N} \mathrm{Tr} \left\{ \gamma_\Psi^{(1)} - \left(\gamma_\Psi^{(1)} \right)^2 \right\} \right)^{\frac{1}{4}}. \tag{6.52}$$

7. Accuracy of the Hartree-Fock Approximation for Large Neutral Atoms

In this section we apply the correlation estimate from Theorem 6.7 to derive a nontrivial error bound on the accuracy of the Hartree-Fock energy of a large neutral atom compared to its actual ground state energy. Focusing on the neutral atom, we abbreviate $H_Z := H_Z(Z, Z, 0)$, $E_{\mathrm{gs}}(Z) := E_{\mathrm{gs}}(Z, Z, 0)$ and $E_{\mathrm{hf}}(Z) := E_{\mathrm{hf}}(Z, Z, 0)$.

To this end we use coherent states as in [13]. We first pick $g \in C_0^\infty(\mathbb{R}_0^+; \mathbb{R}_0^+)$ with $\mathrm{supp}\{g\} \subseteq [0, 1)$ and $4\pi \int 0^\infty |g(r)|^2\, r^2\, dr = 1$. Fixing $\lambda > 0$, we set

$$\forall x \in \mathbb{R}^3 : \quad g_\lambda(x) \ := \ \lambda^{-3/2}\, g(\lambda^{-1}|x|). \tag{7.1}$$

Note that $\mathrm{supp}\{g_\lambda\} \subseteq B(0, \lambda)$ and $\|g_\lambda\|_2 = 1$. Furthermore, for $p, q \in \mathbb{R}^3$, we set

$$\forall x \in \mathbb{R}^3 : \quad f_{pq}(x) \ := \ e^{-ip\cdot x}\, g_\lambda(x - q). \tag{7.2}$$

The family $\{f_{pq}\}_{p,q\in\mathbb{R}^3} \subseteq C_0^\infty(\mathbb{R}^3)$ is called *coherent states* and defines a resolution of the identity in $L^2(\mathbb{R}^3)$ in the following sense.

Lemma 7.1: *Let* $V \in L^p(\mathbb{R}^3)$, *for some* $p \in [1, \infty]$, *and denote by* $d\mu_{pq} := (2\pi)^{-3}\, d^3p\, d^3q$ *the Lebesgue measure on phase space* $\mathbb{R}^3 \times \mathbb{R}^3$ *divided by* $(2\pi)^3$. *Then the following identities hold true in the sense of quadratic forms*

$$\int |f_{pq}\rangle\langle f_{pq}|\, d\mu_{pq} = \mathbb{1}, \tag{7.3}$$

$$\int p^2\, |f_{pq}\rangle\langle f_{pq}|\, d\mu_{pq} = -\Delta + \|\nabla g_\lambda\|_{L^2}^2, \tag{7.4}$$

$$\int V(q)\, |f_{pq}\rangle\langle f_{pq}|\, d\mu_{pq} = (V * |g_\lambda|^2)(q). \tag{7.5}$$

Here, (7.3) *and* (7.5) *hold on* $L^2(\mathbb{R}^3)$, *and* (7.4) *holds on* $H^1(\mathbb{R}^3)$.

Proof: We prove these identities only on $\mathcal{S}(\mathbb{R}^3)$. Then they follow in general by extension by continuity. Given $\varphi \in \mathcal{S}(\mathbb{R}^3)$ and $q, p \in \mathbb{R}^3$, we observe that

$$\langle f_{pq} | \varphi \rangle = \int e^{ip\cdot x} \, g_\lambda(x-q) \, \varphi(x) d^3x = \mathcal{F}\big[(2\pi)^{3/2} \, g_\lambda(\,\cdot\, - q) \, \varphi\big](p), \quad (7.6)$$

since g_λ is real. Hence by Plancherel we have

$$\begin{aligned}
\int d\mu_{pq} \, \langle f_{pq} | \varphi \rangle \, \langle \varphi | f_{pq} \rangle &= \int d^3q \Big(\int d^3p \, \big| \mathcal{F}\big[(2\pi)^{3/2} \, g_\lambda(\,\cdot\, - q) \, \varphi\big](p) \big|^2 \Big) \\
&= \int d^3q \Big(\int d^3x \, |g_\lambda(x-q)|^2 \, |\varphi(x)|^2 \Big) \\
&= \|g_\lambda\|_{L^2}^2 \, \|\varphi\|_{L^2}^2 = \langle \varphi | \varphi \rangle, \qquad (7.7)
\end{aligned}$$

where $\mathcal{F} : \mathcal{S}(\mathbb{R}^3) \to \mathcal{S}(\mathbb{R}^3)$ denotes the Fourier transform. Using the parallelogram identity, this proves (7.3). Similarly, we have that

$$\begin{aligned}
&\int d\mu_{pq} \, p^2 \, \langle f_{pq} | \varphi \rangle \, \langle \varphi | f_{pq} \rangle \\
&= \int d^3q \Big(\int d^3p \, \big| p \, \mathcal{F}\big[(2\pi)^{3/2} \, g_\lambda(\,\cdot\, - q) \, \varphi\big](p) \big|^2 \Big) \\
&= \int d^3q \Big(\int d^3x \, \big| \nabla g_\lambda(x-q) \, \varphi(x) + g_\lambda(x-q) \, \nabla\varphi(x) \big|^2 \Big) \\
&= \int d^3q \Big(\int d^3x \, |\nabla g_\lambda(x-q)|^2 \, |\varphi(x)|^2 + |g_\lambda(x-q)|^2 \, |\nabla\varphi(x)|^2 \Big) \\
&= \big\langle \varphi \big| (-\Delta + \|\nabla g_\lambda\|^2) \, \varphi \big\rangle, \qquad (7.8)
\end{aligned}$$

since

$$\begin{aligned}
&\int d^3q \Big(\int d^3x \, \Big\{ g_\lambda(x-q) \, \varphi(x) \, \nabla g_\lambda(x-q) \cdot \overline{\nabla\varphi(x)} \\
&\qquad\qquad + g_\lambda(x-q) \, \overline{\varphi(x)} \, \nabla g_\lambda(x-q) \cdot \nabla\varphi(x) \Big\} \Big) \\
&= \int d^3x \, \nabla_x\big(|\varphi(x)|^2\big) \cdot \Big(\int d^3q \, \nabla_x\big(g_\lambda(x-q)^2\big) \Big) \\
&= - \int d^3x \, \nabla_x\big(|\varphi(x)|^2\big) \cdot \Big(\int d^3q \, \nabla_q\big(g_\lambda(x-q)^2\big) \Big) = 0. \qquad (7.9)
\end{aligned}$$

This proves (7.4). Finally,

$$\int d\mu_{pq}\, V(q)\, \langle f_{pq}|\varphi\rangle\, \langle\varphi|f_{pq}\rangle = \int d^3x \Big(d^3q\, V(q)\, |g_\lambda(x-q)|^2 \Big)\, |\varphi(x)|^2$$
$$= \int d^3x \big[V * |g_\lambda|^2 \big](x)\, |\varphi(x)|^2$$
$$= \langle\varphi|\, \big[V * |g_\lambda|^2\big]\, \varphi\rangle, \tag{7.10}$$

since $V * |g_\lambda|^2$ is bounded, due to $V \in L^p(\mathbb{R}^3)$, for some $p \in [1, \infty]$. □

Next, we give a simple bound on the kinetic energy of approximate ground states of neutral atoms (see [1, 11]).

Lemma 7.2: *Let $\Psi \in \bigwedge^Z \mathfrak{H}$, with $\|\Psi\| = 1$, have negative energy expectation value, i.e., $\langle\Psi|H_Z\Psi\rangle \le 0$, and denote by ρ_Ψ its one-particle density. Then*

$$\int \rho_\Psi^{5/3}(x)\, d^3x \;\le\; \frac{6}{C_{LT}}\, Z^{7/3}, \tag{7.11}$$

where we recall that C_{LT} is the constant in the Lieb-Thirring inequality (3.11).

Proof: Dropping the electron-electron interaction and using the Lieb-Thirring inequality (3.11), we obtain

$$0 \;\ge\; \langle\Psi|H_Z\Psi\rangle \;\ge\; \frac{C_{LT}}{2} \int \rho_\Psi^{5/3}(x)\, d^3x \;+\; \mathrm{Tr}\Big\{ \Big(-\frac{1}{2}\Delta - \frac{Z}{|x|} \Big)\, \gamma_\Psi \Big\}, \tag{7.12}$$

where γ_Ψ denotes the one-particle density matrix of Ψ. The eigenvalues of the hydrogen-like atom $-\frac{1}{2}\Delta - \frac{Z}{|x|}$ equal $-Z^2 n^{-2}$ and have multiplicity $2n^2$, where $n = 1, 2, 3, \ldots$ is called the principal quantum number. Hence,

$$\mathrm{Tr}\Big\{ \Big(-\frac{1}{2}\Delta - \frac{Z}{|x|} \Big)\, \gamma_\Psi \Big\} \;\ge\; -2\, Z^2\, n_{\max}, \tag{7.13}$$

provided that

$$Z \;=\; \mathrm{Tr}\{\gamma_\Psi\} \;\le\; \sum_{n=1}^{n_{\max}} 2\, n^2. \tag{7.14}$$

Since

$$\sum_{n=1}^{n_{\max}} n^2 \;=\; \frac{n_{\max}^3}{3} + \frac{n_{\max}^2}{2} + \frac{n_{\max}}{6} \;\ge\; \frac{(n_{\max}+1)^3}{9}, \tag{7.15}$$

we may choose $n_{\max} := \lfloor 9^{1/3} Z^{1/3} \rfloor$ and arrive at the asserted bound. □

For the further analysis, we use Thomas-Fermi theory. Let $\rho_{\mathrm{TF}} \in (L^1 \cap L^{5/3})[\mathbb{R}^3; \mathbb{R}^+]$, with $\int \rho_{\mathrm{TF}} = Z$, be the Thomas-Fermi (TF) density of a neutral atom and

$$\Phi_{\mathrm{TF}}(x) \;:=\; \frac{Z}{|x|} - \int \frac{\rho_{\mathrm{TF}}(y)\, d^3y}{|x-y|}, \tag{7.16}$$

the corresponding TF potential. Φ_{TF} scales as $\Phi_{\mathrm{TF}}(x) = Z^{4/3}\Phi_1(Z^{1/3}x)$, where Φ_1 is the TF potential for a neutral atom of unit nuclear charge. Furthermore, Φ_{TF} is the unique positive solution of the TF equation

$$\begin{aligned} \forall x \neq 0: \quad \Delta\Phi(x) \;&=\; \frac{4}{3\pi}\Phi^{3/2}(x), \\ \lim_{|x|\to\infty} \Phi_{\mathrm{TF}}(x) = 0, \qquad & \lim_{x\to 0}\left\{|x|\Phi_{\mathrm{TF}}(x)\right\} = Z. \end{aligned} \tag{7.17}$$

A comparison argument shows that Sommerfeld's solution $(6\pi/|x|)^4$ of the TF equation is a supersolution of this equation and hence,

$$\Phi_{\mathrm{TF}}(x) \;\leq\; \frac{6\pi}{|x|^4}, \tag{7.18}$$

for all $x \in \mathbb{R}^3 \setminus \{0\}$, uniformly in $Z > 0$.

We abbreviate the direct term, i.e., the electrostatic energy $\frac{1}{2}D(\rho,\rho) \geq 0$ of a (not necessarily positive) charge distribution ρ by means of the quadratic form

$$D(f,g) \;:=\; \frac{1}{2}\int \frac{\overline{f(x)}\, g(y)}{|x-y|}\, d^3x\, d^3y, \tag{7.19}$$

which is finite, provided $f, g \in (L^1 \cap L^{5/3})[\mathbb{R}^3]$.

Lemma 7.3: *For* $0 < \varepsilon \leq Z^2/4$, *let* $\Psi \in \bigwedge^Z \mathfrak{H}$, *with* $\|\Psi\| = 1$, *be an* ε*-approximate ground state of a neutral atom, i.e.,* $\langle\Psi|H_Z\Psi\rangle \leq E_{\mathrm{gs}}(Z) + \varepsilon$, *and denote by* γ_Ψ *its one-particle density matrix. Then there exists a universal constant* $C_1 < \infty$ *such that*

$$E_{\mathrm{gs}}(Z) \;\geq\; \int h_{pq}\,\gamma_{pq}\, d\mu_{pq} - \frac{1}{2}D(\rho_{\mathrm{TF}}, \rho_{\mathrm{TF}}) - C_1\left(Z^2 + \lambda^{1/2}Z^{12/5} + \lambda^{-2}Z\right), \tag{7.20}$$

where $h_{pq} := p^2 - \Phi_{\mathrm{TF}}(q)$ *and* $\gamma_{pq} := \langle f_{pq}|\gamma_\Psi f_{pq}\rangle$.

Proof: Using the Lieb-Oxford inequality (2.8), we have that

$$\begin{aligned} E_{\mathrm{gs}}(Z) &\geq \langle\Psi|H_Z\Psi\rangle - \varepsilon \\ &\geq T_\Psi - \int \frac{Z\,\rho_\Psi(x)}{|x|}\, d^3x + \frac{1}{2}D(\rho_\Psi, \rho_\Psi) \\ &\quad - 2\int \rho_\Psi^{4/3}(x)\, d^3x - \varepsilon. \end{aligned} \tag{7.21}$$

The ground state energy is monotonically decreasing in the number of electrons and hence smaller that the ground state energy for one electron which equals $-Z^2/4$. Hence, $\varepsilon \le Z^2/4$ ensures that $\langle\Psi|H_Z\Psi\rangle \le 0$, and hence Lemma 7.2 is applicable. Thus

$$\int \rho_\Psi^{4/3}(x)\,d^3x \le \Big(\int \rho_\Psi(x)\,d^3x\Big)^{1/2}\Big(\int \rho_\Psi^{5/3}(x)\,d^3x\Big)^{1/2} \le \sqrt{6}\,C_{LT}^{-1/2}\,Z^{5/3}. \tag{7.22}$$

Moreover,

$$\frac{1}{2}D(\rho_\Psi,\rho_\Psi) \;\ge\; D(\rho_{\mathrm{TF}},\rho_\Psi) - \frac{1}{2}D(\rho_{\mathrm{TF}},\rho_{\mathrm{TF}}), \tag{7.23}$$

since $D \ge 0$, as a quadratic form. Therefore,

$$\begin{aligned} E_{\mathrm{gs}}(Z) &\ge T_\Psi - \int \Big(\frac{Z}{|x|} - \int \frac{\rho_{\mathrm{TF}}(y)}{|x-y|}\,d^3y\Big)\,\rho_\Psi(x)\,d^3x - \frac{1}{2}D(\rho_{\mathrm{TF}},\rho_{\mathrm{TF}}) \\ &\quad - 2\sqrt{6}\,C_{LT}^{-1/2}\,Z^{5/3} - \varepsilon \qquad\qquad (7.24)\\ &= \mathrm{Tr}\{(-\Delta - \Phi_{\mathrm{TF}}(x))\,\gamma_\Psi\} - \frac{1}{2}D(\rho_{\mathrm{TF}},\rho_{\mathrm{TF}}) - 2\sqrt{6}\,C_{LT}^{-1/2}\,Z^{5/3} - \varepsilon. \end{aligned}$$

By (7.4) and (7.5) we have that

$$\begin{aligned} \mathrm{Tr}\{(-\Delta - \Phi_{\mathrm{TF}}(x))\,\gamma_\Psi\} &= \int h_{pq}\,\gamma_{pq}\,d\mu_{pq} - \|\nabla g_\lambda\|^2\,\mathrm{Tr}\{\gamma_\Psi\} \\ &\quad - \int \Big(\Phi_{\mathrm{TF}}(x) - (\Phi_{\mathrm{TF}} * |g_\lambda|^2)(x)\Big)\,\rho_\Psi(x)\,d^3x \\ &= \int h_{pq}\,\gamma_{pq}\,d\mu_{pq} - \|\nabla g_1\|^2\,\lambda^{-2}\,Z \\ &\quad - \int \Big(\Phi_{\mathrm{TF}}(x) - (\Phi_{\mathrm{TF}} * |g_\lambda|^2)(x)\Big)\,\rho_\Psi(x)\,d^3x. \end{aligned} \tag{7.25}$$

To estimate the last term on the right side of (7.25) we use that $\tilde{\phi} := \rho_{\mathrm{TF}} * |\cdot|^{-1}$ is superharmonic. Since convolution with $|g_\lambda|^2$ is a spherical average, the superharmonicity implies that $\tilde{\phi} \ge \tilde{\phi} * |g_\lambda|^2$ pointwise in $\mathbb{R}^3$. Hence,

$$\Phi_{\mathrm{TF}}(x) - [\Phi_{\mathrm{TF}} * |g_\lambda|^2](x) \;\le\; \frac{Z}{|x|} - \int \frac{Z\,|g_\lambda|^2(y)\,d^3y}{|x-y|} \;\le\; \frac{Z}{|x|}, \tag{7.26}$$

for all $x \in \mathbb{R}^3$. Since g_λ is radially symmetric and supported in the ball of radius λ about the origin, Newton's theorem implies that,

$$\int \frac{Z\,|g_\lambda|^2(y)\,d^3y}{|x-y|} \;=\; \|g_\lambda\|^2\,\frac{Z}{|x|} \;=\; \frac{Z}{|x|}, \tag{7.27}$$

for all $|x| \geq \lambda$. Combining (7.26) and (7.27), we infer that, for all $x \in \mathbb{R}^3$,

$$\Phi_{\mathrm{TF}}(x) - [\Phi_{\mathrm{TF}} * |g_\lambda|^2](x) \;\leq\; \frac{\mathbb{1}(|x| \leq \lambda)\, Z}{|x|}, \tag{7.28}$$

which in turn yields

$$\begin{aligned} &\int \Big(\Phi_{\mathrm{TF}}(x) - (\Phi_{\mathrm{TF}} * |g_\lambda|^2)(x)\Big)\, \rho_\Psi(x)\, d^3x \qquad (7.29)\\ &\leq Z \Big(\int \rho_\Psi^{5/3}(x)\, d^3x\Big)^{3/5} \Big(\int_{|x|\leq\lambda} \frac{d^3x}{|x|^{5/2}}\Big)^{2/5} \\ &\leq 6^{3/5}\, (8\pi)^{2/5}\, C_{LT}^{-3/5}\, \lambda^{1/2}\, Z^{12/5}, \end{aligned}$$

and hence the asserted bound (7.20). □

We complement the lower bound (7.20) on the ground state energy by an upper bound of the same accuracy.

Lemma 7.4: *There exists a universal constant $C_2 < \infty$ such that*

$$E_{\mathrm{gs}}(Z) \;\leq\; \int [h_{pq}]_-\, d\mu_{pq} - \frac{1}{2} D(\rho_{\mathrm{TF}}, \rho_{\mathrm{TF}}) + C_2\, (\lambda^{1/2} Z^{12/5} + \lambda^{-2} Z), \tag{7.30}$$

where $h_{pq} := p^2 - \Phi_{\mathrm{TF}}(q)$ and $[h_{pq}]_- := \min\{h_{pq}, 0\}$ denotes its negative part.

Proof: We first use Lieb's variational principle and the negativity of the exchange term to obtain

$$E_{\mathrm{gs}}(Z) \;\leq\; E_{\mathrm{hf}}(Z) \;\leq\; \mathrm{Tr}\big\{\big(-\Delta - Z\,|x|^{-1}\big)\,\gamma\big\} \;+\; \frac{1}{2} D(\rho_\gamma, \rho_\gamma), \tag{7.31}$$

where γ is any 1-pdm with $\mathrm{Tr}\{\gamma\} = Z$ and ρ_γ is its one-particle density. We choose

$$\gamma \;:=\; \int |f_{pq}\rangle\langle f_{pq}|\, \mathbb{1}[h_{pq} < 0]\, d\mu_{pq} \tag{7.32}$$

which fulfills $0 \leq \gamma \leq \mathbb{1}$ and

$$\mathrm{Tr}\{\gamma\} \;=\; \int \mathbb{1}[h_{pq} < 0]\, d\mu_{pq} \;=\; \frac{1}{6\,\pi^2} \int \Phi_{\mathrm{TF}}^{3/2}(q)\, d^3q \;=\; Z \tag{7.33}$$

by (7.3) and by Thomas-Fermi theory. We observe that

$$\begin{aligned}\mathrm{Tr}\{-\Delta\gamma\} &= \int_{h_{pq}<0} |\nabla f_{pq}(x)|^2\, d^3x\, d\mu_{pq} \\ &= \int_{h_{pq}<0} \big|ip\, g_\lambda(x-q) + \nabla g_\lambda(x-q)\big|^2\, d^3x\, d\mu_{pq} \\ &= \int_{h_{pq}<0} p^2\, d\mu_{pq} \;+\; \|\nabla g_1\|^2\,\lambda^2\, Z,\end{aligned} \tag{7.34}$$

where we use $\int_{h_{pq}<0} d\mu_{pq} = Z$ again. Next, we compute the one-particle density ρ_γ and obtain

$$\begin{aligned}\rho_\gamma(x) = \int_{h_{pq}<0} |f_{pq}(x)|^2\, d\mu_{pq} \;&=\; \frac{1}{6\,\pi^2}\int \Phi_{\mathrm{TF}}^{3/2}(q)\, |g_\lambda(x-q)|^2\, d^3q \\ = \big(\rho_{\mathrm{TF}} * |g_\lambda|^2\big)(x),\end{aligned} \tag{7.35}$$

for all $x \in \mathbb{R}^3$, since Thomas-Fermi theory implies that $6\pi^2\Phi_{\mathrm{TF}}^{3/2} = \rho_{\mathrm{TF}}$. We use superharmonicity as in the proof of Lemma 7.4 again to infer that

$$\begin{aligned}D(\rho_\gamma,\rho_\gamma) &= \int \rho_\gamma(x)\,\big(|\cdot|^{-1} * \rho_{\mathrm{TF}} * |g_\lambda|^2\big)(x)\, d^3x \\ &\le \int \rho_\gamma(x)\,\big(|\cdot|^{-1} * \rho_{\mathrm{TF}}\big)(x)\, d^3x \\ &= \int \big(|\cdot|^{-1} * \rho_\gamma\big)(x)\, \rho_{\mathrm{TF}}(x)\, d^3x \\ &= \int \big(|\cdot|^{-1} * \rho_{\mathrm{TF}} * |g_\lambda|^2\big)(x)\, \rho_{\mathrm{TF}}(x)\, d^3x \\ &\le \int \big(|\cdot|^{-1} * \rho_{\mathrm{TF}} * |g_\lambda|^2\big)(x)\, \rho_{\mathrm{TF}}(x)\, d^3x \\ &= D(\rho_{\mathrm{TF}},\rho_{\mathrm{TF}}).\end{aligned} \tag{7.36}$$

Again similar to the proof of Lemma 7.4, we use Newton's theorem and obtain

$$\begin{aligned}\mathrm{Tr}\{|x|^{-1}\gamma\} D(\rho_\gamma,\rho_\gamma) &= \int \frac{\rho_\gamma(y)\, d^3y}{|y|} \\ &= \int \bigg(\int \frac{|g_\lambda(y-x)|^2\, d^3y}{|y|}\bigg) \rho_{\mathrm{TF}}(x)\, d^3x \\ &\ge \int_{|x|\ge\lambda} \frac{\rho_{\mathrm{TF}}(x)\, d^3x}{|x|} \\ &= \int \frac{\rho_{\mathrm{TF}}(x)\, d^3x}{|x|} \;-\; \int_{|x|\le\lambda} \frac{\rho_{\mathrm{TF}}(x)\, d^3x}{|x|}.\end{aligned} \tag{7.37}$$

By Thomas-Fermi theory there exists a constant $C' >$ such that $\int \rho_{\mathrm{TF}}^{5/3} = C' \, Z^{7/3}$, and hence we have that

$$\int_{|x|\leq\lambda} \frac{\rho_{\mathrm{TF}}(x)\, d^3x}{|x|} \leq \Big(\int \rho_{\mathrm{TF}}^{5/3}(x)\, d^3x \Big)^{3/5} \Big(\int_{|x|\leq\lambda} \frac{d^3x}{|x|^{5/2}} \Big)^{2/5} \leq (C')^{3/5}\, (8\pi)^{2/5}\, \lambda^{1/2}\, Z^{12/5}. \tag{7.38}$$

Inserting this estimate into (7.37) and adding up (7.37), (7.36), and (7.34) we arrive at the asserted estimate (7.30). □

As a consequence of Lemmata 7.3 and 7.4, we obtain the important result that the ground state energy and the TF energy of a neutral atom agree to leading order.

Corollary 7.5: *For some constant $C < \infty$,*

$$\left| \frac{E_{\mathrm{gs}}(Z) - E_{\mathrm{TF}}(Z)}{E_{\mathrm{TF}}(Z)} \right| \leq C\, Z^{-(16/175)}. \tag{7.39}$$

Proof: We first choose $\lambda := Z^{-14/25}$ and apply Lemmata 7.3 and 7.4. This yields

$$\left| E_{\mathrm{gs}}(Z) - \int [h_{pq}]_-\, d\mu_{pq} + \frac{1}{2} D(\rho_{\mathrm{TF}}, \rho_{\mathrm{TF}}) \right| \leq 2\,(C_1 + C_2)\, Z^{53/25}. \tag{7.40}$$

On the other hand,

$$\int [h_{pq}]_-\, d\mu_{pq} - \frac{1}{2} D(\rho_{\mathrm{TF}}, \rho_{\mathrm{TF}}) = E_{\mathrm{TF}}(Z) = -C_{\mathrm{TF}}\, Z^{7/3}, \tag{7.41}$$

for some universal constant $C_{\mathrm{TF}} > 0$, which is a fact from Thomas-Fermi theory. Eqs. (7.40) and (7.41) imply the assertion. □

Before we come to the main result of this section we need to derive a final bound that uses Thomas-Fermi theory.

Lemma 7.6: *There exists a universal constant $C_3 < \infty$ such that, for all $E > 0$*

$$\int \mathbb{1}[-E \leq h_{pq} < 0]\, d\mu_{pq} \leq C_3\, E^{3/4}. \tag{7.42}$$

Proof: We observe that

$$\int \mathbb{1}[h_{pq} < -E]\, d\mu_{pq} = \frac{1}{6\,\pi^2} \int [\Phi_{\mathrm{TF}}(q) - E]_+^{3/2}\, d^3q, \tag{7.43}$$

where $[s]_+ := \max\{s, 0\}$ is the positive part of $s \in \mathbb{R}$. By (7.18), $\Phi_{\mathrm{TF}}(q) \leq C_S\,|q|^{-4}$, for all $q \neq 0$, where Sommerfeld's constant C_S equals $C_S := (6\pi)^4$ (and is, in particular, independent of Z). Thus

$$\begin{aligned}
&\int \mathbb{1}[-E \leq h_{pq} < 0]\, d\mu_{pq} \qquad\qquad (7.44)\\
&= \frac{1}{6\,\pi^2} \int \Big(\Phi_{\mathrm{TF}}^{3/2}(q) - \big[\Phi_{\mathrm{TF}}(q) - E\big]_+^{3/2} \Big)\, d^3q \\
&\leq \frac{1}{6\,\pi^2} \int \Big(C_S^{3/2} |q|^{-6} \;-\; \big[C_S |q|^{-4} - E\big]_+^{3/2} \Big)\, d^3q \\
&\leq \frac{1}{6\,\pi^2} \bigg\{ \int_{C_S|q|^{-4} < E} C_S^{3/2} \frac{d^3q}{|q|^6} + E \int_{C_S|q|^{-4} > E} C_S^{1/2} |q|^{-2}\, d^3q \bigg\} \\
&= \frac{4\, C_S^{3/4}}{9\,\pi}\, E^{3/4}
\end{aligned}$$

using that $s \mapsto s^{3/2} - [s - E]_+^{3/2}$ is monotonically increasing, for $s \geq 0$. □

We come to the main result of this section

Theorem 7.7: *Let $\Psi \in \bigwedge^Z \mathfrak{H}$, with $\|\Psi\| = 1$, be an $Z^2/4$-approximate ground state of a neutral atom, i.e., $\langle \Psi | H_Z \Psi \rangle \leq E_{\mathrm{gs}}(Z) + Z^2/4$, and denote by γ_Ψ its one-particle density matrix. Then there exists a universal constant $C < \infty$ such that*

$$\mathrm{Tr}\{\gamma_\Psi - \gamma_\Psi^2\} \;\leq\; C\, Z^{159/175}. \qquad (7.45)$$

Proof: We first choose $\lambda := Z^{-14/25}$ and apply Lemmata 7.3 and 7.4. This yields

$$\begin{aligned}
E \int_{h_{pq} < -E} (1 - \gamma_{pq})\, d\mu_{pq} &\leq \int_{h_{pq} < -E} \big(- [h_{pq}]_- \big)\, (1 - \gamma_{pq})\, d\mu_{pq} \\
&\leq \int \big(- [h_{pq}]_- \big)\, (1 - \gamma_{pq})\, d\mu_{pq} \\
&\leq \int h_{pq}\, \gamma_{pq}\, d\mu_{pq} \;-\; \int [h_{pq}]_-\, d\mu_{pq} \\
&\leq 3\,(C_1 + C_2)\, Z^{53/25}, \qquad (7.46)
\end{aligned}$$

for any $E > 0$, using that $1 - \gamma_{pq} \geq 0$. Furthermore, this estimate and Lemma 7.6 imply that

$$\int_{h_{pq}\geq -E} \gamma_{pq}\, d\mu_{pq} \tag{7.47}$$
$$= Z - \int_{h_{pq}<-E} \gamma_{pq}\, d\mu_{pq} = \int_{h_{pq}<0} d\mu_{pq} - \int_{h_{pq}<-E} \gamma_{pq}\, d\mu_{pq}$$
$$= \int_{h_{pq}<-E} (1-\gamma_{pq})\, d\mu_{pq} + \int_{-E\leq h_{pq}<0} d\mu_{pq}$$
$$\leq 3\,(C_1+C_2)\,\frac{Z^{53/25}}{E} + C_3\, E^{3/4}.$$

We choose $E := Z^{212/175}$ and obtain from (7.46) and (7.47) that

$$\int \min\left\{1-\gamma_{pq}\,,\,\gamma_{pq}\right\} d\mu_{pq} \;\leq\; 6\,(C_1+C_2+C_3)\, Z^{159/175}. \tag{7.48}$$

Finally, $0 \leq \gamma_\Psi \leq 1$ implies that $\gamma_\Psi - \gamma_\Psi^2 \leq \mathbb{1} - \gamma_\Psi$, as well as, $\gamma_\Psi - \gamma_\Psi^2 \leq \gamma_\Psi$. Hence,

$$\left\langle f_{pq}\right| (\gamma_\Psi - \gamma_\Psi^2)\, f_{pq}\big\rangle \;\leq\; \min\left\{1-\gamma_{pq}\,,\,\gamma_{pq}\right\}, \tag{7.49}$$

and we arrive at

$$\mathrm{Tr}\{\gamma_\Psi - \gamma_\Psi^2\} \;=\; \int \left\langle f_{pq}\right| (\gamma_\Psi - \gamma_\Psi^2)\, f_{pq}\big\rangle\, d\mu_{pq} \;\leq\; 6\,(C_1+C_2+C_3)\, Z^{159/175} \tag{7.50}$$

using (7.3). □

The estimate on the accuracy of the Hartree-Fock approximation for large neutral atoms is now an immediate consequence of Theorems 6.7 and 7.7.

Corollary 7.8: *For some constant $C < \infty$,*

$$|E_{\mathrm{gs}}(Z) - E_{\mathrm{hf}}(Z)| \;\leq\; C\, Z^{(5/3)-(4/175)}. \tag{7.51}$$

Proof: Since $E_{\mathrm{hf}}(Z) \geq E_{\mathrm{gs}}(Z)$, we only need to show that $E_{\mathrm{gs}}(Z) \geq E_{\mathrm{hf}}(Z) - CZ^{(5/3)-(4/175)}$. By Theorem 6.7, we have that

$$E_{\mathrm{gs}}(Z) - E_{\mathrm{hf}}(Z) \;\geq\; \left\langle \Psi_{\mathrm{gs}}\right| H(Z)\Psi_{\mathrm{gs}}\big\rangle - \mathcal{E}_{\mathrm{hf}}(\gamma_{\mathrm{gs}}) \tag{7.52}$$
$$= \left\langle \Psi \,\middle|\, \sum_{1\leq m<n\leq N} \frac{1}{|x_m - x_n|}\Psi \right\rangle$$
$$- \frac{1}{2}\int \left(\rho_\psi(x)\,\rho_\psi(y) - \left|\gamma_\Psi^{(1)}(x,y)\right|^2 \right) \frac{d^3x\,d^3y}{|x-y|}$$
$$\geq -C\left(\int \rho_\psi^{5/3}(x)\, d^3x\right)^{\frac{1}{2}} Z^{1/2} \left(\frac{1}{Z}\mathrm{Tr}\left\{\gamma_\Psi^{(1)} - \left(\gamma_\Psi^{(1)}\right)^2\right\}\right)^{\frac{1}{4}},$$

where $\Psi \in \bigwedge^Z \mathfrak{H}$ is a(n approximate) ground state of H. Since $\int \rho_\psi^{5/3}(x)\, d^3x \le C Z^{7/3}$, thanks to Lemma 7.2, we hence get

$$E_{\mathrm{gs}}(Z) - E_{\mathrm{hf}}(Z) \;\ge\; C\, Z^{5/3} \left(\frac{\mathrm{Tr}\{\gamma_\Psi - \gamma_\psi^2\}}{Z} \right)^{1/4}, \tag{7.53}$$

and Corollary 7.8 follows from (7.45). □

8. *N*-Representability

We rewrite the ground state energy of an atom ($K = 1$) or molecule ($K \ge 2$) with N electrons using the notation of Section 5,

$$E_{\mathrm{gs}}(N) = \inf\Big\{ \langle H \rangle_\rho \Big| \rho \text{ is a density matrix },\ \rho \check{\mathbb{N}} = \check{\mathbb{N}} \rho = N\rho \Big\}, \tag{8.1}$$

where

$$H = \sum_{k,m} h_{km} a_k^* a_m + \sum_{k,\ell,m,n} V_{k,\ell;m,n} a_\ell^* a_k^* a_m a_n, \tag{8.2}$$

with

$$h = -\Delta - \sum_{k=1}^{K} \frac{Z_k}{|x - R_k|}, \quad V = \frac{1}{|x - y|}. \tag{8.3}$$

By means of the 1-pdm and 2-pdm we can reformulate this as

$$E_{\mathrm{gs}}(N) = \inf\{ \mathcal{E}_N(\Gamma) \mid \Gamma \in D_N^{(2)} \}, \tag{8.4}$$

where the energy functional is given by

$$\mathcal{E}_N(\Gamma) := \mathrm{Tr}\left\{ \left(\frac{1}{N-1} (1 \otimes h) + V \right) \Gamma \right\} \tag{8.5}$$

and

$$D_N^{(2)} := \left\{ \gamma_\rho^{(2)} \mid \rho \text{ is a density matrix },\ \check{\mathbb{N}}\rho = \rho\check{\mathbb{N}} = N\rho \right\} \tag{8.6}$$

denotes the set of N-representable 2-pdm. Here we use that

$$\begin{aligned} \mathrm{Tr}\left\{ (1 \otimes h)\, \gamma_\rho^{(2)} \right\} &= \sum_{k,\ell,m,n} \delta_{k,m} h_{\ell n} \langle a_\ell^* a_k^* a_m a_n \rangle_\rho \\ &= \sum_{\ell,n} h_{\ell n} \langle a_\ell^* \check{\mathbb{N}} a_n \rangle_\rho = \sum_{\ell,n} h_{\ell n} \langle (N-1)\, a_\ell^* a_n \rangle_\rho \\ &= (N-1) \sum_{\ell,n} h_{\ell n} \langle a_\ell^* a_n \rangle_\rho = (N-1)\, \mathrm{Tr}\left\{ h \gamma_\rho^{(1)} \right\} \end{aligned} \tag{8.7}$$

holds true. Comparing to (8.1), in which one varies over N degrees of freedom, we have reduced the variation to 2 degrees of freedom in (8.4). This comes, however, at a high price: The set $D_N^{(2)}$ of N-representable 2-pdm has a complicated and essentially unknown structure. It seems that nothing has been gained by (8.4), but surprisingly this is not the case. Replacing $D_N^{(2)}$ by any superset, we obtain from (8.4) a lower bound to $E_{\text{gs}}(N)$. For instance, by Theorem 5.2

$$\begin{aligned} \forall \Gamma \in D_N^{(2)} : 0 \le \gamma \le 1, \operatorname{Tr}\{\gamma\} = N \\ 0 \le \Gamma \le N, \operatorname{Tr}\{\Gamma\} = N(N-1), \end{aligned} \tag{8.8}$$

where γ results from Γ by taking a partial trace, we obtain

$$E_{\text{gs}}(N) \ge \inf\left\{\mathcal{E}_N(\Gamma) \,\middle|\, 0 \le \Gamma \le N, \operatorname{Tr}\{\Gamma\} = N(N-1)\right\}. \tag{8.9}$$

This lower bound is not accurate, and this fact has discouraged research in this approach for many years. In the recent past, however, a refined version of this approximation has been demonstrated to yield very accurate bounds. To see this, we first reformulate Theorem 5.2. We call $\rho \in \mathscr{L}_+^1(\mathcal{F}_f[\mathfrak{H}])$ an **N-particle density matrix (N-pdm).**

$$:\Longleftrightarrow \quad \rho \ge 0, \;\operatorname{Tr}(\rho) = 1, \;\rho\check{\mathbb{N}} = \check{\mathbb{N}}\rho = N\rho. \tag{8.10}$$

Lemma 8.1: *If ρ is an N-pdm with $N \ge 2$ then*

$$\begin{aligned} \left\{0 \le \gamma_\rho^{(1)} \le 1\right\} &\Longleftrightarrow \left\{\forall f, g \in \mathfrak{H} : \left\langle \left(a(f) + a^*(g)\right)\left(a^*(f) + a(g)\right)\right\rangle_\rho \ge 0\right\} \\ &\Longleftrightarrow \left\{\forall f, g \in \mathfrak{H} : \langle a(f)\, a^*(f)\rangle_\rho, \langle a^*(g)\, a(g)\rangle_\rho \ge 0\right\}. \end{aligned} \tag{8.11}$$

Proof: We have

$$\forall \varphi, \psi \in \mathfrak{H} : \left\langle \varphi \,\middle|\, \gamma_\rho^{(1)} \psi \right\rangle = \langle a^*(\psi)\, a(\varphi)\rangle_\rho. \tag{8.12}$$

Since ρ preserves the particle number, we have that $\langle a(f)\, a(g)\rangle_\rho = \langle a^*(f)\, a^*(g)\rangle_\rho = 0$ and hence

$$\begin{aligned} &\left\langle \left(a(f) + a^*(g)\right)\left(a^*(f) + a(g)\right)\right\rangle_\rho \\ &= \left\langle \left(a(f)\, a^*(f) + a^*(g)\, a(g)\right)\right\rangle_\rho = \|f\|^2 - \left\langle f \,\middle|\, \gamma_\rho^{(1)} f\right\rangle + \left\langle g \,\middle|\, \gamma_\rho^{(1)} g\right\rangle. \end{aligned} \tag{8.13}$$

Note that $\left\langle g \middle| \gamma_\rho^{(1)} g \right\rangle \geq 0$ and $\|f\|^2 \geq \left\langle f \middle| \gamma_\rho^{(1)} f \right\rangle$, because $\gamma_\rho^{(1)} \geq 0$ and $\gamma_\rho^{(1)} \leq 1$, respectively. Inserting this into (8.13) we obtain

$$\left\langle \big(a(f) + a^*(g)\big)\big(a^*(f) + a(g)\big) \right\rangle_\rho \geq 0. \tag{8.14}$$

Conversely, if $\left\langle \big(a(f) + a^*(g)\big)\big(a^*(f) + a(g)\big) \right\rangle_\rho \geq 0$, for all $f, g \in \mathfrak{H}$, then $\left\langle a(f)a^*(f) \right\rangle_\rho, \left\langle a(g)a^*(g) \right\rangle_\rho \geq 0$ imply that

$$\forall g: \ \langle g | \gamma_\rho^{(1)} g \rangle \geq 0, \qquad \forall f: \ \|f\|^2 \geq \langle f | \gamma_\rho^{(1)} f \rangle, \tag{8.15}$$

and $0 \leq \gamma_\rho^{(1)} \leq \mathbb{1}$. □

Now, we perform some computations. To ease the reading, we abbreviate $\gamma := \gamma_\rho^{(1)}$ and $\Gamma := \gamma_\rho^{(2)}$. Furthermore, we fix an orthonormal basis $\{\varphi_n\}_{n\in\mathbb{N}} \subseteq \mathfrak{H}$ and write

$$\gamma_{km} := \langle \varphi_k | \gamma \varphi_m \rangle \quad \text{and} \quad \Gamma_{k\ell;mn} := \langle \varphi_k \otimes \varphi_\ell | \Gamma \varphi_m \otimes \varphi_n \rangle. \tag{8.16}$$

(i) Suppose that

$$\mathbb{A} := \sum_{k,\ell} A_{k,\ell} a_k a_\ell \quad \text{and} \quad \Psi_A := \sum_{k,\ell} \overline{A_{k,\ell}} \, \varphi_k \otimes \varphi_\ell, \tag{8.17}$$

where we assume $(A_{k,\ell})_{k\ell}$ to be antisymmetric, $A_{k,\ell} = -A_{\ell,k}$, or equivalently

$$A = -A^T. \tag{8.18}$$

Then

$$\begin{aligned} \left\langle \mathbb{A}^* \mathbb{A} \right\rangle_\rho &= \sum_{k,\ell,m,n} \overline{A_{k,\ell}} A_{m,n} \left\langle a_\ell^* a_k^* a_m a_n \right\rangle_\rho \\ &=: \sum_{k,\ell,m,n} \overline{A_{k,\ell}} A_{m,n} \, \Gamma_{mn;k\ell} \\ &= \langle \Psi_A \mid \Gamma \Psi_A \rangle . \end{aligned} \tag{8.19}$$

(ii) Introducing $\mathbb{B} := \sum_{k,\ell} B_{k\ell} a_k^* a_\ell$, we observe that

$$a_\ell^* a_k a_m^* a_n = \delta_{km} a_\ell^* a_n - a_\ell^* a_m^* a_k a_n = \delta_{km} a_\ell^* a_n + a_m^* a_\ell^* a_k a_n \tag{8.20}$$

and hence

$$\begin{aligned} \left\langle \mathbb{B}^* \mathbb{B} \right\rangle_\rho &= \sum_{k,\ell,m,n} \overline{B_{k\ell}} B_{mn} \left\langle a_\ell^* a_k a_m^* a_n \right\rangle_\rho \\ &= \sum_{k,\ell,m,n} \overline{B_{k\ell}} B_{mn} \left(\delta_{km} \gamma_{n,\ell} + \Gamma_{kn;\ell m} \right) \\ &= \operatorname{Tr}\{B^* B \gamma\} + \operatorname{Tr}\{(B^* \otimes B)\Gamma\} . \end{aligned} \tag{8.21}$$

(iii) We furthermore introduce

$$\mathbb{D} := \sum_{k,\ell} D_{k\ell} a_\ell a_k^* = \mathrm{Tr}\{D\} - \sum_{k,\ell} D_{k\ell} a_k^* a_\ell \tag{8.22}$$

and observe that

$$\begin{aligned}
&\Big\langle (\mathbb{B}+\mathbb{D})^* (\mathbb{B}+\mathbb{D}) \Big\rangle_\rho \\
&= \Big\langle \Big(\mathrm{Tr}[D] + \sum_{k,\ell} (B-D)_{k\ell}\, a_k^* a_\ell\Big)^* \Big(\mathrm{Tr}[D] + \sum_{k,\ell} (B-D)_{k\ell}\, a_k^* a_\ell\Big) \Big\rangle_\rho \\
&= |\mathrm{Tr}[D]|^2 + 2\mathrm{Re}\Big(\overline{\mathrm{Tr}[D]}\ \mathrm{Tr}\{(B-D)\gamma\}\Big) \\
&\quad + \mathrm{Tr}\big\{(B-D)^*(B-D)\gamma\big\} + \mathrm{Tr}\big\{\big((B-D)^* \otimes (B-D)\big)\Gamma\big\}.
\end{aligned} \tag{8.23}$$

(iv) From

$$\begin{aligned}
a_m a_n a_\ell^* a_k^* &= \delta_{\ell n} a_m a_k^* - \delta_{\ell m} a_n a_k^* + a_l^* a_m a_n a_k^* \\
&= \delta_{\ell n}\delta_{km} - \delta_{\ell n} a_k^* a_m - \delta_{\ell m}\delta_{kn} + \delta_{\ell m} a_k^* a_n + a_\ell^* a_m a_n a_k^* \\
&= \delta_{\ell n}\delta_{km} - \delta_{\ell n} a_k^* a_m - \delta_{\ell m}\delta_{kn} + \delta_{\ell m} a_k^* a_n \\
&\quad + \delta_{kn} a_\ell^* a_m - \delta_{km} a_\ell^* a_n + a_\ell^* a_k^* a_m a_n
\end{aligned} \tag{8.24}$$

and $\mathbb{E} := \sum_{k,\ell} E_{k\ell} a_k^* a_\ell^*$, with $E = -E^T$ it follows that

$$\begin{aligned}
\big\langle \mathbb{E}^*\mathbb{E} \big\rangle_\rho &= \sum_{k,\ell,m,n} \overline{E_{nm}} E_{\ell k} \langle a_m a_n a_\ell^* a_k^* \rangle_\rho \\
&= 2\mathrm{Tr}\{EE^*\} + \mathrm{Tr}\big\{(E^T E^* + E^T E^* + E\overline{E} - EE^*)\gamma\big\} \\
&\qquad + \langle \Psi_{\overline{E}} \mid \Gamma \Psi_{\overline{E}} \rangle
\end{aligned} \tag{8.25}$$

thus

$$\big\langle \mathbb{E}^*\mathbb{E} \big\rangle_\rho = 2\mathrm{Tr}\{EE^*\} - 4\mathrm{Tr}\{EE^*\gamma\} + \langle \Psi_{\overline{E}} \mid \Gamma \Psi_{\overline{E}} \rangle. \tag{8.26}$$

Putting (i)-(iv) together, we arrive at

$$\begin{aligned}
&\Big\langle (\mathbb{A}+\mathbb{B}+\mathbb{D}+\mathbb{E})^* (\mathbb{A}+\mathbb{B}+\mathbb{D}+\mathbb{E}) \Big\rangle_\rho \\
&= \Big\langle \big(\mathbb{A}^*\mathbb{A} + (\mathbb{B}+\mathbb{D})^*(\mathbb{B}+\mathbb{D}) + \mathbb{E}^*\mathbb{E}\big) \Big\rangle_\rho \\
&= \langle \Psi_A \mid \Gamma \Psi_A \rangle + \langle \Psi_{\overline{E}} \mid \Gamma \Psi_{\overline{E}} \rangle + 2\mathrm{Tr}\{EE^*\} \\
&\quad - 4\mathrm{Tr}\{EE^*\gamma\} + |\mathrm{Tr}[D]|^2 + 2\mathrm{Re}(\mathrm{Tr}[D]\ \mathrm{Tr}\{(B-D)\gamma\}) \\
&\quad + \mathrm{Tr}\big\{(B-D)^*(B-D)\gamma\big\} + \mathrm{Tr}\big\{\big[(B-D)^* \otimes (B-D)\big]\Gamma\big\},
\end{aligned} \tag{8.27}$$

where we use that ρ is particle number preserving. From this the following theorem derives.

Theorem 8.2: *Let ρ be an N-pdm and $\gamma = \gamma_\rho^{(1)}$ and $\Gamma = \gamma_\rho^{(2)}$ be its 1-pdm and 2-pdm, respectively. Then the following statements are equivalent:*

For all polynomials $P(a^*, a)$ of degree ≤ 2:

$$\left\langle P(a^*, a)^* \, P(a^*, a) \right\rangle_\rho \geq 0. \tag{8.28}$$

$$\Longleftrightarrow$$

For all $A = -A^T$ and B of finite rank:

$$0 \leq \mathrm{Tr}\{A^* A \gamma\} \leq \mathrm{Tr}\{A^* A\}, \tag{8.29}$$

$$\langle \Psi_A \mid \Gamma \Psi_A \rangle \geq 0, \tag{8.30}$$

$$2\mathrm{Tr}(A^* A) - 4\mathrm{Tr}\{A^* A \gamma\} + \langle \Psi_A \mid \Gamma \Psi_A \rangle \geq 0, \tag{8.31}$$

$$\mathrm{Tr}\{B^* B \gamma\} + \mathrm{Tr}\{(B^* \otimes B)\Gamma\} - |\mathrm{Tr}\{B\gamma\}|^2 \geq 0. \tag{8.32}$$

Proof: The assertion follows directly from (8.27) and the observation that

$$\min_D \left\{ |\mathrm{Tr}(D)|^2 + 2\mathrm{Re}\left(\overline{\mathrm{Tr}(D)} \, \mathrm{Tr}\{B\gamma\} \right) \right\} = -|\mathrm{Tr}\{B\gamma\}|^2. \tag{8.33}$$

Note that

$$\begin{aligned} &2\mathrm{Tr}\{A^* A\} - 4\mathrm{Tr}\{A^* A \gamma\} + \langle \Psi_A \mid \Gamma \Psi_A \rangle \\ &= \langle \Psi_A \mid [2 - 2(1 \otimes \gamma) - 2(\gamma \otimes 1) + \Gamma] \, \Psi_A \rangle . \end{aligned} \tag{8.34}$$

□

Remarks:

- Conditions (8.30)-(8.32) are known as P-Condition, Q-Condition, and G-Condition in quantum chemistry. More precisely,

  ***P*-Condition:** $\Longleftrightarrow$

  $$\forall A = -A^T : \quad \langle \Psi_A \mid \Gamma \Psi_A \rangle \geq 0, \tag{8.35}$$

  ***Q*-Condition:** $\Longleftrightarrow$

  $$\forall A = A^T : \quad 2\mathrm{Tr}\{A^* A\} - 4\mathrm{Tr}\{A^* A \gamma\} + \langle \Psi_A \mid \Gamma \Psi_A \rangle \geq 0, \tag{8.36}$$

  ***G*-Condition:** $\Longleftrightarrow$

  $$\forall B : \quad \mathrm{Tr}\{B^* B \gamma\} + \mathrm{Tr}\{(B^* \otimes B)\Gamma\} \geq \left| \mathrm{Tr}\{B\gamma\} \right|^2. \tag{8.37}$$

- It is remarkable that Theorem 6.5 is implied by the P- and the G-Condition. More precisely, if $0 \leq \gamma \leq 1$, $\gamma \in \mathscr{L}^1(\mathfrak{H})$, $\Gamma \in \mathscr{L}^1(\mathfrak{H} \otimes \mathfrak{H})$,

$\mathrm{Ex}\,\Gamma = \Gamma\,\mathrm{Ex}$, and (γ, Γ) fulfills (8.35) and (8.37) then

$$\begin{aligned} &\mathrm{Tr}\Big\{(X \otimes X)\Big(\Gamma - (1-\mathrm{Ex})(\gamma \otimes \gamma)\Big)\Big\} \\ &\geq -\mathrm{Tr}\{X\gamma\} \min\Big[1,\ 10\sqrt{\mathrm{Tr}\big\{X(\gamma - \gamma^2)\big\}}\Big], \end{aligned} \tag{8.38}$$

for all orthogonal projections $X = X^* = X^2$ on $\mathfrak{H}$ [5]. Hence, also Theorem 6.7 is implied by the P- and G-Condition. In particular,

$$\begin{aligned} E_{\mathrm{GPQ}}(N) :=& \inf\Big\{\mathcal{E}(\Gamma)\Big|\ \Gamma \in \mathscr{L}^1(\mathfrak{H} \otimes \mathfrak{H}),\ \mathrm{Tr}(\Gamma) = N(N-1), \\ &\mathrm{Ex}\,\Gamma = \Gamma\,\mathrm{Ex}, \Gamma \text{ fulfills the } P, Q, G\text{-Conditions}\Big\} \\ \geq& E_{\mathrm{gs}} - C\,N^{\frac{5}{3}-\varepsilon}, \end{aligned} \tag{8.39}$$

for neutral atoms or molecules, where $\varepsilon > 0$ is a positive number.

- For small N, the approximation of $E_{\mathrm{gs}}(N)$ by $E_{\mathrm{GPQ}}(N)$ is numerically not very accurate, which discouraged scientists to use N-representability methods for a long time. Fukuda et al [20], Cances et al [7], and Mazziotti [17] have shown in the past decade, however, that the addition of two further conditions, namely, Erdahl's T_1-Condition and T_2-Condition yield an excellent approximation of the form (8.39), i.e.,

$$\begin{aligned} E_{\mathrm{gs}}(N) \geq& E_{\mathrm{GPQT1T2}}(N) \\ :=& \inf\Big\{\mathcal{E}(\Gamma)\Big|\ \Gamma \in \mathscr{L}^1(\mathfrak{H} \otimes \mathfrak{H}),\ \mathrm{Tr}(\Gamma) = N(N-1), \\ &\mathrm{Ex}\,\Gamma = \Gamma\,\mathrm{Ex}, \Gamma \text{ fulfills the } P, Q, G, T_1, T_2\text{-Conditions}\Big\}. \end{aligned} \tag{8.40}$$

We briefly present Erdahl's T_1-Condition and T_2-Condition. We start with a computation,

$$\begin{aligned} & a_k a_\ell a_m a_{\hat m}^* a_{\hat\ell}^* a_{\hat k}^* + a_{\hat m}^* a_{\hat\ell}^* a_{\hat k}^* a_k a_\ell a_m \\ &= \delta_{m\hat m} a_k a_\ell a_{\hat\ell}^* a_{\hat k}^* - \delta_{\ell\hat m} a_k a_m a_{\hat\ell}^* a_{\hat k}^* + \delta_{k\hat m} a_\ell a_m a_{\hat\ell}^* a_{\hat k}^* \\ &\quad - a_{\hat m}^* a_k a_\ell a_m a_{\hat\ell}^* a_{\hat k}^* + a_{\hat m}^* a_{\hat\ell}^* a_{\hat k}^* a_k a_\ell a_m \\ &= \delta_{m\hat m}\left(\delta_{\ell\hat\ell} a_k a_{\hat k}^* - \delta_{k\hat\ell} a_\ell a_{\hat k}^* + a_{\hat\ell}^* a_k a_\ell a_{\hat k}^*\right) \\ &\quad - \delta_{\ell\hat m}\left(\delta_{m\hat\ell} a_k a_{\hat k}^* - \delta_{k\hat\ell} a_m a_{\hat k}^* + a_{\hat\ell}^* a_k a_m a_{\hat k}^*\right) \\ &\quad + \delta_{k\hat m}\left(\delta_{m\hat\ell} a_\ell a_{\hat k}^* - \delta_{\ell\hat\ell} a_m a_{\hat k}^* + a_{\hat\ell}^* a_\ell a_m a_{\hat k}^*\right) \\ &\quad - \delta_{m\hat\ell} a_{\hat m}^* a_k a_\ell a_{\hat k}^* + \delta_{\ell\hat\ell} a_{\hat m}^* a_k a_m a_{\hat k}^* - \delta_{k\hat\ell} a_{\hat m}^* a_\ell a_m a_{\hat k}^* \\ &\quad + a_{\hat m}^* a_{\hat\ell}^* a_k a_\ell a_m a_{\hat k}^* + a_{\hat m}^* a_{\hat\ell}^* a_{\hat k}^* a_k a_\ell a_m \end{aligned}$$

hence

$$
\begin{aligned}
&a_k a_\ell a_m a^*_{\hat{m}} a^*_{\hat{\ell}} a^*_{\hat{k}} + a^*_{\hat{m}} a^*_{\hat{\ell}} a^*_{\hat{k}} a_k a_\ell a_m \\
&\quad = \delta_{m\hat{m}} \Big(\delta_{\ell\hat{\ell}} \Big[\delta_{k\hat{k}} - a^*_{\hat{k}} a_k \Big] - \delta_{k\hat{\ell}} \Big[\delta_{\ell\hat{k}} - a^*_{\hat{k}} a_\ell \Big] \\
&\qquad\qquad + \delta_{\ell\hat{k}} a^*_{\hat{\ell}} a_k - \delta_{k\hat{k}} a^*_{\hat{\ell}} a_\ell + a^*_{\hat{\ell}} a^*_{\hat{k}} a_k a_\ell \Big) \\
&\qquad - \delta_{\ell\hat{m}} \Big(\delta_{m\hat{\ell}} \Big[\delta_{k\hat{k}} - a^*_{\hat{k}} a_k \Big] - \delta_{k\hat{\ell}} \Big[\delta_{m\hat{k}} - a^*_{\hat{k}} a_m \Big] \\
&\qquad\qquad + \delta_{m\hat{k}} a^*_{\hat{\ell}} a_k - \delta_{k\hat{k}} a^*_{\hat{\ell}} a_m + a^*_{\hat{\ell}} a^*_{\hat{k}} a_k a_m \Big) \\
&\qquad + \delta_{k\hat{m}} \Big(\delta_{m\hat{\ell}} \Big[\delta_{\ell\hat{k}} - a^*_{\hat{k}} a_\ell \Big] - \delta_{\ell\hat{\ell}} \Big[\delta_{m\hat{k}} - a^*_{\hat{k}} a_m \Big] \\
&\qquad\qquad + \delta_{m\hat{k}} a^*_{\hat{\ell}} a_\ell - \delta_{\ell\hat{k}} a^*_{\hat{\ell}} a_m + a^*_{\hat{\ell}} a^*_{\hat{k}} a_\ell a_m \Big) \\
&\qquad - \delta_{m\hat{\ell}} \Big(\delta_{\ell\hat{k}} a^*_{\hat{m}} a_k - \delta_{k\hat{k}} a^*_{\hat{m}} a_\ell + a^*_{\hat{m}} a^*_{\hat{k}} a_k a_\ell \Big) \\
&\qquad + \delta_{\ell\hat{\ell}} \Big(\delta_{m\hat{k}} a^*_{\hat{m}} a_k - \delta_{k\hat{k}} a^*_{\hat{m}} a_m + a^*_{\hat{m}} a^*_{\hat{k}} a_k a_m \Big) \\
&\qquad - \delta_{k\hat{\ell}} \Big(\delta_{m\hat{k}} a^*_{\hat{m}} a_\ell - \delta_{\ell\hat{k}} a^*_{\hat{m}} a_m + a^*_{\hat{m}} a^*_{\hat{k}} a_\ell a_m \Big) \\
&\qquad + \delta_{m\hat{k}} a^*_{\hat{m}} a^*_{\hat{\ell}} a_k a_\ell - \delta_{\ell\hat{k}} a^*_{\hat{m}} a^*_{\hat{\ell}} a_k a_m + \delta_{k\hat{k}} a^*_{\hat{m}} a^*_{\hat{\ell}} a_\ell a_m
\end{aligned}
$$

and thus

$$
\begin{aligned}
&a_k a_\ell a_m a^*_{\hat{m}} a^*_{\hat{\ell}} a^*_{\hat{k}} + a^*_{\hat{m}} a^*_{\hat{\ell}} a^*_{\hat{k}} a_k a_\ell a_m \\
&\quad = \Big\{ \delta_{m\hat{m}} \delta_{\ell\hat{\ell}} \delta_{k\hat{k}} - \delta_{m\hat{m}} \delta_{k\hat{\ell}} \delta_{\ell\hat{k}} - \delta_{\ell\hat{m}} \delta_{m\hat{\ell}} \delta_{k\hat{k}} \\
&\qquad + \delta_{\ell\hat{m}} \delta_{k\hat{\ell}} \delta_{m\hat{k}} + \delta_{k\hat{m}} \delta_{m\hat{\ell}} \delta_{\ell\hat{k}} - \delta_{k\hat{m}} \delta_{\ell\hat{\ell}} \delta_{m\hat{k}} \Big\} \\
&\qquad - \Big\{ \delta_{m\hat{m}} \delta_{\ell\hat{\ell}} a^*_{\hat{k}} a_k - \delta_{m\hat{m}} \delta_{k\hat{\ell}} a^*_{\hat{k}} a_\ell - \delta_{m\hat{m}} \delta_{\ell\hat{k}} a^*_{\hat{\ell}} a_k \\
&\qquad + \delta_{m\hat{m}} \delta_{k\hat{k}} a^*_{\hat{\ell}} a_\ell - \delta_{\ell\hat{m}} \delta_{m\hat{\ell}} a^*_{\hat{k}} a_k + \delta_{\ell\hat{m}} \delta_{k\hat{\ell}} a^*_{\hat{k}} a_m \\
&\qquad + \delta_{\ell\hat{m}} \delta_{m\hat{k}} a^*_{\hat{\ell}} a_k - \delta_{\ell\hat{m}} \delta_{k\hat{k}} a^*_{\hat{\ell}} a_m + \delta_{k\hat{m}} \delta_{m\hat{\ell}} a^*_{\hat{k}} a_\ell \\
&\qquad - \delta_{k\hat{m}} \delta_{\ell\hat{\ell}} a^*_{\hat{k}} a_m - \delta_{k\hat{m}} \delta_{m\hat{k}} a^*_{\hat{\ell}} a_\ell + \delta_{k\hat{m}} \delta_{\ell\hat{k}} a^*_{\hat{\ell}} a_m \\
&\qquad + \delta_{m\hat{\ell}} \delta_{\ell\hat{k}} a^*_{\hat{m}} a_k - \delta_{m\hat{\ell}} \delta_{k\hat{k}} a^*_{\hat{m}} a_\ell - \delta_{\ell\hat{\ell}} \delta_{m\hat{k}} a^*_{\hat{m}} a_k \\
&\qquad + \delta_{\ell\hat{\ell}} \delta_{k\hat{k}} a^*_{\hat{m}} a_m + \delta_{k\hat{\ell}} \delta_{m\hat{k}} a^*_{\hat{m}} a_\ell - \delta_{k\hat{\ell}} \delta_{\ell\hat{k}} a^*_{\hat{m}} a_m \Big\} \\
&\qquad + \Big\{ \delta_{m\hat{m}} a^*_{\hat{\ell}} a^*_{\hat{k}} a_k a_\ell - \delta_{\ell\hat{m}} a^*_{\hat{\ell}} a^*_{\hat{k}} a_k a_m + \delta_{k\hat{m}} a^*_{\hat{\ell}} a^*_{\hat{k}} a_\ell a_m \\
&\qquad - \delta_{m\hat{\ell}} a^*_{\hat{m}} a^*_{\hat{k}} a_k a_\ell + \delta_{\ell\hat{\ell}} a^*_{\hat{m}} a^*_{\hat{k}} a_k a_m - \delta_{k\hat{\ell}} a^*_{\hat{m}} a^*_{\hat{k}} a_l a_m \\
&\qquad + \delta_{m\hat{k}} a^*_{\hat{m}} a^*_{\hat{\ell}} a_k a_\ell - \delta_{\ell\hat{k}} a^*_{\hat{m}} a^*_{\hat{\ell}} a_k a_m + \delta_{k\hat{k}} a^*_{\hat{m}} a^*_{\hat{\ell}} a_\ell a_m \Big\}.
\end{aligned}
\tag{8.41}
$$

Now, if $M_{k_1,k_2,k_3} = (-1)^{\pi} M_{k_{\pi(1)},k_{\pi(2)},k_{\pi(3)}}$ is totally antisymmetric then we obtain

$$\begin{aligned}\sum_{k,\ell,m,\hat{k},\hat{\ell},\hat{m}} & M_{k,\ell,m}\overline{M_{\hat{k},\hat{\ell},\hat{m}}}\left\langle a_k a_\ell a_m a_{\hat{m}}^* a_{\hat{\ell}}^* a_{\hat{k}}^* + a_{\hat{m}}^* a_{\hat{\ell}}^* a_{\hat{k}}^* a_k a_\ell a_m \right\rangle_\rho \\ &= 6\sum_{k,\ell,m} |M_{k,\ell,m}|^2 - 18\sum_{\ell,m}\sum_{k,\hat{k}} M_{k,\ell,m}\overline{M_{\hat{k},\ell,m}}\gamma_{k,\hat{k}} \\ &\quad + 9\sum_m \sum_{k,\ell,\hat{k},\hat{\ell}} M_{k,\ell,m}\overline{M_{\hat{k},\hat{\ell},m}}\Gamma_{k\ell,\hat{k}\hat{\ell}}. \end{aligned} \tag{8.42}$$

We set $M_m(k,\ell) := M_{k,\ell,m}$ and $\Psi_{M_m} := \sum_{k,\ell}\overline{M_m(k,\ell)}\varphi_k \otimes \varphi_\ell$ and obtain the

T_1-Condition: $\Longleftrightarrow$

$$\sum_m \left(2\mathrm{Tr}\left\{M_m^* M_m\right\} - 6\mathrm{Tr}\left\{M_m^* M_m \gamma\right\} + 3\left\langle \Psi_{M_m} \mid \Gamma \Psi_{M_m}\right\rangle\right) \geq 0. \tag{8.43}$$

Analogously, we consider

$$\mathbb{D} := \sum_{k,\ell,m=1}^{\infty} D_m\big(k,\ell\big)\; a_m^* a_\ell^* a_k, \tag{8.44}$$

where $D_m\big(k,\ell\big) = -D_m\big(\ell,k\big) \in \mathbb{C}$ is antisymmetric in two indices, and use that

$$\mathrm{Tr}\Big(\rho\big\{\mathbb{D}^*,\mathbb{D}\big\}\Big) = \mathrm{Tr}\Big(\rho\big\{\mathbb{D}^*\mathbb{D} + \mathbb{D}\mathbb{D}^*\big\}\Big) \geq 0, \tag{8.45}$$

from which we obtain the $\underline{T_2\text{-Condition}}$ $:\Longleftrightarrow$

$$\sum_\alpha \left(\langle \Psi_{\widetilde{D}_\alpha} \mid \Gamma \Psi_{\widetilde{D}_\alpha} \rangle + 4\mathrm{Tr}\Big\{ \Big(D_\alpha^* \otimes D_\alpha\Big)\Gamma \Big\} + 2\mathrm{Tr}\Big\{ D_\alpha^* D_\alpha \gamma \Big\} \right) \geq 0, \tag{8.46}$$

where

$$\Psi_{\widetilde{D}_\alpha} := \sum_{k,m} D_m\big(k,\alpha\big)\varphi_k \otimes \varphi_m. \tag{8.47}$$

Acknowledgments

It is a pleasure to thank Mohammed Amayri for typing and Mohammed Amayri, Hans Konrad Knörr, Edmund Menge, and especially Andreas Groh for careful proofreading. Further thanks go to Edmund Menge for computing the T_2-Condition. Finally I thank the anonymous referee for her or his useful critique that encouraged me to improve on the first version of the manuscript. These notes are based on a course held at Mainz University in Spring 2008 and the papers [1, 2, 3, 6, 4, 5].

References

1. V. Bach. Error bound for the Hartree-Fock energy of atoms and molecules. *Commun. Math. Phys.*, 147:527–548, 1992.
2. V. Bach. Accuracy of mean field approximations for atoms and molecules. *Commun. Math. Phys.*, 155:295–310, 1993.
3. V. Bach. Approximative theories for large Coulomb systems. In J. Rauch and B. Simon, editors, *Quasiclassical Methods*, volume 95 of *IMA Volumes in Mathematics and its Applications*. Springer, 1997.
4. V. Bach, J. Fröhlich, and L. Jonsson. Bogolubov-Hartree-Fock mean field theory for neutron stars and other systems with attractive interactions. *J. Math. Phys.*, 50:102102, 2009. doi:10.1063/1.3225565.
5. V. Bach, H. K. Knörr, and E. Menge. Fermion correlation inequalites derived from g- and p-conditions. *In preparation*, 2012.
6. V. Bach, E. H. Lieb, and J. P. Solovej. Generalized Hartree-Fock theory and the Hubbard model. *J. Stat. Phys.*, 76:3–90, 1994.
7. E. Cancès, M. Lewin, and G. Stoltz. The electronic ground state energy problem: a new reduced density matrix approach. *J. Chem. Phys.*, 125:064101–064106, 2006.
8. P. A. M. Dirac. Note on exchange phenomena in the Thomas-Fermi atom. *Proc. Cambridge Philos. Soc.*, 26:376–385, 1931.
9. F. Dyson. Ground-state energy of a finite system of charged particles. *J. Math. Phys.*, 8:1538–1545, 1967.
10. C. Fefferman and R. de la Llave. Relativistic stability of matter – I. *Revista Matematica Iberoamericana*, 2(1, 2):119–161, 1986.
11. E. H. Lieb. The stability of matter. *Rev. Mod. Phys.*, 48:653–669, 1976.
12. E. H. Lieb. A lower bound for Coulomb energies. *Phys. Lett.*, 70A:444–446, 1979.
13. E. H. Lieb. Thomas-Fermi and related theories of atoms and molecules. *Rev. Mod. Phys.*, 53:603–641, 1981.
14. E. H. Lieb and S. Oxford. An improved lower bound on the indirect Coulomb energy. *Int. J. Quantum Chem.*, 19:427–439, 1981.
15. Elliott H. Lieb and Walter E. Thirring. – erratum – bound for the kinetic energy of Fermions which proves the stability of matter. *Phys. Rev. Lett.*, 35(16):1116, October 1975.
16. Elliott H. Lieb and Walter E. Thirring. Bound for the kinetic energy of Fermions which proves the stability of matter. *Phys. Rev. Lett.*, 35(11):687–689, September 1975.
17. D. Mazziotti. Variational two-electron reduced-density-matrix theory. In D. Mazziotti, editor, *Reduced-Density-Matrix Mechanics*, volume 134 of *Advances in Chemical Physics*, chapter 3, pages 21–59. Wiley-Interscience, 2007.
18. E. M. Stein and G. Weiss. *Introduction to Fourier Analysis on Euclidean Spaces.* Princeton University Press, Princeton, New Jersey, 2nd edition, 1971.

19. W. Thirring. *Lehrbuch der Mathematischen Physik 4: Quantenmechanik großer Systeme*. Springer-Verlag, Wien, New York, 1st edition, 1980.
20. Z. Zhao, B. J. Braams, M. Fukuda, M. L. Overton, and J. K. Percus. The reduced density matrix method for electronic structure calculations and the role of three-index representability conditions. *J. Chem. Phys.*, 120:2095–2104, 2004.

ON THE DYNAMICS OF A FERMI GAS IN A RANDOM MEDIUM WITH DYNAMICAL HARTREE–FOCK INTERACTIONS

Thomas Chen

Department of Mathematics
University of Texas at Austin
Austin, TX 78712, USA
tc@math.utexas.edu

This article addresses some aspects of the dynamics of a Fermi gas in a random medium, at weak disorders. We present some joint results with I. Sasaki on the Boltzmann limit for the thermal momentum distribution function, and on the persistence of quasifreeness, for the case of a free Fermi gas in a random medium. Subsequently, we present some recent joint results with I. Rodnianski on the derivation of the Boltzmann limit for a Fermi gas in a random medium with nonlinear self-interactions modeled in dynamical Hartree-Fock theory.

1. Introduction

We survey some recent results in [15, 16] addressing the dynamics of a gas of fermions (electrons) in a random medium. A prominent example of materials described by such models are semiconductors. Some of the main questions raised in this context address how electric conductivity or the absence thereof can be explained from first principles in quantum mechanics.

A widely used model to study aspects of such systems is the Anderson model which describes a single electron in a random medium. It is defined by the Hamiltonian $H_\omega = -\Delta + \eta\omega_x$ acting on $\ell^2(\mathbb{Z}^d)$ where $(\Delta f)(x) = \sum_{|y-x|=1} f(y)$ is the nearest neighbor Laplacian on $\mathbb{Z}^d$, and $\{\omega_x\}$ is an i.i.d. random field of random variables which act as multiplication operators.

The absence of electron transport (*Anderson localization*, electric insulators) for large $|\eta| \gg 1$, is nowadays mathematically well-understood, [2, 32]. On the other hand, the weak disorder regime, $|\eta| \ll 1$, poses some very prominent open problems. Only in dimension $d = 1$, it is known that Anderson localization occurs for all values of $|\eta| > 0$, [11]. In $d = 2$, it is conjectured that even for small $|\eta| \ll 1$, Anderson localization persists. In $d \geq 3$, it is conjectured that there is a component of absolutely continuous spectrum, thus implying the existence of delocalized states and electric conduction.

In order to investigate the long time dynamics of the weakly disordered Anderson model, it has proven to be a remarkably powerful strategy to study kinetic and diffusive scaling limits of the associated Schrödinger dynamics. In the seminal works [25, 27, 33, 50], it is proven for a kinetic scaling determined by macroscopic time and space coordinates $(T, X) = \eta^2(t, x)$ that, as $\eta \to 0$, the semiclassical dynamics is determined by a linear Boltzmann equation, see also [12, 13, 14]. Moreover, it has been established in the breakthrough work [27] that for $d \geq 3$ and a diffusive time scale, the dynamics predicted by the Anderson model is determined by a heat equation. It is conjectured that this result holds for all times, which would consequently explain electric conductivity, and the delocalization of the electron wave function in the relevant energy regimes. We also refer to the important related works [3, 8, 9, 10, 20, 22, 24, 36, 37, 39, 38, 42, 43, 44, 46, 47, 48, 49].

The Anderson model neglects the repulsion between the electrons due to Coulomb interactions, and the Pauli principle. The work addressed here focuses on the dynamics of systems of many electrons in a random medium, at positive temperature, [15, 16]. One of our main motivations is to investigate the extent to which manybody effects influence the predictions of the Anderson model at small disorders.

The presentation is structured as follows. In Section 2, we discuss the dynamics of an ideal Fermi gas in a random medium, based on joint work with I. Sasaki, [15]. We analyze the dynamics of an ideal homogenous Fermi gas in a weak random potential, and derive the kinetic scaling limit for the momentum distribution function with a translation invariant initial state and prove that it is determined by a linear Boltzmann equation. Moreover, we prove that if the initial state is quasifree, then the time evolved state, averaged over the randomness is not quasifree, but has a quasifree kinetic limit. We show that the momentum distributions determined by the Gibbs states of a free fermion field are stationary solutions of the linear Boltzmann equation; this includes the limit of zero temperature. We note that recently, important results on the persistence of localization in fermionic manybody models at strong disorders (a topic which is not addressed here) have been established in [4, 17, 18]. We also refer to [1, 5, 6] for works related to the topics discussed here.

In section 4, we discuss the joint work [16] with I. Rodnianski, which investigates the dynamics of a Fermi gas in a random medium where the particle interactions between the fermions are modeled in dynamical Hartree-Fock theory. We derive Boltzmann equations in kinetic scaling limits for scaling regimes determined by different ratios between the strengths of the randomness, and of the particle interactions. Central to this work is the development of methods to control the *nonlinear* self-interactions of the quantum field, combined with Feynman graph expansion methods to govern the randomness.

We note that for the translation invariant model without the random potential (i.e., $\eta = 0$), but including the full repulsive particle pair interaction (without Hartree-Fock approximation), the related problem of deriving the Boltzmann-Uhlenbeck-Uehling equation from the microscopic quantum dynamics is an extremely challenging open problem; for some work in this direction, see [7, 29, 34, 40, 51]. Some very interesting recent progress was recently established in [41]. The contextually related question of the stability of the Fermi sea for a gas of interacting fermions is a quintessential problem in mathematical physics which has in recent years received much attention, especially due to the landmark works of Feldman, Knörrer, and Trubowitz summarized in [30].

Acknowledgment

It is a pleasure to thank Prof. Heinz Siedentop for his kind invitation to participate at the program on Complex Quantum Systems at the National University of Singapore. I also thank the Institute for Mathematical Sciences and the Centre for Quantum Technologies at NUS for their generous hospitality. I am grateful to the referee for detailed and helpful comments. The work presented in this review was supported by the US NSF grants DMS-0407644 / DMS-0524909 and DMS-0704031 / DMS-0940145.

2. Fermi Gas in a Random Medium

Starting with this section, we address some results concerning the dynamics of an ideal Fermi gas in a weakly disordered random medium, at positive temperature, based on joint work of the author with I. Sasaki, [15]. One of our main goals is to compare the dynamics of manybody models of this type with dynamical properties of the weakly disordered Anderson model.

We consider a fermion gas in a finite box $\Lambda_L := [-\frac{L}{2}, \frac{L}{2}]^d \cap \mathbb{Z}^d$ of side length $L \gg 1$, with periodic boundary conditions, in dimensions $d \geq 3$. We denote its dual lattice by $\Lambda_L^* := \Lambda_L / L \subset \mathbb{T}^d$. Throughout this text, we are using the notation

$$\int dp \equiv \frac{1}{|\Lambda_L|} \sum_{p \in \Lambda_L^*} \tag{2.1}$$

for brevity. We denote the Fourier transform and its inverse transform by

$$\widehat{h}(p) = \sum_{x \in \Lambda_L} h(x) e^{-2\pi i p x} \quad , \quad f^{\vee}(x) = \int dp\, f(p) e^{2\pi i p x} \,,$$

respectively. We let

$$\mathfrak{F}(\Lambda_L) = \bigoplus_{n \geq 0} \mathfrak{F}_n(\Lambda_L) \,, \tag{2.2}$$

denote the fermionic Fock space of scalar electrons where

$$\mathfrak{F}_0(\Lambda_L) = \mathbb{C} \quad , \quad \mathfrak{F}_n(\Lambda_L) = \bigwedge_1^n \ell^2(\Lambda_L) \,, \; n \geq 1 \,. \tag{2.3}$$

Moreover, we introduce creation- and annihilation operators a_p^+, a_q, for p, $q \in \Lambda_L^*$, satisfying the canonical anticommutation relations

$$a_p^+ \, a_q \, + \, a_q \, a_p^+ \; = \; \delta(p - q) \; := \; \begin{cases} L^d & \text{if } p = q \\ 0 & \text{otherwise.} \end{cases} \tag{2.4}$$

We denote by $\Omega \in \mathfrak{F}(\Lambda_L)$ the Fock vacuum vector, satisfying $a_p \Omega = 0$ for all $p \in \Lambda_L^*$. For $x \in \Lambda_L$, we define $a_x := \int dp\, e^{2\pi i p x} a_p$ and $a_x^* = (a_x)^*$.

We consider the fermionic manybody Hamiltonian

$$H_\omega \; := \; T \, + \, \eta \, V_\omega \tag{2.5}$$

where

$$T \; = \; \int dp\, E(p) \, a_p^+ \, a_p \tag{2.6}$$

is the kinetic energy operator. We assume that T is the second quantization of the centered nearest neighbor Laplacian $(\Delta f)(x) = \sum_{|y-x|=1} f(y)$ on $\mathbb{Z}^d$. Accordingly,

$$E(p) \; = \; \sum_{j=1}^{d} \cos(2\pi p_j) \,, \tag{2.7}$$

is the symbol of Δ. The operator

$$V_\omega \; := \; \sum_{x \in \Lambda_L} \omega_x \, a_x^+ a_x \tag{2.8}$$

couples the fermions to a static random potential. Similarly to the case of the Anderson model, $\{\omega_x\}_{x\in\Lambda_L}$ is a field of i.i.d. real-valued random variables which we assume to be centered, normalized, and Gaussian for simplicity. Accordingly,

$$\mathbb{E}[\,\omega_x\,] \;=\; 0\,, \;\; \mathbb{E}[\,\omega_x^2\,] \;=\; 1 \tag{2.9}$$

for all $x \in \Lambda_L$, and higher correlations, $\mathbb{E}[\prod_{\ell=1}^m \omega_{x_\ell}]$ decompose into the sums of products of all possible pair correlations (Wick's theorem).

Let

$$N \;:=\; \sum_{x\in\Lambda_L} a_x^+ a_x \tag{2.10}$$

denote the particle number operator. The fact that

$$[H_\omega, N] \;=\; 0 \tag{2.11}$$

holds can be easily verified.

Furthermore, we define a C^*-algebra $\mathfrak{A}$ by the direct limit

$$\mathfrak{A} \;=\; \overline{\bigvee_{L>0} \mathcal{B}(\mathfrak{F}(\Lambda_L))}^{\|\,\|_{op}}\,, \tag{2.12}$$

as $L \to \infty$, where $\overline{(\,\cdot\,)}^{\|\,\|_{op}}$ denotes closure with respect to the operator norm, and $B(\mathfrak{F}(\Lambda_L))$ stands for the C^*-algebra of bounded operators on $\mathfrak{F}(\Lambda_L)$. Our goal is to study the dynamics on $\mathfrak{A}$ given by

$$\alpha_t(A) \;=\; e^{itH_\omega}\, A\, e^{-itH_\omega} \tag{2.13}$$

for $A \in \mathfrak{A}$, generated by the random Hamiltonian H_ω.

2.1. *Statement of the main results*

We consider a normalized, translation-invariant, deterministic state

$$\rho_0 : \mathfrak{A} \longrightarrow \mathbb{C}\,. \tag{2.14}$$

Here, deterministic shall mean that it does not depend on $\{\omega_x\}_x$. Accordingly, we define the associated time-evolved state

$$\rho_t(A) \;:=\; \rho_0(\,e^{itH_\omega}\, A\, e^{-itH_\omega}\,)\,, \tag{2.15}$$

with $t \in \mathbb{R}$, and initial condition given by ρ_0. We particularly focus on the dynamics of the averaged two-point functions

$$\mathbb{E}[\,\rho_t(\,a_p^+ a_q\,)\,]\,, \tag{2.16}$$

where $p, q \in \Lambda_L^*$. Clearly,

$$\mathbb{E}[\,\rho_0(\,a_p^+ a_q\,)\,] \;=\; \rho_0(\,a_p^+ a_q\,) \;=\; \delta(p-q)\,\frac{1}{L^d}\,\rho_0(\,a_p^+ a_p\,)\,, \tag{2.17}$$

where

$$\delta(k) \;:=\; L^d \delta_k\,, \tag{2.18}$$

and where

$$\delta_k \;=\; \begin{cases} 1 & \text{if } k \equiv 0 \bmod \Lambda_L^* \\ 0 & \text{otherwise} \end{cases} \tag{2.19}$$

denotes the Kronecker delta on the lattice Λ_L^* (mod $\mathbb{T}^d$). We remark that for fermions,

$$0 \;\leq\; \frac{1}{L^d}\,\rho_0(\,a_p^+ a_p\,) \;\leq\; 1\,, \tag{2.20}$$

since $\|a_p^{(+)}\| = L^{d/2}$ in operator norm, $\forall p \in \Lambda_L^*$.

2.2. *Boltzmann limit of the momentum distribution function*

We denote the microscopic time, position, and velocity variables by (t, x, v), and the corresponding macroscopic variables by $(T, X, V) = (\eta^2 t, \eta^2 x, v)$. We prove that the momentum distribution $f_t(q)$ converges to a solution of a linear Boltzmann equation in the limit $\eta \to 0$.

Theorem 2.1: (T. Chen and I. Sasaki, [15]) *Let ρ_0 be a translation invariant state on $\mathfrak{A}$. Then, the averaged two-point functions are translation invariant (i.e., diagonal in a_p^+, a_p),*

$$\mathbb{E}[\rho_t(\,a^+(f) a(g)\,)] \;=\; \int dp\, f(p)\, \overline{g(p)}\, \mathbb{E}[\rho_t(\,a_p^+ a_p\,)]\,, \tag{2.21}$$

for any $f, g \in \mathcal{S}(\mathbb{T}^d)$ of Schwartz class (which have rapidly decaying Fourier coefficients), with $a(g) = \int dp\, \overline{g}(p) a_p$ and $a^+(f) = (a(f))^*$. Moreover, the thermodynamic limit

$$\Omega_T^{(2;\eta)}(f; g) \;:=\; \lim_{L\to\infty} \mathbb{E}[\rho_{T/\eta^2}(\,a^+(f)\,a(g)\,)] \tag{2.22}$$

exists for all $f, g \in \mathcal{S}(\mathbb{T}^d)$, and $T > 0$.

For any $T > 0$ and all $f, g \in \mathcal{S}(\mathbb{T}^d)$, the limit

$$\Omega_T^{(2)}(f;g) := \lim_{\eta \to 0} \Omega_T^{(2;\eta)}(f;g) \tag{2.23}$$

exists, and is the inner product of f, g with respect to a Borel measure $F_T(p)dp$,

$$\Omega_T^{(2)}(f;g) = \int dp\, F_T(p)\, f(p)\, \overline{g(p)}\,, \tag{2.24}$$

where $F_T(V)$ satisfies the linear Boltzmann equation

$$\partial_T F_T(V) = 2\pi \int_{\mathbb{T}^d} dU\, \delta(\, E(U) - E(V)\,)\,(\, F_T(U) - F_T(V)\,)\,, \tag{2.25}$$

with initial condition

$$F_0(p) = \lim_{L \to \infty} \frac{1}{L^d}\, \rho_0(\, a_p^+\, a_p\,) \tag{2.26}$$

for $p \in \mathbb{T}^d$.

2.3. *Outline of the proof*

We give a sketchy outline of the proof of Theorem 2.1. The strategy employed applies results in the works [12, 25, 33] which address the Anderson model at weak disorders. To begin with, we consider the Heisenberg evolution of the creation- and annihilation operators,

$$a(f,t) := e^{itH_\omega} a(f) e^{-itH_\omega}\,, \tag{2.27}$$

where f denotes an arbitrary test function (more precisely, f is a Schwartz class function on the unit torus $\mathbb{T}^d$ in frequency space, to accommodate the thermodynamic limit under which $\Lambda_L^* \to \mathbb{T}^d$ which will eventually be taken).

We make the key observation that since H_ω is bilinear in a^+, a, it follows that $a(f,t)$ is a linear superposition of annihilation operators. Therefore, there exists a function f_t such that

$$a(f,t) = a(f_t) = \int dp\, \overline{f_t(p)}\, a_p = (a^+(f_t))^*\,. \tag{2.28}$$

In particular,

$$\begin{aligned} i\partial_t a(f_t) &= [\, H_\omega\,,\, a(f_t)\,] \\ &= \int dp\, f_t(p)\, E(p)\, a_p + \eta \int dp \int du\, f_t(p)\, \widehat{V}_\omega(u-p)\, a_u \\ &= a(\,\Delta f_t^\vee\,) + a(\,\eta\, V_\omega^{(1)} f_t^\vee\,)\,, \end{aligned} \tag{2.29}$$

and moreover, it is clear that $a(f,0) = a(f_0) = a(f)$. We recall that Δ denotes the nearest neighbor Laplacian on Λ_L. Here, $H_\omega^{(1)} = H_\omega|_{\mathfrak{F}_1}$ stands for the 1-particle Anderson Hamiltonian, and $V_\omega^{(1)} = V_\omega|_{\mathfrak{F}_1}$ is the 1-particle multiplication operator $(V_\omega^{(1)} f^\vee)(x) = \omega_x f^\vee(x)$.

Thus, f_t is the solution of the 1-particle random Schrödinger equation generated by the weakly disordered Anderson model,

$$i\partial_t f_t^\vee = H_\omega^{(1)} f_t^\vee := \Delta f_t^\vee + \eta V_\omega^{(1)} f_t^\vee , \tag{2.30}$$

with initial condition given by the test function

$$f_0^\vee = f^\vee . \tag{2.31}$$

Accordingly,

$$\begin{aligned} \rho_t(a^+(f) a(g)) &= \rho_0(a^+(f_t) a(g_t)) \\ &= \int dp\, dq\, \rho_0(a_p^+ a_q) f_t(p) \overline{g_t(q)} \\ &= \int dp\, J(p) f_t(p) \overline{g_t(p)} . \end{aligned} \tag{2.32}$$

We note here that the momentum conservation condition

$$\rho_0(a_p^+ a_q) = \delta(p-q) J(p) \tag{2.33}$$

follows from translation invariance of ρ_0, with

$$0 \leq J(p) = \frac{1}{L^d} \rho_0(a_p^+ a_p) = \frac{1}{1+e^{h(p)}} \leq 1 , \tag{2.34}$$

where $J(p) \leq 1$ follows from the Pauli principle.

We pick some $N \in \mathbb{N}$ which remains to be optimized later, and expand f_t, g_t into the truncated Duhamel series,

$$f_t = f_t^{(\leq N)} + f_t^{(>N)} , \tag{2.35}$$

with

$$f_t^{(\leq N)} := \sum_{n=0}^{N} f_t^{(n)} . \tag{2.36}$$

The Duhamel term of n-th order (in powers of η) is given by

$$\begin{aligned} f_t^{(n)}(p) := \eta^n e^{\epsilon t} \int d\alpha\, e^{it\alpha} \int dk_0 \cdots dk_n\, \delta(p-k_0) \\ \Big(\prod_{j=0}^{n} \frac{1}{E(k_j) - \alpha - i\epsilon} \Big) \Big(\prod_{j=1}^{n} \widehat{V}_\omega(k_j - k_{j-1}) \Big) f(k_n) . \end{aligned} \tag{2.37}$$

Similarly to the Boltzmann limit for the weakly disordered Anderson model in [25, 33, 12], we choose

$$\epsilon = \frac{1}{t} \tag{2.38}$$

so that the factor $e^{\epsilon t}$ remains bounded for all t. By

$$f_t^{\vee(>N)} = i\eta \int_0^t ds\, e^{i(t-s)H_\omega} V_\omega^{(1)} f_t^{\vee(N)}(s) \,, \tag{2.39}$$

we account for the Duhamel remainder term.

Accordingly,

$$\rho_t(\, a^+(f)\, a(g)\,) = \rho_0(\, a^+(f_t)\, a(g_t)\,) = \sum_{n,\widetilde{n}\in\mathcal{I}_N} \rho_t^{(n,\widetilde{n})}(f;g) \tag{2.40}$$

where

$$\rho_t^{(n,\widetilde{n})}(f;g) := \rho_0(\, a^+(f_t^{(n)})\, a(g_t^{(\widetilde{n})})\,) \tag{2.41}$$

for $\mathcal{I}_N := \{1, \ldots, N, > N\}$.

Next, we use the following notation. If $n, \widetilde{n} \leq N$, and $n + \widetilde{n}$ is odd, $\mathbb{E}[\rho_t^{(n,\widetilde{n})}(p,q)] = 0$. Thus, let

$$\bar{n} := \frac{n + \widetilde{n}}{2} \in \mathbb{N} \,, \tag{2.42}$$

and we define $\{u_j\}_{j=0}^{2\bar{n}+1}$ by

$$u_j := \begin{cases} k_{n-j} & \text{if } j \leq n \\ \widetilde{k}_{j-n-1} & \text{if } j \geq n+1\,. \end{cases} \tag{2.43}$$

Thus, for $n, \widetilde{n} \leq N$ (and $\widehat{V}_\omega(u)^* = \widehat{V}_\omega(-u)$),

$$\begin{aligned} \mathbb{E}[\rho_t^{(n,\widetilde{n})}(f;g)] &= \eta^{2\bar{n}}\, e^{2\epsilon t} \int d\alpha\, d\widetilde{\alpha}\, e^{it(\alpha-\widetilde{\alpha})} \\ &\int du_0 \cdots du_{2\bar{n}+1}\, f(u_0)\, \overline{g(u_{2\bar{n}+1})}\, J(u_n)\, \delta(u_n - u_{n+1}) \\ &\prod_{j=0}^{n} \frac{1}{E(u_j) - \alpha - i\epsilon} \prod_{\ell=n+1}^{2\bar{n}+1} \frac{1}{E(u_\ell) - \widetilde{\alpha} + i\epsilon} \\ &\mathbb{E}\Big[\prod_{j=1}^{n} \widehat{V}_\omega(u_j - u_{j-1}) \prod_{j=n+2}^{2\bar{n}+1} \widehat{V}_\omega(u_j - u_{j-1}) \Big]. \end{aligned} \tag{2.44}$$

This expression is analogous to those occurring in the analysis of the Anderson model in [25, 33, 12].

2.4. *Feynman graph expansion*

We take the expectation with respect to the random potential, and compute all correlations explicitly. To this end, we introduce the set of *Feynman graphs* $\Gamma_{n,\widetilde{n}}$, with $n+\widetilde{n}\in 2\mathbb{N}$, as follows.

We consider two horizontal solid lines, which we refer to as *particle lines*, joined by a distinguished vertex which we refer to as the ρ_0-vertex (corresponding to the term $\rho_0(\,a^+_{u_n}\,a_{u_{n+1}}\,)$, determined by J). On the line on its left, we introduce n vertices, and on the line on its right, we insert $\widetilde{n}$ vertices. We refer to those vertices as *interaction vertices*, and enumerate them from 1 to $2\bar{n}$ starting from the left. The edges between the interaction vertices are referred to as *propagator lines*. We label them by the momentum variables u_0, ..., $u_{2\bar{n}+1}$, increasingly indexed starting from the left. To the j-th propagator line, we associate the resolvent $\frac{1}{E(u_j)-\alpha-i\epsilon}$ if $0\le j\le n$, and $\frac{1}{E(u_j)-\widetilde{\alpha}+i\epsilon}$ if $n+1\le j\le 2\bar{n}+1$. To the ℓ-th interaction vertex (adjacent to the edges labeled by $u_{\ell-1}$ and u_ℓ), we associate the random potential $\widehat{V}_\omega(u_\ell-u_{\ell-1})$, where $1\le\ell\le 2\bar{n}+1$.

A *contraction graph* associated to the above pair of particle lines joined by the ρ_0-vertex, and decorated by $n+\widetilde{n}$ interaction vertices, is the graph obtained by pairwise connecting interaction vertices by dashed *contraction lines*. We denote the set of all such contraction graphs by $\Gamma_{n,\widetilde{n}}$; it contains

$$|\Gamma_{n,\widetilde{n}}| \;=\; (2\bar{n}-1)(2\bar{n}-3)\cdots 3\cdot 1 \;=\; \frac{(2\bar{n})!}{\bar{n}!2^{\bar{n}}} \;=\; O(\bar{n}!) \tag{2.45}$$

elements.

If in a given graph $\pi\in\Gamma_{n,\widetilde{n}}$, the ℓ-th and the ℓ'-th vertex are joined by a contraction line, we write

$$\ell\sim_\pi\ell'\,, \tag{2.46}$$

and we associate the delta distribution

$$\delta(u_\ell-u_{\ell-1}-(u_{\ell'}-u_{\ell'-1})) \;=\; \mathbb{E}[\,\widehat{V}_\omega(\,u_\ell-u_{\ell-1}\,)\,\widehat{V}_\omega(\,u_{\ell'}-u_{\ell'-1}\,)\,] \tag{2.47}$$

to this contraction line.

2.5. *Classification of graphs*

Next, we classify Feynman graphs as follows; see [12, 33], and Figure 1.

- A subgraph consisting of one propagator line adjacent to a pair of vertices ℓ and $\ell+1$, and a contraction line connecting them, i.e., $\ell\sim_\pi\ell+1$, where both ℓ, $\ell+1$ are either $\le n$ or $\ge n+1$, is called an *immediate recollision*.

- The graph $\pi \in \Gamma_{n,n}$ (i.e., $n = \widetilde{n} = \bar{n}$) with $\ell \sim_\pi 2n+1-\ell$ for all $\ell = 1, \ldots, n$, is called a *basic ladder* diagram. The contraction lines are called *rungs* of the ladder. We note that a rung contraction always has the form $\ell \sim_\pi \ell'$ with $\ell \le n$ and $\ell' \ge n+1$. Moreover, in a basic ladder diagram one always has that if $\ell_1 \sim_\pi \ell'_1$ and $\ell_2 \sim_\pi \ell'_2$ with $\ell_1 < \ell_2$, then $\ell'_2 < \ell'_1$.
- A diagram $\pi \in \Gamma_{n,\widetilde{n}}$ is called a *decorated ladder* if any contraction is either an immediate recollision, or a rung contraction $\ell_j \sim_\pi \ell'_j$ with $\ell_j \le n$ and $\ell'_j \ge n$ for $j = 1, \ldots, k$, and $\ell_1 < \cdots < \ell_k$, $\ell'_1 > \cdots > \ell'_k$. Evidently, a basic ladder diagram is the special case of a decorated ladder which contains no immediate recollisions (so that necessarily, $n = \widetilde{n}$).
- A diagram $\pi \in \Gamma_{n,\widetilde{n}}$ is called *crossing* if there is a pair of contractions $\ell \sim_\pi \ell'$, $j \sim_\pi j'$, with $\ell < j < \ell' < j'$.
- A diagram $\pi \in \Gamma_{n,\widetilde{n}}$ is called *nesting* if there is a subdiagram with $\ell \sim_\pi \ell + 2k$, with $k \ge 1$, and either $\ell \ge n+1$ or $\ell + 2k \le n$, with $j \sim_\pi j+1$ for $j = \ell+1, \ell+3, \ldots, \ell+2k-1$. The latter corresponds to a progression of $k-1$ immediate recollisions.

We note that any diagram that is not a decorated ladder contains at least a crossing or a nesting subdiagram.

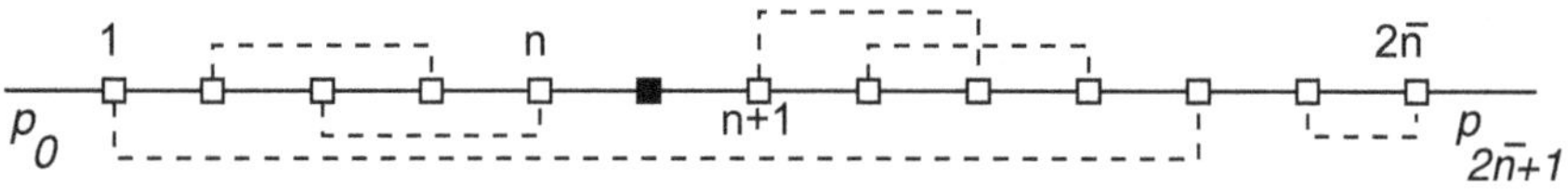

Fig. 1. An example of a Feynman graph, $\pi \in \Gamma_{n,\widetilde{n}}$, with $n = 4$, $\widetilde{n} = 6$. The distinguished vertex is the ρ_0-vertex.

Accordingly, to prove the theorem, we first control the thermodynamic limit $L \to \infty$, for an arbitrary but fixed $\eta > 0$. Subsequently, we show that the Feynman amplitudes of crossing and nesting diagrams yield small error terms that vanish as $\eta \searrow 0$, while that the amplitudes of decorated ladder diagrams are dominant. Finally, the sum of Feynman amplitudes associated to decorated ladder diagrams is verified to solve the linear Boltzmann equation, as asserted in the theorem. □

2.6. *Discussion of the result*

An example of an initial condition of particular physical interest is given by a Gibbs state (with inverse temperature β and chemical potential μ) for a non-interacting fermion gas,

$$\rho_0(A) = \frac{1}{Z_{\beta,\mu}} \operatorname{Tr}(e^{-\beta(T-\mu N)} A) \tag{2.48}$$

where $Z_{\beta,\mu} := \operatorname{Tr}(e^{-\beta(T-\mu N)})$.

The momentum distribution in the free Gibbs state is the *Fermi-Dirac distribution*

$$F_0(p) = \lim_{L\to\infty} \rho_0(\frac{1}{L^d} a_p^+ a_p) = \frac{1}{1+e^{\beta(E(p)-\mu)}} .$$

The probability of having a plane wave with momentum p then is $\frac{F_0(p)}{\int dp\, F_0(p)}$.

We make the key observation that for all $0 < \beta \leq \infty$,

$$F_0(p) = \frac{1}{1+e^{\beta(E(p)-\mu)}}$$

is a *stationary solution* of the Boltzmann equation. This remains true in zero temperature limit $\beta \to \infty$ where (in the weak sense)

$$\frac{1}{1+e^{\beta(E(p)-\mu)}} \to \chi[E(p) < \mu]$$

which is nontrivial whenever $\mu > 0$.

3. Persistence of Quasifreeness in the Boltzmann Limit

We shall next address the persistence of the property of *quasifreeness* under the dynamics considered above. A state ρ_0 is quasifree (determinantal) if

$$\begin{aligned} &\rho_0(a^+(f_1) \cdots a^+(f_r) a(g_1) \cdots a(g_s)) \\ &\qquad = \delta_{r,s} \det \left[\rho_0(a^+(f_i) a(g_j)) \right]_{1\leq i,j\leq r} . \end{aligned}$$

In a quasifree state, the particles are uncorrelated, and the Pauli principle is made evident by the explicit determinantal structure. Here we are particularly interested in the influence of the random potential on the property of quasirandomness.

We observe that since H_ω is bilinear in the creation- and annihilation operators, it follows that for any operator of the form $K = \int dpdq\, \kappa(p,q) a_p^+ a_q$ bilinear in a^+, a (with suitable kernel κ),

$$K(t) := e^{itH_\omega} K e^{-itH_\omega} \tag{3.1}$$

is also bilinear in a^+, a. Therefore, if K is deterministic,

$$\rho_t(A) = \frac{1}{Z_K} \operatorname{Tr}(e^{-K(t)} A) \tag{3.2}$$

is quasifree with probability 1 (where $Z_K := \operatorname{Tr}(e^{-K})$).

However, although quasifreeness holds almost surely, the *average* with respect to the random potential, $\mathbb{E}[\rho_t(\,\cdot\,)]$, is *not quasifree*, for any $\eta > 0$, and $t > 0$. Notably, quasifreeness is a *nonlinear* condition on determinants. Nevertheless, the kinetic scaling limit produces a quasifree limiting state.

Theorem 3.1: (T. Chen and I. Sasaki, [15]) *Let ρ_0 be a number conserving, quasifree, and translation invariant state on $\mathfrak{A}$. Then, the following holds. For any normal ordered monomial in creation- and annihilation operators,*

$$a^+(f_1)\cdots a^+(f_r)\,a(g_1)\cdots a(g_r)\,, \tag{3.3}$$

with $r, s \in \mathbb{N}$ and Schwartz class test functions $f_j, g_\ell \in \mathcal{S}(\mathbb{T}^d)$, and any $T > 0$, the macroscopic $2r$-point function

$$\begin{aligned}&\Omega_T^{(2r)}(\,f_1,\ldots,f_r\,;\,g_1,\ldots,g_r\,)\\&\quad:= \lim_{\eta\to 0}\lim_{L\to\infty}\mathbb{E}[\rho_{T/\eta^2}(\,a^+(f_1)\cdots a^+(f_r)\,a(g_1)\cdots a(g_r)\,)]\end{aligned} \tag{3.4}$$

exists and is quasifree,

$$\Omega_T^{(2r)}(\,f_1,\ldots,f_r\,;\,g_1,\ldots,g_r\,) = \det\big[\,\Omega_T^{(2)}(\,f_i\,,\,g_j\,)\,\big]_{1\le i,j\le r}\,. \tag{3.5}$$

The macroscopic 2-point function is the same as in Theorem 2.1,

$$\Omega_T^{(2)}(\,f\,;\,g\,) = \int dp\,F_T(p)\,f(p)\,\overline{g(p)}\,, \tag{3.6}$$

and $F_T(p)$ solves the linear Boltzmann equation (2.25) with initial condition (2.26).

For the proof, we employ the fact that the main estimate

$$\lim_{\eta\to 0}\lim_{L\to\infty}\Big|\,\mathbb{E}[\rho_{T/\eta^2}(\,a^+(f_1)\cdots a^+(f_r)\,a(g_1)\cdots a(g_r)\,)] - \det\big[\,\Omega_T^{(2)}(\,f_i\,;\,g_j\,)\,\big]_{1\le i,j\le r}\Big| = 0$$

can be interpreted as a corollary of results in [13]. □

3.1. *Outline of the proof of Theorem 3.1*

The result obtained in Theorem 3.1 follows from an application of results in [13] where we refer the reader for details. We shall here give a rough outline of the strategy. The expectation

$$\lim_{L\to\infty}\mathbb{E}\Big[\,\prod_{j=1}^{r}\rho_t(\,a^+(f_j)\,a(g_{s(j)})\,)\,\Big] \tag{3.7}$$

can be represented by a graph expansion as follows. We expand each of the factors

$$\rho_t(\, a^+(f_j)a(g_{s(j)})\,) \;=\; \sum_{n,\widetilde{n}=1}^{N+1} \int dp\, J(p)\, f_{j,t}^{(n)}(p)\, \overline{g_{s(j),t}^{(\widetilde{n})}(p)} \tag{3.8}$$

separately into a truncated Duhamel series of level N, using the same definitions as in (2.40). The treatment of the remainder term (where at least one of the indices $n, \widetilde{n}$ equals $N+1$) is more involved, and we refer to [12, 25, 33] for details.

For the analysis of the expectation (3.7), we further develop the Feynman graph expansions in the previous section. For $r > 1$, we consider r particle lines parallel to one another, each containing a distinguished ρ_0-vertex separating it into a left and a right part. Enumerating them from 1 to r, the j-th particle line contains n_j interaction vertices on the left of the ρ_0-vertex, and $\widetilde{n}_j$ interaction vertices on its right. We note that for $r > 1$, only $\sum_{j=1}^{r}(n_j + \widetilde{n}_j)$ has to be an even number, but not each individual

$$\widehat{n}_j \;:=\; n_j \,+\, \widetilde{n}_j\,. \tag{3.9}$$

On the j-th interaction line, we label the propagator lines by momentum variables $u_0^{(j)}, \dots, u_{\widehat{n}_j+1}^{(j)}$, with indices increasing from the left, see also Figure 2.

A *contraction graph* of degree $\{(n_j, \widetilde{n}_j)\}_{j=1}^r$ is obtained by connecting pairs of interaction vertices by contraction lines. We denote the set of contraction graphs of degree $\{(n_j, \widetilde{n}_j)\}_{j=1}^r$ by $\Gamma_{\{(n_j,\widetilde{n}_j)\}_{j=1}^r}$. If the ℓ-th vertex

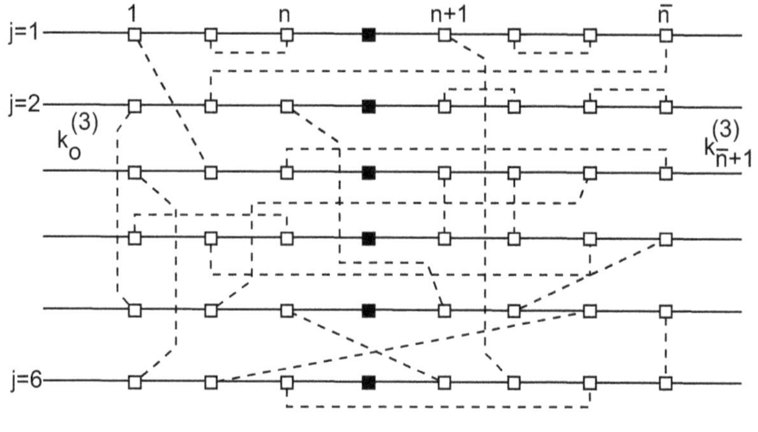

Fig. 2. Order r Feynman graph. The particle line indexed by $j = 3$ is disconnected.

on the j-th particle line is connected by a contraction line to the ℓ'-th vertex on the j'-th particle line, we write

$$(j;\ell) \sim_\pi (j';\ell'). \tag{3.10}$$

To a graph $\pi \in \Gamma_{\{(n_j,\widetilde{n}_j)\}_{j=1}^r}$, we associate the *Feynman amplitude*

$$\begin{aligned}\mathrm{Amp}_\pi(\{f_j, g_{s(j)}\};\eta;T) &:= \eta^{2\sum_{1\le j\le r}(n_j+\widetilde{n}_j)}\, e^{2r\epsilon t} \prod_{j=1}^r \int d\alpha_j\, d\widetilde{\alpha}_j\, e^{it(\alpha_j-\widetilde{\alpha}_j)} \\ &\int du_0^{(j)} \cdots du_{\widetilde{n}_j+1}^{(j)}\, f_j(u_0^{(j)})\, \overline{g_{s(j)}(u_{\widetilde{n}_j+1}^{(j)})}\, J(u_{n_j}^{(j)})\, \delta(u_{n_j}^{(j)} - u_{n_j+1}^{(j)}) \\ &\delta_\pi(\{u_i^{(j)}\}_{i=0}^{\widetilde{n}_j+1}) \prod_{\ell=0}^{n_j} \frac{1}{E(u_\ell^{(j)})-\alpha_j-i\epsilon} \prod_{\ell'=n_j+2}^{\widetilde{n}_j} \frac{1}{E(u_{\ell'}^{(j)})-\widetilde{\alpha}_j+i\epsilon},\end{aligned} \tag{3.11}$$

where

$$\epsilon = \frac{1}{t} = \frac{\eta^2}{T} \tag{3.12}$$

for $T > 0$. The delta distribution

$$\delta_\pi(\{u_j^{(j)}\}_{j=0}^{\widetilde{n}_j+1}) = \prod_{(j;\ell)\sim_\pi(j';\ell')} \delta(u_\ell^{(j)} - u_{\ell-1}^{(j)} - (u_{\ell'}^{(j')} - u_{\ell'-1}^{(j')})) \tag{3.13}$$

is the product of delta distributions associated to all contraction lines in π.

3.1.1. *Completely disconnected graphs*

The subclass

$$\Gamma^{disc}_{\{(n_j,\widetilde{n}_j)\}_{j=1}^r} \subset \Gamma_{\{(n_j,\widetilde{n}_j)\}_{j=1}^r} \tag{3.14}$$

of *completely disconnected* graphs of degree $\{(n_j,\widetilde{n}_j)\}_{j=1}^r$ consists of those graphs in which contraction lines only connect interaction vertices on the same particle line.

It follows from the proof of Theorem 2.1 that

$$\lim_{L\to\infty} \sum_{\substack{0\le n_j,\widetilde{n}_j\le N\\ j=1,\dots,r}} \sum_{\pi\in\Gamma^{disc}_{\{(n_j,\widetilde{n}_j)\}_{j=1}^r}} \mathrm{Amp}_\pi(\{f_j, g_{s(j)}\};\eta;T) \tag{3.15}$$

$$\begin{aligned}&= \lim_{L\to\infty} \prod_{j=1}^r \sum_{n_j,\widetilde{n}_j=1}^N \mathbb{E}[\int dp\, J(p)\, f_{j,T/\eta^2}(p)\, \overline{g_{s(j)},T/\eta^2(p)}\,] \\ &= \lim_{L\to\infty} \prod_{j=1}^r \left(\mathbb{E}[\rho_{T/\eta^2}(a^+(f_j)a(g_{s(j)}))] + O(\eta^\delta)\right).\end{aligned} \tag{3.16}$$

The term of order $O(\eta^\delta)$ accounts for the remainder term associated to the j-th particle line (i.e., the terms involving $\mathbb{E}[\,\rho^{(n_j,\widetilde{n}_j)}_{T/\eta^2}(p,q)\,]$ where at least one of the indices $n_j, \widetilde{n}_j$ equals N). Thus, for any fixed $r \in \mathbb{N}$, we obtain

$$\lim_{\eta\to 0}\lim_{L\to\infty} \sum_{\substack{0\le n_j,\widetilde{n}_j\le N\\ j=1,\dots,r}} \sum_{\pi\in\Gamma^{disc}_{\{(n_j,\widetilde{n}_j)\}_{j=1}^r}} \mathrm{Amp}_\pi(\{f_j, g_{s(j)}\};\eta;T)$$
$$= \prod_{j=1}^r \Omega^{(2)}_T(\,f_j\,;\,g_{s(j)}\,)\,. \tag{3.17}$$

That is, the sum over completely disconnected graphs yields the product of averaged 2-point functions in the kinetic scaling limit.

3.1.2. *Non-disconnected graphs*

Following [13], we refer to the complement of the set of completely disconnected graphs in $\Gamma_{\{(n_j,\widetilde{n}_j)\}_{j=1}^r}$,

$$\Gamma^{n-d}_{\{(n_j,\widetilde{n}_j)\}_{j=1}^r} := \Gamma_{\{(n_j,\widetilde{n}_j)\}_{j=1}^r} \setminus \Gamma^{disc}_{\{(n_j,\widetilde{n}_j)\}_{j=1}^r}\,, \tag{3.18}$$

as the set of *non-disconnected graphs*. The remaining part of the proof of Theorem 3.1 consists of showing that the sum over non-disconnected graphs, combined with the remainder terms, can be bounded by $O(\eta^\delta)$, for L sufficiently large, and some $\delta > 0$.

The following condition is required in [13] (which addresses the weakly disordered Anderson model) for the validity of the estimate analogous to (3.22) below: The initial condition ϕ_0 (corresponding to the test functions f_j, g_ℓ in our case) of the random Schrödinger evolution studied in [13] is assumed to satisfy a "concentration of singularity condition" (that is, singularities in momentum space are not too much "spread out" in the limit $\eta \to 0$). It states that in frequency space $\mathbb{T}^d$,

$$\widehat{\phi}_0 = \widehat{\phi}_0^{(reg)} + \widehat{\phi}_0^{(sing)}\,, \tag{3.19}$$

where

$$\|\,\widehat{\phi}_0^{(reg)}\,\|_\infty < c \tag{3.20}$$

and

$$\|\,|\widehat{\phi}_0^{(sing)}| * |\widehat{\phi}_0^{(sing)}|\,\|_2 < c'\,\eta^{3/2} \tag{3.21}$$

are satisfied uniformly in L, as $L \to \infty$.

In the case at hand, this means that we have to require that f_j, g_ℓ satisfy the concentration of singularity condition. This is, however, naturally fulfilled since f_j, g_ℓ are η-independent Schwartz class functions (in contrast, the initial states considered in [13] are of WKB type, and scale non-trivially with η.) Hence, all results in [13] can indeed be adopted straightforwardly.

In [13], it is established that the amplitude of every non-disconnected graph with $n_j, \widetilde{n}_j \leq N$ for $j = 1, \dots, r$, is bounded by

$$\sup_{\pi \in \Gamma^{n-d}_{\{(n_j,\widetilde{n}_j)\}_{j=1}^r}} \left| \mathrm{Amp}_\pi(\{f_j, g_{s(j)}\}; \eta; T) \right| \tag{3.22}$$

$$< \epsilon^{1/5} \, (c\,\eta^2 \, \epsilon^{-1} \, \log \frac{1}{\epsilon})^{\frac{r}{2}\sum_{j=1}^r \widehat{n}_j} (\log \frac{1}{\epsilon})^{4r} \,, \tag{3.23}$$

where we recall that $\epsilon = \frac{1}{t} = \frac{\eta^2}{T}$ for $T > 0$. The main property of this key estimate is a factor $\epsilon^{1/5}$ smaller than the sum over amplitudes of disconnected graphs; this improvement is obtained from exploiting the structure of momentum space constraints implied by the existence of at least one contraction line connecting two different particle lines; see [13].

The number of non-disconnected graphs is bounded by

$$\left| \Gamma^{n-d}_{\{(n_j,\widetilde{n}_j)\}_{j=1}^r} \right| \leq (\sum_{j=1}^r \widehat{n}_j)! \leq (2rN)! \tag{3.24}$$

where $\widehat{n}_j = n_j + \widetilde{n}_j$. Therefore, the sum of amplitudes of all non-disconnected graphs with $0 \leq n_j, \widetilde{n}_j \leq N$ is bounded by

$$\sum_{1\leq j\leq r} \; \sum_{0\leq n_j,\widetilde{n}_j\leq N} \; \sum_{\pi \in \Gamma^{n-d}_{\{(n_j,\widetilde{n}_j)\}_{j=1}^r}} \left| \mathrm{Amp}_\pi(\{f_j, g_{s(j)}\}; \eta; T) \right| \tag{3.25}$$

$$\leq ((2rN)!)^2 \epsilon^{1/5} \, (c\,\eta^2 \, \epsilon^{-1} \, \log \frac{1}{\epsilon})^{rN} (\log \frac{1}{\epsilon})^{4r}$$

$$\leq \eta^{c_0 r} \,, \tag{3.26}$$

for some $c_0 > 0$, given the choice of parameters $t = \frac{1}{\epsilon} = \frac{T}{\eta^2}$ (with T denoting the macroscopic time) and $N = C \frac{|\log \eta|}{\log |\log \eta|}$, see also [25, 33, 12]. Here we have estimated the sum over pairs $0 \leq n_j, \widetilde{n}_j \leq N$, $1 \leq j \leq r$, by a factor $(2rN)!$, since $\#\{(n_j, \widetilde{n}_j)\}_{j=1}^r \,|\, \sum_j \widehat{n}_j = m \} \leq m!$.

The discussion of Duhamel remainder terms is more technical, but similarly as in [13], it is shown in [15] that all expressions in the expansion comprising at least one Duhamel remainder term (with index $N + 1$) can be bounded by $C^r \, \eta^{c_0 r}$ for a constant C independent of η and r. Combined with (3.25) for the given choice of parameters, this implies that in the limit $\eta \to 0$, only the contributions associated to completely disconnected graphs survive, thus verifying quasifreeness in the kinetic scaling limit. This completes the proof of Theorem 4.4. For more details addressing the arguments outlined here, we refer to [13, 15].

4. Fermi Gas with Dynamical Hartree-Fock Interactions

In our next step, we include particle interactions between the fermions, modeled in dynamical Hartree-Fock theory. As a consequence, the main task is to control both randomness and the nonlinearities arising from the self-interaction of the field. The results presented here are based on joint work with I. Rodnianski, [16].

We consider the time-dependent Hamiltonian

$$H(t) = T + \eta V_\omega + \lambda W(t) \tag{4.1}$$

where the fermion-fermion interaction is modeled by

$$W(t) = \sum_{x,y} v(x-y) \{ \mathbb{E}[\rho_t(a_x^+ a_x)] a_y^+ a_y - \mathbb{E}[\rho_t(a_y^+ a_x)] a_x^+ a_y \}. \tag{4.2}$$

The terms on the rhs correspond to the Hartree-Fock direct and exchange term, respectively. The coupling constant λ accounts for the strength of interaction between the fermions. The kinetic energy operator T and the operator V_ω which describes the interaction of each fermion with the static random potential are as in the previous section. For technical reasons that we will not further address here, we assume that $\|\widehat{v}\|_{H^{3/2+\delta}(\mathbb{T}^3)} < C$ for $\delta > 0$ arbitrary but fixed.

We are interested in the dynamics of the two-point function, which is determined by

$$\begin{aligned} & i\partial_t \rho_t(a_p^+ a_q) \\ & = (E(p) - E(q)) \rho_t(a_p^+ a_q) \\ & \quad + \lambda \int du \, \mathbb{E}[\rho_t(\frac{1}{L^d} a_u^+ a_u)] (\widehat{v}(u-p) \rho_t(a_u^+ a_q) - \widehat{v}(q-u) \rho_t(a_p^+ a_u)) \\ & \quad + \eta \int du \, \widehat{V}_\omega(u-p) \rho_t(a_u^+ a_q) - \widehat{V}_\omega(q-u) \rho_t(a_p^+ a_u) \end{aligned} \tag{4.3}$$

for any realization of the random potential, where

$$\widehat{V}_\omega(u) := \sum_{x \in \Lambda_L} \omega_x e^{-2\pi i u \cdot x} \tag{4.4}$$

is well-defined, almost surely (where we ultimately let $L \to \infty$).

We make the following key observations:

- For a generic realization of the random potential, the problem is *not* translation invariant.
- The equation (4.3) for the momentum distribution function $\rho_t(a_p^+ a_p)$ does *not* close.

However, we can close the equation for the momentum distribution function by taking the expectation, $\mathbb{E}$, with respect to the random potential. In particular, the $\mathbb{E}$-average is translation invariant, due to the homogeneity of the randomness. Accordingly, the average state $\mathbb{E}[\rho_t(\,\cdot\,)] : \mathfrak{A} \to \mathbb{C}$ solves

$$\begin{aligned} i\partial_t \, \mathbb{E}[\, \rho_t(\, A\,)\,] &= \mathbb{E}[\, \rho_t(\,[\,H(t)\,,\,A\,]\,)\,] \\ \mathbb{E}[\rho_0] &= \rho_0\,. \end{aligned} \tag{4.5}$$

This is a self-consistent nonlinear initial value problem determining $\mathbb{E}[\rho_t(\,\cdot\,)]$.

We note that for almost every realization of V_ω, we have

$$\rho_t(\,A\,) \;=\; \rho_0(\,\mathcal{U}_t^*\,A\,\mathcal{U}_t\,)$$

for $A \in \mathfrak{A}$, with $\mathcal{U}_t$ unitary,

$$i\partial_t\,\mathcal{U}_t \;=\; H(t)\,\mathcal{U}_t\,,$$

and $\mathcal{U}_0 = \mathbf{1}$. Notably, the Hamiltonian $H(t)$ itself depends on $\mathbb{E}[\rho_t(\,\cdot\,)]$. In particular, we note that

$$\rho_t(\,a^+(f)\,a(g)\,) \;=\; \rho_0(\,a^+(f,t)\,a(g,t)\,)\,. \tag{4.6}$$

The Heisenberg evolution of the creation- and annihilation operators is determined by

$$a(f,t) \;:=\; \mathcal{U}_t^*\,a(f)\,\mathcal{U}_t\,. \tag{4.7}$$

Similarly as in the case discussed for the ideal Fermi gas, there exists a function f_t such that

$$a(f,t) \;=\; a(f_t)\,, \tag{4.8}$$

where f_t is the solution of the 1-particle random Schrödinger equation

$$i\partial_t f_t(p) \;=\; E(p)f_t(p) \,+\, \eta\,(\,\widehat{V}_\omega * f_t\,)(p) \,-\, \lambda\,(\,\widehat{v} * \mu_t\,)(p)f_t(p) \tag{4.9}$$

with initial condition

$$f_0 \;=\; f\,. \tag{4.10}$$

Noting that the Hamiltonian $H(t)$ itself depends on the unknown quantity

$$\mu_t(p) := \frac{1}{L^3}\,\mathbb{E}[\,\rho_t(\,a_p^+ a_p\,)\,]\,, \tag{4.11}$$

we determine μ_t by writing the solution to (4.5) in integral form, as an expansion in powers of η.

For arbitrary test functions f and g, we consider the pair correlation function

$$\begin{aligned}\rho_t(\,a^+(f)\,a(g)\,) &= \rho_0(\,a^+(f_t)\,a(g_t)\,)\\ &= \int dp\,dq\,\rho_0(\,a_p^+\,a_q)\,f_t(p)\,\overline{g_t(q)}\\ &= \int dp\,J(p)\,f_t(p)\,\overline{g_t(p)}\,.\end{aligned} \tag{4.12}$$

Passing to the last line, we have used the momentum conservation condition

$$\rho_0(\,a_p^+\,a_q\,) \;=\; J(p)\,\delta(p-q) \tag{4.13}$$

obtained from the translation invariance of the initial state ρ_0, where

$$0 \le J(p) \;=\; \frac{1}{L^3}\rho_0(\,a_p^+ a_p\,) \le 1\,, \tag{4.14}$$

similarly as in the case of the ideal Fermi gas.

The solution f_t of (4.9), (4.10), satisfies the Duhamel formula

$$f_t(p) \;=\; U_{0,t}(p)\,f(p) \,+\, i\,\eta\int_0^t ds\,U_{s,t}(p)\,(\,\widehat{V}_\omega * f_s\,)(p) \tag{4.15}$$

where

$$U_{s,t}(p) := e^{i\int_s^t ds'\,(\,E(p)-\lambda\kappa_{s'}(p)\,)} \tag{4.16}$$

and

$$\kappa_s(u) := (\,\widehat{v} * \mu_s\,)(u)\,. \tag{4.17}$$

We note that the term $U_{0,t}(p)f(p)$ solves (4.9) for $\eta = 0$ (no random potential) with initial condition (4.10).

Let $N \in \mathbb{N}$, which remains to be optimized. The N-fold iterate of (4.15) is given by the truncated Duhamel expansion with remainder term,

$$f_t \;=\; f_t^{(\le N)} \,+\, f_t^{(>N)}\,, \tag{4.18}$$

where

$$f_t^{(\le N)} := \sum_{n=0}^{N} f_t^{(n)}\,, \tag{4.19}$$

and $f_t^{(>N)}$ is the Duhamel remainder term of order N. We define

$$t_{-1} := 0\,, \;\; t_j = s_0 + \cdots + s_j\,, \tag{4.20}$$

for $j = 0, \ldots, n$, and

$$\begin{aligned}&\mathcal{R}(k_0, \ldots, k_n; z)\\ &:= \int_{\mathbb{R}_+^{n+1}} ds_0 \cdots ds_n \Big(\prod_{j=0}^{n} e^{-is_j(E(k_j)-z)} e^{i\lambda \int_{t_{j-1}}^{t_j} ds'\, \kappa_{s'}(k_j)} \Big)\,,\end{aligned} \tag{4.21}$$

for $z \in \mathbb{C}$.

The n-th order term in the Duhamel expansion is given by

$$\begin{aligned} f_t^{(n)}(p) := (i\eta)^n \int_0^t dt_n \cdots \int_0^{t_2} dt_1 \int dk_0 \cdots dk_n\, \delta(p - k_0) \\ \Big[\prod_{j=0}^{n} U_{t_{j-1},t_j}(k_j) \Big] \Big[\prod_{\ell=1}^{n} \widehat{V}_\omega(k_\ell - k_{\ell-1}) \Big] f(k_n)\,. \end{aligned} \tag{4.22}$$

Expressed in terms of the time increments $s_j := t_j - t_{j-1}$,

$$\begin{aligned} f_t^{(n)}(p) = (i\eta)^n \int ds_0 \cdots ds_n\, \delta(t - \sum_{j=0}^{n} s_j) \int dk_0 \cdots dk_n\, \delta(p - k_0) \\ \Big[\prod_{j=0}^{n} e^{-i \int_{t_{j-1}}^{t_j} ds' (E(k_j) - \lambda \kappa_{s'}(k_j))} \Big] \Big[\prod_{\ell=1}^{n} \widehat{V}_\omega(k_\ell - k_{\ell-1}) \Big] f(k_n)\,. \end{aligned} \tag{4.23}$$

Expressing the delta distribution $\delta(t - \sum_{j=0}^{n} s_j)$ in terms of its Fourier transform, we find

$$\begin{aligned} f_t^{(n)}(p) = (i\eta)^n\, e^{\epsilon t} \int d\alpha\, e^{-it\alpha} \int dk_0 \cdots dk_n\, \delta(p - k_0) \\ \mathcal{R}(k_0, \ldots, k_n; \alpha + i\epsilon) \Big[\prod_{j=1}^{n} \widehat{V}_\omega(k_j - k_{j-1}) \Big] f(k_n)\,. \end{aligned} \tag{4.24}$$

The above three equivalent expressions for $f_t^{(n)}(p)$ have different advantages in different contexts, and will all be used in the sequel.

The Duhamel remainder term of order N is given by

$$f_t^{(>N)} = i\eta \int_0^t ds\, \mathcal{U}_{s,t}\, V_\omega^{(1)}\, f_s^{(N)}\,. \tag{4.25}$$

We choose

$$\epsilon = \frac{1}{t} \tag{4.26}$$

so that the factor $e^{\epsilon t}$ in (4.24) remains bounded for all t.

Substituting the truncated Duhamel expansion for $a^+(f_t)$, $a(g_t)$ in (4.12), one obtains

$$\rho_t(\,a^+(f)\,a(g)\,) = \rho_0(\,a^+(f_t)\,a(g_t)\,) = \sum_{n,\widetilde{n}=0}^{N+1} \rho_t^{(n,\widetilde{n})}(f,g) \tag{4.27}$$

where

$$\rho_t^{(n,\widetilde{n})}(f,g) := \rho_0(\,a^+(f_t^{(n)})\,a(g_t^{(\widetilde{n})})\,)\,. \tag{4.28}$$

If $n, \widetilde{n} \leq N$, we have

$$\begin{aligned}\mathbb{E}[\rho_t^{(n,\widetilde{n})}(f,g)] &= \eta^{2\bar{n}} \sum_{\pi\in\Gamma_{n,\widetilde{n}}} \int_0^t dt_q \cdots \int_0^{t_2} dt_1 \int_0^t d\theta_q \cdots \int_0^{\theta_2} d\theta_1 \\ &\int du_0 \cdots du_{2\bar{n}+1}\, f(u_0)\, \overline{g(u_{2\bar{n}+1})}\, J(u_n)\, \delta(u_n - u_{n+1}) \\ &\qquad \Big[\prod_{j=0}^{n} U_{t_{j-1},t_j}(u_j)\Big]\Big[\prod_{j=n+1}^{2\bar{n}+1} \overline{U_{\theta_{j-1},\theta_j}(u_j)}\Big] \\ &\qquad \mathbb{E}\Big[\prod_{j=1}^{n} \widehat{V}_\omega(u_j - u_{j-1}) \prod_{j=n+2}^{2\bar{n}+1} \widehat{V}_\omega(u_j - u_{j-1})\Big]\end{aligned} \tag{4.29}$$

and using (4.24), this is equivalent to

$$\begin{aligned}\mathbb{E}[\rho_t^{(n,\widetilde{n})}(f,g)] &= \eta^{2\bar{n}}\, e^{2\epsilon t} \sum_{\pi\in\Gamma_{n,\widetilde{n}}} \int d\alpha\, d\widetilde{\alpha}\, e^{it(\alpha-\widetilde{\alpha})} \\ &\int du_0 \cdots du_{2\bar{n}+1}\, f(u_0)\, \overline{g(u_{2\bar{n}+1})}\, J(u_n)\, \delta(u_n - u_{n+1}) \\ &\qquad \mathcal{R}(u_0,\dots,u_n;\alpha+i\epsilon)\,\mathcal{R}(u_{n+1},\dots,u_{2\bar{n}+1};\widetilde{\alpha}-i\epsilon) \\ &\qquad \mathbb{E}\Big[\prod_{j=1}^{n} \widehat{V}_\omega(u_j - u_{j-1}) \prod_{j=n+2}^{2\bar{n}+1} \widehat{V}_\omega(u_j - u_{j-1})\Big]\end{aligned} \tag{4.30}$$

where $t_{-1}, \theta_{-1} := 0$ in (4.29).

4.1. *Statement of main results*

We introduce macroscopic variables (T, X), related to the microscopic variables (t, x) by

$$(T, X) = (\zeta t, \zeta x), \tag{4.31}$$

with ζ a real parameter. We will study kinetic scaling limits associated to different scaling ratios between ζ, η and λ.

The random potential has an average effect on the dynamics of μ_t by an amount proportional to its variance, $O(\eta^2 t)$, in the time interval $[0, t]$. Since the strength of the fermion pair interactions is $O(\lambda)$, both effects are comparable if $\lambda = O(\eta^2)$. Accordingly, we distinguish the following scaling regimes.

4.1.1. *The regime* $\lambda \leq C\eta^2$

In this regime, the interaction between electrons and the effect of the random potential per time unit is comparable.

Theorem 4.1: (T. Chen and I. Rodnianski, [16]) *Assume that* $\lambda \leq O(\eta^2)$*. Then, for any fixed, finite* $T > 0$*, and any choice of test functions* f*,* g*,*

$$\lim_{\eta \to 0} \lim_{L \to \infty} \mathbb{E}[\, \rho_{T/\eta^2}(\, a^+(f) a(g)\,)\,] = \int dp\, f(p)\, \overline{g(p)}\, F_T(p) \tag{4.32}$$

holds, where $F_T(p)$ *satisfies the linear Boltzmann equation*

$$\partial_T F_T(p) = 2\pi \int du\, \delta(\, E(u) - E(p)\,)\, (\, F_T(u) - F_T(p)\,) \tag{4.33}$$

with initial condition $F_0 = \mu_0$.

The Boltzmann equations obtained in the kinetic scaling limit are linear because the Hartree-Fock interactions cancel, due to translation invariance.

Remarks about the Proof. For the proof, we use the nonlinear evolution

$$U_{s,t}(p) := e^{i \int_s^t ds' \, (\, E(p) - \lambda \widehat{v} * \mu_{s'}\,)} \tag{4.34}$$

as the reference dynamics, instead of free evolution $e^{i(t-s)E(p)}$ as in previous sections, and we invoke the Feynman graph expansion in powers of η.

Since the free evolution operator depends on the unknown $\mu_t(p)$, and satisfies a *nonlinear* evolution equation, the resolvent calculus used for the problems discussed previously is not available. Accordingly, a majority of arguments in [16] are based on *stationary phase estimates.*

The recombination of contributions associated to decorated ladders is much more difficult for the problem at hand than for the linear problems discussed previously. Our approach involves a very careful analysis of phase cancellations and stationary phase effects. □

4.1.2. *The regime* $\eta = o(\sqrt{\lambda})$

In the regime $\eta = o(\sqrt{\lambda})$, the limiting distribution is *stationary* for any initial momentum distribution.

Theorem 4.2: (T. Chen and I. Rodnianski, [16]) *Assume that* $\eta^2 = O(\lambda^{1+\delta})$ *for* $\delta > 0$ *arbitrary. Then, for any fixed, finite* $T > 0$,

$$\lim_{\lambda\to 0}\lim_{L\to\infty} \mathbb{E}[\,\rho_{T/\lambda}(\,a^+(f)\,a(g)\,)\,] = \int dp\, f(p)\,\overline{g(p)}\,F_T(p)\,, \tag{4.35}$$

for arbitrary test functions f, g, *and*

$$\partial_T F_T(p) = 0\,, \tag{4.36}$$

for $F_0 = \mu_0$. *Accordingly,* $F_T = F_0$ *is stationary.*

4.1.3. *The regime* $t = T/\eta^2$ *and* $\lambda = O_\eta(1)$

This regime is much more difficult to control than those discussed above, and poses challenging open problems. In [16], we prove a partial result that emphasizes some interesting properties the kinetic scaling limit determined by $T = \eta^2 t$ and $\eta \to 0$, with λ small but independent of η. We are considering, for $\lambda = O(1)$, the rescaled, formal fixed point equation

$$\begin{aligned}\int dp\, f(p)\,\overline{g(p)}\,\mu_{T/\eta^2}(p) &= \mathcal{G}^{(L)}[\,\mu_\bullet(\,\bullet\,);\eta;\lambda;T;f,g\,]\\ &:= \mathbb{E}[\,\rho_{T/\eta^2}(\,a^+(f)a(g)\,)\,]\end{aligned} \tag{4.37}$$

for $\mu_\bullet(\,\bullet\,)$. The existence and uniqueness of solutions for this fixed point equation is currently an open problem. Below, we will make the assumption that there exist limiting stationary solutions, and determine their form under this hypothesis.

We base our discussion on the following hypotheses for the case $\lambda = O(1)$:

(H1) There exist solutions $F^{(\eta)}(T) := \lim_{L\to\infty} \mu_{T/\eta^2}$ of (4.37), such that the limit $w - \lim_{\eta\to 0} F^{(\eta)}(T) =: F(T) = F(0)$ exists and is stationary.

(H2) The stationary fixed point solution in (H1) satisfies

$$\begin{aligned} F(T) &= \lim_{\eta\to 0}\lim_{L\to\infty} \mathcal{G}^{(L)}[\,F^{(\eta)};\eta;\lambda;T;f,g\,] \\ &= \lim_{\eta\to 0}\lim_{L\to\infty} \mathcal{G}^{(L)}[\,F;\eta;\lambda;T;f,g\,]\,. \end{aligned} \tag{4.38}$$

The first equality sign here is equivalent to *(H1)*, while the second equality sign accounts for the assumption that $F^{(\eta)}$ can be replaced by the limiting fixed point F before letting $\eta \to 0$, to produce the same result.

Based on the analysis given in [16], we are able to prove hypothesis *(H2)* if $F^{(\eta)} = F + O(\eta^2)$. Error bounds of order $O(\eta^2)$ require more precise estimates of "crossing" and "nesting" terms in the Feynman graph expansion than considered in [16], but are available from [26, 27, 28, 29].

Proposition 4.3: *Let λ be small but independent of η, and assume that $F \in L^\infty(\mathbb{T}^3)$ independent of t. Then, the thermodynamic limit*

$$\mathcal{G}[\,F\,;\eta;\lambda;T;f,g\,] := \lim_{L\to\infty} \mathcal{G}^{(L)}[\,F;\eta;\lambda;T;f,g\,] \tag{4.39}$$

exists.

The proof of this proposition follows straightforwardly from results established in [12, 13, 15, 33].

Theorem 4.4: (T. Chen and I. Rodnianski, [16]) *Assume that $\lambda \leq O_\eta(1)$, and let*

$$\widetilde{E}_\lambda(u) := E(u) + \lambda(\,\widehat{v} * F\,)(u)\,. \tag{4.40}$$

We assume that $F \in L^\infty(\mathbb{T}^3)$ admits the bounds

$$\sup_\alpha \int dp \,\frac{1}{|\widetilde{E}_\lambda(q) - \alpha - i\epsilon|}\,,\ \sup_q \int d\alpha \,\frac{1}{|\widetilde{E}_\lambda(q) - \alpha - i\epsilon|} \leq C\,\log\frac{1}{\epsilon}\,, \tag{4.41}$$

and

$$\sup_{\alpha_i} \sup_{u\in\mathbb{T}^3} \int dq\, dp \frac{1}{|\widetilde{E}_\lambda(q) - \alpha_1 - i\epsilon|} \frac{1}{|\widetilde{E}_\lambda(p) - \alpha_2 - i\epsilon|} \frac{1}{|\widetilde{E}_\lambda(p \pm q + u) - \alpha_3 - i\epsilon|} \leq \epsilon^{-b} \quad (4.42)$$

for some $0 < b < 1$.

Then, F *satisfies*

$$\int dp\, f(p)\, \overline{g(p)}\, F(p) = \lim_{\eta\to 0} \mathcal{G}[\, F; \eta; \lambda; T; f, g\,], \quad (4.43)$$

independent of T*, if and only if it satisfies*

$$F(p) = \mu_0(p) = \frac{1}{\widetilde{m}_\lambda(p)} \int du\, \delta(\, \widetilde{E}_\lambda(u) - \widetilde{E}_\lambda(p)\,)\, F(u), \quad (4.44)$$

where

$$\widetilde{m}_\lambda(p) := 2\pi \int du\, \delta(\, \widetilde{E}_\lambda(u) - \widetilde{E}_\lambda(p)\,) \quad (4.45)$$

is the (normalized) measure of the level surface of $\widetilde{E}_\lambda$ *for the value* $\widetilde{E}_\lambda(p)$.

We point out the following noteworthy observations related to Theorem 4.4.

- The solution of (4.40) corresponds to a renormalized kinetic energy which is shifted by the average interaction energy for fermion pairs.
- The fixed point equation (4.44) for F shows that the stationary kinetic limits of μ_t are concentrated and equidistributed on level surfaces of the renormalized kinetic energy function $\widetilde{E}_\lambda(\,\cdot\,)$.
- The bounds (4.41) and (4.42) correspond to the "crossing estimates" in [12, 33, 26, 38]. They ensure sufficient non-degeneracy of the renormalized energy level surfaces so that the Feynman graph expansions introduced below are convergent. However, they do not seem sufficient to prove hypothesis *(H2)* under the assumption that *(H1)* holds.
- We note that if $\lambda \leq o_\eta(1)$, the stationary solutions found in Theorem 4.4 reduce to those of the linear Boltzmann equation derived in Theorem 4.1.

The problem of deriving the time dependent kinetic scaling limit for the model in the regime $t = T/\eta^2$ and $\lambda = O_\eta(1)$ is currently open.

References

1. R. Adami, G. Golse, A. Teta, *Rigorous derivation of the cubic NLS in dimension one*, J. Stat. Phys. **127**, no. 6, 1194–1220 (2007).
2. M. Aizenman, S. Molchanov, *Localization at large disorder and at extreme energies: an elementary derivation*, Commun. Math. Phys. **157**, 245–278 (1993).
3. M. Aizenman, R. Sims, S. Warzel, *Absolutely continuous spectra of quantum tree graphs with weak disorder*, Comm. Math. Phys. **264**, no. 2, 371–389 (2006).
4. M. Aizenman, S. Warzel, *Localization bounds for multiparticle systems*, Comm. Math. Phys. **290**, no. 3, 903–934 (2009).
5. W. Aschbacher, V. Jaksic, Y. Pautrat, C.-A. Pillet, *Transport properties of quasi-free fermions*, J. Math. Phys. **48**, no. 3 (2007).
6. V. Bach, E.H. Lieb, J.P. Solovej, *Generalized Hartree-Fock theory and the Hubbard model*, J. Stat. Phys. **76** (1-2), 3–89 (1994).
7. D. Benedetto, F. Castella, R. Esposito, M. Pulvirenti, *Some considerations on the derivation of the nonlinear quantum Boltzmann equation*, J. Stat. Phys. **116**, no. 1-4, 381–410 (2004).
8. J. Bellissard, *Random matrix theory and the Anderson model*, J. Statist. Phys. **116**, no. 1-4, 739–754 (2004).
9. J. Bourgain, *On random Schrödinger operators on* $\mathbb{Z}^2$, Discrete and Continuous Dynamical Systems **8**, no 1, 1-15 (2002).
10. J. Bourgain, *Random lattice Schrödinger operators with decaying potential: Some higher dimensional phenomena*, Springer LNM, Vol 1807 (2003), 70-98.
11. R. Carmona, J. Lacroix, *Spectral theory of random Schrödinger operators*, Birkhäuser Boston (1990).
12. T. Chen, *Localization lengths and Boltzmann limit for the Anderson model at small disorders in dimension 3*, J. Stat. Phys., **120** (1-2), 279-337 (2005).
13. T. Chen, *Convergence in higher mean of a random Schrödinger to a linear Boltzmann evolution*, Comm. Math. Phys., **267**, 355-392 (2006).
14. T. Chen, *Localization lengths for Schrödinger operators on* Z^2 *with decaying random potentials*, Int. Math. Res. Notices, **2005:54**, 3341-3373 (2005).
15. T. Chen, I. Sasaki, *Boltzmann limit and quasifreeness for a homogenous Fermi gas in a weakly disordered random medium*, J. Stat. Phys., **132** (2), 329-353, 2008.
16. T. Chen, I. Rodnianski, *Boltzmann limit for a homogenous Fermi gas with dynamical Hartree-Fock interactions in a random medium*, J. Stat. Phys., **142** (5), 1000–1051 (2011).
17. V. Chulaevsky, Y. Suhov, *Eigenfunctions in a two-particle Anderson tight binding model*, Comm. Math. Phys. **289**, no. 2, 701–723 (2009).
18. V. Chulaevsky, Y. Suhov, *Multi-particle Anderson localisation: induction on the number of particles*, Math. Phys. Anal. Geom. **12**, no. 2, 117–139 (2009).
19. Denissov, S. *Absolutely continuous spectrum of multidimensional Schrödinger operator*, Int. Math. Res. Notices 74, 3963–3982 (2004).
20. J.-M. Combes, P. Hislop, *Localization for some continuous, random Hamiltonians in d-dimensions*, J. Funct. Anal., **124**, 149-180 (1994).

21. H.L. Cycon, R.G. Froese, W. Kirsch, B. Simon, *Schrödinger operators*, Springer Verlag (1987).
22. W. De Roeck, J. Fröhlich, *Diffusion of a massive quantum particle coupled to a quasi-free thermal medium*, Comm. Math. Phys. **303** (3), 613–707 (2011).
23. S. Denisov, *Absolutely continuous spectrum of multidimensional Schrödinger operator*, Int. Math. Res. Notices **74**, 3963–3982 (2004).
24. M. Disertori, H. Pinson, T. Spencer, *Density of states for random band matrices*, Comm. Math. Phys. **232**, no. 1, 83–124 (2002).
25. L. Erdös, *Linear Boltzmann equation as the scaling limit of the Schrödinger evolution coupled to a phonon bath*, J. Stat. Phys. **107** (5), 1043-1127 (2002).
26. L. Erdös, M. Salmhofer, *Decay of the Fourier transform of surfaces with vanishing curvature*, Math. Z. **257** (2), 261–294 (2007).
27. L. Erdös, M. Salmhofer, H.-T. Yau, *Quantum diffusion for the Anderson model in the scaling limit*, Ann. Henri Poincaré, **8** (4), 621–685 (2007).
28. L. Erdös, M. Salmhofer, H.-T. Yau, *Quantum diffusion of the random Schrödinger evolution in the scaling limit. II. The recollision diagrams.* Comm. Math. Phys. **271** (1), 1–53 (2007).
29. L. Erdös, M. Salmhofer, H.-T. Yau, *On the quantum Boltzmann equation*, J. Stat. Phys., **116** (114), 367–380 (2004).
30. J. Feldman, H. Knörrer, E. Trubowitz, *A two dimensional Fermi liquid, Part 1: Overview*, Comm. Math. Phys, **247**, 1-47 (2004).
31. A. Figotin, F. Germinet, A. Klein, P. Müller, *Persistence of Anderson localization in Schrödinger operators with decaying random potentials*, Ark. Mat. **45** (1), 15–30 (2007).
32. J. Fröhlich, T. Spencer, *Absence of diffusion in the Anderson tight binding model for large disorder or low energy*, Comm. Math. Phys. **88**, 151–184 (1983).
33. L. Erdös, H.-T. Yau, *Linear Boltzmann equation as the weak coupling limit of a random Schrödinger equation*, Comm. Pure Appl. Math. **53** (6), 667–735 (2000).
34. N.T. Ho and L.J. Landau, *Fermi gas on a lattice in the van Hove limit*, J. Stat. Phys., **87**, 821–845 (1997).
35. N.M. Hugenholtz, *Derivation of the Boltzmann equation for a Fermi gas*, J. Stat. Phys., **32**, 231–254 (1983).
36. A. Klein, *Extended states in the Anderson model on the Bethe lattice*, Adv. Math. **133**, no. 1, 163–184 (1998).
37. P. Kuchment, *Quantum graphs: An introduction and a brief survey*, Analysis on graphs and its applications, 291–312, Proc. Sympos. Pure Math., **77**, Amer. Math. Soc., Providence, RI, 2008.
38. J. Lukkarinen, *Asymptotics of resolvent integrals: The suppression of crossings for analytic lattice dispersion relations*, J. Math. Pures Appl. (9) 87, no. 2, 193–225 (2007).
39. J. Lukkarinen, H. Spohn, *Kinetic limit for wave propagation in a random medium*, Arch. Ration. Mech. Anal. **183**, no. 1, 93–162 (2007).
40. J. Lukkarinen, H. Spohn, *Not to normal order—notes on the kinetic limit*

for weakly interacting quantum fluids, J. Stat. Phys. **134**, no. 5-6, 1133–1172 (2009).
41. J. Lukkarinen, H. Spohn, *Weakly nonlinear Schrödinger equation with random initial data*, Invent. Math. **183** (1), 79–188 (2011).
42. J. Magnen, G. Poirot, V. Rivasseau, *Renormalization group methods and applications: First results for the weakly coupled Anderson model*, Phys. A **263**, no. 1-4, 131-140 (1999).
43. J. Magnen, G. Poirot, V. Rivasseau, *Ward type identities for the 2d Anderson model at weak disorder*, J. Stat. Phys., **93**, 1/2, 331-358 (1998).
44. J. Magnen, G. Poirot, V. Rivasseau, *The Anderson model as a Matrix model*, Proc. Adv. QFT Conf., 1996.
45. Poirot, G., *Mean Green's function of the Anderson model at weak disorder with an infra-red cut-off*, Ann. Inst. H. Poincaré Phys. Théor. 70, no. 1, 101-146 (1999).
46. I. Rodnianski, W. Schlag, *Classical and quantum scattering for a class of long range random potentials*, Int. Math. Res. Notices, **2003:5**, 243-300 (2003).
47. J. Schenker, *Hölder equicontinuity of the integrated density of states at weak disorder*, Lett. Math. Phys., **70**, no. 3, 195–209 (2004).
48. W. Schlag, C. Shubin, T. Wolff, *Frequency concentration and localization lengths for the Anderson model at small disorders*, J. Anal. Math., **88**, 173–220 (2002).
49. T. Spencer, M.R. Zirnbauer, *Spontaneous symmetry breaking of a hyperbolic sigma model in three dimensions*, Comm. Math. Phys. **252**, no. 1-3, 167–187 (2004).
50. H. Spohn, *Derivation of the transport equation for electrons moving through random impurities*, J. Statist. Phys., 17, no. 6, 385-412 (1977).
51. H. Spohn, *The phonon Boltzmann equation, properties and link to weakly anharmonic lattice dynamics*, J. Stat. Phys. **124**, no. 2-4, 1041–1104 (2006).
52. Stein, E. *Harmonic Analysis*, Princeton University Press (1993).

ON THE MINIMIZATION OF HAMILTONIANS OVER PURE GAUSSIAN STATES

Jan Dereziński

Department of Mathematical Methods in Physics
Faculty of Physics, University of Warsaw
Hoża 74, 00-682 Warszawa, Poland
Jan.Derezinski@fuw.edu.pl

Marcin Napiórkowski

Department of Mathematical Methods in Physics
Faculty of Physics, University of Warsaw
Hoża 74, 00-682 Warszawa, Poland
Marcin.Napiorkowski@fuw.edu.pl

Jan Philip Solovej

Department of Mathematics, University of Copenhagen
Universitetsparken 5, 2100 Copenhagen, Denmark
solovej@math.ku.dk

A Hamiltonian defined as a polynomial in creation and annihilation operators is considered. After a minimization of its expectation value over pure Gaussian states, the Hamiltonian is Wick-ordered in creation and annihilation operators adapted to the minimizing state. It is shown that this procedure eliminates from the Hamiltonian terms of degrees 1 and 2 that do not preserve the particle number, and leaves only the terms that can be interpreted as quasiparticles excitations. We propose to call this fact *Beliaev's Theorem*, since to our knowledge it was mentioned for the first time in a paper by Beliaev from 1959.

1. Introduction

Various phenomena in many-body quantum physics are explained with help of *quasiparticles*. Unfortunately, we are not aware of a rigorous definition of this concept, except for some very special cases.

A typical situation when one speaks about quasiparticles seems to be the following: Suppose that the Hamiltonian of a system can be written as $H = H_0 + V$, where H_0 is in some sense dominant and V is a perturbation that in first approximation can be neglected. Suppose also that

$$H_0 = B + \sum_i \omega_i b_i^* b_i, \tag{1.1}$$

where B is a number, operators b_i^*/b_i satisfy the standard canonical commutation/anticommutation relations (CCR/CAR) and the Hilbert space contains a state annihilated by b_i (the *Fock vacuum for* b_i). We then say that the operators b_i^*/b_i *create/annihilate a quasiparticle.*

Of course, the above definition is very vague.

In our paper we describe a simple theorem that for many Hamiltonians gives a natural decomposition $H = H_0 + V$ with H_0 of the form (1.1), and thus suggests a possible definition of a quasiparticle. Our starting point is a fairly general Hamiltonian H defined on a bosonic or fermionic Fock space. For simplicity we assume that the 1-particle space is finite dimensional. With some technical assumptions, the whole picture should be easy to generalize to the infinite dimensional case. We assume that the Hamiltonian is a polynomial in creation and annihilation operators a_i^*/a_i, $i = 1, \dots, n$. (This is a typical assumption in Many Body Quantum Physics and Quantum Field Theory.)

An important role in Many Body Quantum Physics is played by the so-called *Gaussian states*, called also *quasi-free states*. Gaussian states can be *pure* or *mixed*. The former are typical for the zero temperature, whereas the latter for positive temperatures. In our paper we do not consider mixed Gaussian states.

Pure Gaussian states are obtained by applying Bogoliubov transformations to the Fock vacuum state (given by the vector Ω annihilated by a_i's). Pure Gaussian states are especially convenient for computations.

We minimize the expectation value of the Hamiltonian H with respect to pure Gaussian states, obtaining a state given by a vector $\tilde{\Omega}$. By applying an appropriate Bogoliubov transformation, we can replace the old creation and annihilation operators a_i^*, a_i by new ones b_i^*, b_i, which are adapted to the "new vacuum" $\tilde{\Omega}$, i.e., that satisfy $b_i\tilde{\Omega} = 0$. We can rewrite the Hamiltonian H in the new operators and Wick order them, that is, put b_i^* on the left and b_i on the right. The theorem that we prove says that

$$H = B + \sum_{ij} D_{ij} b_i^* b_j + V,$$

where V has only terms of the order greater than 2. In particular, H does not contain terms of the type b_i^*, b_i, $b_i^*b_j^*$, or b_ib_j. It is thus natural to set $H_0 := B + \sum_{ij} D_{ij}b_i^*b_j$. D_{ij} is a hermitian matrix. Clearly, it can be diagonalized, so that H_0 acquires the form of (1.1).

We present several versions of this theorem. First we assume that the Hamiltonian is even. In this case it is natural to restrict the minimization to even pure Gaussian states. In the fermionic case, we can also minimize over odd pure Gaussian states. In the bosonic case, we consider also Hamiltonians without the evenness assumption, and then we minimize with respect to all pure Gaussian states.

The procedure of minimizing over Gaussian states is widely applied in practical computations and is known under many names. In the fermionic case in the contex of nuclear physics it often goes under the name of the *Hartree-Fock-Bogoliubov method* [11]. It is closely related to the *Bardeen-Cooper-Schrieffer approximation* used in superconductivity [1] and the *Fermi liquid theory* developed by Landau [10]. In the bosonic case it is closely related to the *Bogoliubov approximation* used in the theory of superfluidity [4], see also [12, 5]. In both bosonic and fermionic cases it is often called the *mean-field approach* [8].

The fact that we describe in our paper is probably very well known, at least on the intuitive level, to many physicists, especially in condensed matter theory. One can probably say that it summarizes in abstract terms one of the most widely used methods of contemporary quantum physics. The earliest reference that we know to a statement similar to our main result is formulated in a paper of Beliaev [2]. Beliaev studied fairly general fermionic Hamiltonians by what we would nowadays call the Hartree-Fock-Bogoliubov approximation. In a footnote on page 10 he writes:

The condition $H_{20} = 0$ may be easily shown to be exactly equivalent to the requirement of a minimum "vacuum" energy U. Therefore, the ground state of the system in terms of new particles is a "vacuum" state. The excited states are characterized by definite numbers of new particles, elementary excitations.

Therefore, we propose to call the main result of our paper *Beliaev's Theorem.*

The proof of Beliaev's Theorem is not difficult, especially when it is formulated in an abstract way, as we do. Nevertheless, in concrete situations, when similar computations are performed, consequences of this result may often appear somewhat miraculous. The authors of this work witnessed it several times: the authors themselves, or their colleagues, after tedious computations and numerous mistakes watched the unwanted terms disappear [5, 6]. As we show, these terms have to disappear by a general argument.

Acknowledgments

J. D. thanks V. Zagrebnov for useful discussions. J. D. and J. P. S. thank the Danish Council for Independent Research for support during a visit in the Fall of 2010 of J. D. to the Department of Mathematics, University of Copenhagen. The research of J. D. and M. N was supported in part by the National Science Center (NCN) grant No. 2011/01/B/ST1/04929. The work of M. N. was also supported by the Foundation for Polish Science International PhD Projects Programme co-financed by the EU within the Regional Development Fund.

2. Preliminaries

2.1. *2nd quantization*

We will consider in parallel the bosonic and fermionic case.

Let us describe our notation concerning the 2nd quantization. We will always assume that the 1-particle space is $\mathbb{C}^n$. (It is easy to extend our analysis to the infinite dimensional case.) The *bosonic Fock space* will be denoted $\Gamma_{\mathrm{s}}(\mathbb{C}^n)$ and the *fermionic Fock space* $\Gamma_{\mathrm{a}}(\mathbb{C}^n)$. We use the notation $\Gamma_{\mathrm{s/a}}(\mathbb{C}^n)$ for either the bosonic or fermionic Fock space. $\Omega \in \Gamma_{\mathrm{s/a}}(\mathbb{C}^n)$ stands for the *Fock vacuum.* If r is an operator on $\mathbb{C}^n$, then $\Gamma(r)$ stands for its *2nd quantization*, that is

$$\Gamma(r) := \left(\overset{\infty}{\underset{n=0}{\oplus}} r^{\otimes n} \right) \Big|_{\Gamma_{\mathrm{s/a}}(\mathbb{C}^n)}.$$

a_i^*, a_i denote the standard *creation* and *annihilation operators* on $\Gamma_{\mathrm{s/a}}(\mathbb{C}^n)$, satisfying the usual canonical commutation/anticommutation relations.

2.2. *Wick quantization*

Consider an arbitrary polynomial on $\mathbb{C}^n$, that is a function of the form

$$h(\overline{z}, z) := \sum_{\alpha,\beta} h_{\alpha,\beta} \overline{z}^{\alpha} z^{\beta}, \tag{2.1}$$

where $z = (z_1, \ldots, z_n) \in \mathbb{C}^n$, $\overline{z}$ denotes the complex conjugate of z and $\alpha = (\alpha_1, \ldots, \alpha_n) \in (\mathbb{N} \cup \{0\})^n$ represent multiindices. In the bosonic/fermionic case we always assume that the coefficients $h_{\alpha,\beta}$ are symmetric/antisymmetric separately in the indices of $\overline{z}$ and z.

We write $|\alpha| = \alpha_1 + \cdots + \alpha_n$. We say that h is *even* if the sum in (2.1) is restricted to even $|\alpha| + |\beta|$.

The Wick quantization of (2.1) is the operator on $\Gamma_{\mathrm{s/a}}(\mathbb{C}^n)$ defined as

$$h(a^*, a) := \sum_{\alpha,\beta} h_{\alpha,\beta}(a^*)^\alpha a^\beta. \tag{2.2}$$

In the fermionic case, (2.2) defines a bounded operator on $\Gamma_{\mathrm{a}}(\mathbb{C}^n)$. In the bosonic case, (2.2) can be viewed as an operator on $\bigcap_{n>0} \mathrm{Dom} N^n \subset \Gamma_{\mathrm{s}}(\mathbb{C}^n)$, where

$$N = \sum_{i=1}^{n} a_i^* a_i$$

is the number operator.

2.3. *Bogoliubov transformations*

We will now present some basic well known facts about Bogoliubov transformations. For proofs and additional information we refer to [3] (see also [7], [9]). We will often use the summation convention of summing with respect to repeated indices.

Operators of the form

$$Q = \theta_{ij} a_i^* a_j^* + h_{kl} a_k^* a_l + \overline{\theta}_{ij} a_j a_i \pm \frac{1}{2} h_{kk}, \tag{2.3}$$

where h is a self-adjoint matrix, will be called *quadratic Hamiltonians.* In the bosonic/fermionic case we can always assume that θ is symmetric/antisymmetric. (The term $\pm\frac{1}{2}h_{kk}$, with the sign depending on the bosonic/fermionic case, means that Q is the Weyl quantization of the corresponding quadratic expression.) The group generated by operators of the form $\mathrm{e}^{\mathrm{i}Q}$, where Q is a quadratic Hamiltonian, is called the *metaplectic (Mp)* group in the bosonic case and the *Spin* group in the fermionic case.

In the bosonic case, the group generated by Mp together with $\mathrm{e}^{\mathrm{i}(y_i a_i^* + \overline{y}_i a_i)}$, $y_i \in \mathbb{C}$, $i = 1, \ldots, n$, is called the *affine mataplectic (AMp)* group.

In the fermionic case, the group generated by operators $y_i a_i^* + \overline{y}_i a_i$ with $\sum |y_i|^2 = 1$ (which are unitary) is called the *Pin* group. Note that *Spin* is a subgroup of *Pin* of index 2.

In the bosonic case, consider $U \in AMp$. It is well known that

$$Ua_iU^* = p_{ij}a_j + q_{ij}a_j^* + \xi_i, \quad Ua_i^*U^* = \overline{p}_{ij}a_j^* + \overline{q}_{ij}a_j + \overline{\xi}_i \tag{2.4}$$

for some matrices p and q and a vector ξ.

In the fermionic case, consider $U \in Pin$. Then

$$Ua_iU^* = p_{ij}a_j + q_{ij}a_j^*, \quad Ua_i^*U^* = \overline{p}_{ij}a_j^* + \overline{q}_{ij}a_j \tag{2.5}$$

for some matrices p and q.

The maps (2.4) and (2.5) are often called *Bogoliubov transformations*. Bogoliubov transformations can be interpreted as automorphism of the corresponding *classical phase space*. Let us describe briefly this interpretation.

Consider the space $\mathbb{C}^n \oplus \mathbb{C}^n$. It has a distinguished $2n$-dimensional real subspace consisting of vectors $(z, \overline{z}) = ((z_i)_{i=1,\ldots,n}, (\overline{z}_i)_{i=1,\ldots,n})$, which we will call *the real part of* $\mathbb{C}^n \oplus \mathbb{C}^n$, and which can be interpreted as the classical phase space. The real part of $\mathbb{C}^n \oplus \mathbb{C}^n$ is equipped with a symplectic form

$$(z, \overline{z})\omega(z', \overline{z}') := \mathrm{Im}(z|z'), \tag{2.6}$$

and a scalar product

$$(z, \overline{z}) \cdot (z', \overline{z}') := \mathrm{Re}(z|z'). \tag{2.7}$$

Consider the bosonic case. Note that the transformation (2.4), viewed as a map on the real part of $\mathbb{C}^n \oplus \mathbb{C}^n$ given by the matrix $\begin{bmatrix} p & q \\ \overline{q} & \overline{p} \end{bmatrix}$ and the vector $\begin{bmatrix} \xi \\ \overline{\xi} \end{bmatrix}$, preserves the symplectic form (2.6) – in other words, it belongs to ASp, the *affine symplectic group*. More precisely, it is easily checked that in this way we obtain a 2-fold covering homomorphism of AMp onto ASp.

In the fermionic case there is an analogous situation. The transformation (2.5), viewed as a map on the real part of $\mathbb{C}^n \oplus \mathbb{C}^n$ given by the matrix $\begin{bmatrix} p & q \\ \overline{q} & \overline{p} \end{bmatrix}$, preserves the scalar product (2.7) – in other words, it belongs to O, the *orthogonal group*. More precisely, it is easily checked that in this way we obtain a 2-fold covering homomorphism of Pin onto O.

2.4. *Pure Gaussian states*

We will use the term *pure state* to denote a normalized vector modulo a phase factor. In particular, we will distinguish between a pure state and its *vector representative*.

On Fock spaces we have a distinguished pure state called the *(Fock) vacuum state*, corresponding to Ω. States given by vectors of the form $U\Omega$, where $U \in Mp$ or $U \in Spin$, will be called *even pure Gaussian states.* The family of even pure Gaussian states will be denoted by $\mathfrak{G}_{\mathrm{s/a},0}$.

In the bosonic case, states given by vectors of the form $U\Omega$ where $U \in AMp$ will be called *Gaussian pure states.* The family of bosonic pure Gaussian states will be denoted by $\mathfrak{G}_{\mathrm{s}}$.

In the fermionic case, states given by vectors of the form $U\Omega$, where $U \in Pin$ will be called *fermionic pure Gaussian states.* The family of fermionic pure Gaussian states is denoted $\mathfrak{G}_{\mathrm{a}}$.

Fermionic pure Gaussian states that are not even will be called *odd fermionic pure Gaussian states.* The family of odd fermionic pure Gaussian states is denoted $\mathfrak{G}_{\mathrm{a},1}$.

One can ask whether pure Gaussian states have *natural* vector representatives (that is, whether one can naturally fix the phase factor of their vector representatives). In the bosonic case this is indeed always possible. If $c = [c_{ij}]$ is a symmetric matrix satisfying $\|c\| < 1$, then the vector

$$\det(1 - c^*c)^{1/4}\mathrm{e}^{\frac{1}{2}c_{ij}a_i^*a_j^*}\Omega \tag{2.8}$$

defines a state in $\mathfrak{G}_{\mathrm{s},0}$ (see [13]). If $\theta = [\theta_{ij}]$ is a symmetric matrix satisfying $c = \mathrm{i}\frac{\tanh\sqrt{\theta\theta^*}}{\sqrt{\theta\theta^*}}\theta$, then (2.8) equals

$$\mathrm{e}^{\mathrm{i}X_\theta}\Omega \tag{2.9}$$

with

$$X_\theta := \theta_{ij}a_i^*a_j^* + \overline{\theta}_{ij}a_ja_i. \tag{2.10}$$

Each state in $\mathfrak{G}_{\mathrm{s},0}$ is represented uniquely as (2.8) (or equivalently as (2.9)). In particular, (2.9) provides a smooth parametrization of $\mathfrak{G}_{\mathrm{s},0}$ by symmetric matrices.

The manifold of fermionic even pure Gaussian states is more complicated. We will say that a fermionic even pure Gaussian state given by Ψ is *nondegenerate* if $(\Omega|\Psi) \neq 0$ (if it has a nonzero overlap with the vacuum). Every nondegenerate fermionic even pure Gaussian state can be represented by a vector

$$\det(1 + c^*c)^{-1/4}\mathrm{e}^{\frac{1}{2}c_{ij}a_i^*a_j^*}\Omega, \tag{2.11}$$

where $c = [c_{ij}]$ is an antisymmetric matrix. If $\theta = [\theta_{ij}]$ is an antisymmetric matrix satisfying $c = \mathrm{i}\frac{\tan\sqrt{\theta\theta^*}}{\sqrt{\theta\theta^*}}\theta$, $\|\theta\| < \pi/2$, then (2.11) equals

$$\mathrm{e}^{\mathrm{i}X_\theta}\Omega \tag{2.12}$$

with

$$X_\theta := \theta_{ij} a_i^* a_j^* + \overline{\theta}_{ij} a_j a_i. \tag{2.13}$$

Vectors (2.11) are natural representatives of their states. It is easy to see that only nondegenerate fermionic pure Gaussian states possess natural vector representatives.

Not all even fermionic pure Gaussian states are nondegenerate. *Slater determinants* with an even nonzero number of particles are examples of even Gaussian pure states that are not nondegenerate.

Nondegenerate pure Gaussian states form an open dense subset of $\mathfrak{G}_{\mathrm{a},0}$ containing the Fock state (corresponding to $c = \theta = 0$). In particular, (2.11) provides a smooth parametrization of a neighborhood of the Fock state in $\mathfrak{G}_{\mathrm{a},0}$ by antisymmetric matrices.

The fact that each even bosonic/nondegenerate fermionic pure Gaussian state can be represented by a vector of the form (2.8)/(2.11) goes under the name of the *Thouless Theorem.* (See [14]; this name is used eg. in the monograph by Ring and Schuck [11].) The closely related fact saying that these vectors can be represented in the form (2.9)/(2.12) is sometimes called the *Ring-Schuck Theorem.*

By definition, the group AMp/Pin acts transitively on $\mathfrak{G}_{\mathrm{s/a}}$. In other words, for any $\tilde{\Omega} \in \mathfrak{G}_{\mathrm{s/a}}$ we can find $U \in AMp/Pin$ such that $\tilde{\Omega} = U\Omega$. Such a U is not defined uniquely – it can be replaced by $U\Gamma(r)$, where r is unitary on $\mathbb{C}^n$.

Clearly, if we set

$$b_i := U a_i U^*, \quad b_i^* := U a_i^* U^*, \tag{2.14}$$

then $b_i \tilde{\Omega} = 0$, $i = 1, \ldots, n$, and they satisfy the same CCR/CAR as a_i, $i = 1, \ldots, n$. If h is a polynomial of the form (2.1), then we can Wick quantize it using the transformed operators:

$$h(b^*, b) = \sum_{\alpha,\beta} h_{\alpha,\beta} (b^*)^\alpha b^\beta.$$

Obviously, $U h(a^*, a) U^* = h(b^*, b)$.

3. Main Result

As explained in the introduction, we think that the following result should be called *Beliaev's Theorem.*

Theorem 3.1: *Let h be a polynomial on $\mathbb{C}^n$ and $H := h(a^*, a)$ its Wick quantization. We consider the following functions:*

(1) *(bosonic case, even pure Gaussian states)* $\mathfrak{G}_{\mathrm{s},0} \ni \Phi \mapsto (\Phi|H\Phi)$;
(2) *(bosonic case, arbitrary pure Gaussian states)* $\mathfrak{G}_{\mathrm{s}} \ni \Phi \mapsto (\Phi|H\Phi)$;
(3) *(fermionic case, even pure Gaussian states)* $\mathfrak{G}_{\mathrm{a},0} \ni \Phi \mapsto (\Phi|H\Phi)$;
(4) *(fermionic case, odd pure Gaussian states)* $\mathfrak{G}_{\mathrm{a},1} \ni \Phi \mapsto (\Phi|H\Phi)$.

In (1), (3) and (4) we assume in addition that the polynomial h is even.

For a vector $\tilde{\Omega}$ representing a pure Gaussian state, let $U \in AMp/Pin$ satisfy $\tilde{\Omega} = U\Omega$. Set $b_i = Ua_iU^*$ and suppose that $\tilde{h}$ is the polynomial satisfying $H = \tilde{h}(b^*, b)$. Then the following statements are equivalent:

(A) $\tilde{\Omega}$ represents a stationary point of the function defined in (1)–(4).

(B)

$$\tilde{h}(b^*, b) = B + D_{ij}b_i^*b_j + \text{terms of higher order in } b\text{'s.}$$

Proof: Let us prove the case (2), which is a little more complicated than the remaining cases. Let us fix $U \in AMp$ so that $\tilde{\Omega} = U\Omega$. Clearly, we can write

$$H = \tilde{h}(b^*, b) = B + \overline{K}_i b_i + K_i b_i^* + O_{ij}b_j^*b_i^* + \overline{O}_{ij}b_ib_j + D_{ij}b_i^*b_j$$
$$+ \text{terms of higher order in } b\text{'s.} \qquad (3.1)$$

We know that in a neighborhood of $\tilde{\Omega}$ arbitrary pure Gaussian states are parametrized by a symmetric matrix θ and a vector y:

$$\theta \mapsto U\mathrm{e}^{\mathrm{i}\phi(y)}\mathrm{e}^{\mathrm{i}X_\theta}\Omega,$$

where $X_\theta := \theta_{ij}a_i^*a_j^* + \overline{\theta}_{ij}a_ja_i$ and $\phi(y) = y_ia_i^* + \overline{y}_ia_i$. We get

$$(U\mathrm{e}^{\mathrm{i}\phi(y)}\mathrm{e}^{\mathrm{i}X_\theta}\Omega|HU\mathrm{e}^{\mathrm{i}\phi(y)}\mathrm{e}^{\mathrm{i}X_\theta}\Omega) = (\mathrm{e}^{\mathrm{i}\phi(y)}\mathrm{e}^{\mathrm{i}X_\theta}\Omega|U^*\tilde{h}(b^*, b)U\mathrm{e}^{\mathrm{i}\phi(y)}\mathrm{e}^{\mathrm{i}X_\theta}\Omega)$$
$$= (\Omega|\mathrm{e}^{-\mathrm{i}X_\theta}\mathrm{e}^{-\mathrm{i}\phi(y)}\tilde{h}(a^*, a)\mathrm{e}^{\mathrm{i}\phi(y)}\mathrm{e}^{\mathrm{i}X_\theta}\Omega). \qquad (3.2)$$

Now

$$\mathrm{e}^{-\mathrm{i}X_\theta}\mathrm{e}^{-\mathrm{i}\phi(y)}\tilde{h}(a^*, a)\mathrm{e}^{\mathrm{i}\phi(y)}\mathrm{e}^{\mathrm{i}X_\theta} = B - \mathrm{i}(\overline{\theta}_{ij}O_{ij} - \theta_{ij}\overline{O}_{ij}) - \mathrm{i}(\overline{y}_iK_i - y_i\overline{K}_i)$$
$$+ \text{terms containing } a_i \text{ or } a_i^* + O(\|\theta\|^2, \|y\|^2).$$

Therefore, (3.2) equals

$$B - \mathrm{i}(\overline{\theta}_{ij}O_{ij} - \theta_{ij}\overline{O}_{ij}) - \mathrm{i}(\overline{y}_iK_i - y_i\overline{K}_i) + O(\|\theta\|^2, \|y\|^2). \qquad (3.3)$$

Since vectors y and matrices θ are independent variables, (3.3) is stationary at $\tilde{\Omega}$ if and only if $[O_{ij}]$ is a zero matrix and $[K_i]$ is a zero vector. This ends the proof of part (2).

To prove (3) and (4) we note that, for $U \in Pin$, the neighborhood of $\tilde{\Omega} = U\Omega$ in the set of fermionic pure Gaussian states is parametrized by antisymmetric matrices θ:

$$\theta \mapsto U\mathrm{e}^{iX_\theta}\Omega,$$

where again $X_\theta := \theta_{ij}a_i^* a_j^* + \overline{\theta}_{ij}a_j a_i$. Therefore, it suffices to repeat the above proof with $y_i = K_i = 0$, $i = 1, \dots, n$.

The proof of (1) is similar. □

Proposition 3.2: *In addition to the assumptions of Theorem 3.1 (2), suppose that $\tilde{\Omega}$ corresponds to a minimum. Then the matrix $[D_{ij}]$ is positive.*

Proof: Using that O and K are zero, we obtain

$$\mathrm{e}^{-\mathrm{i}\phi(y)}\tilde{h}(a^*, a)\mathrm{e}^{\mathrm{i}\phi(y)} = B + \overline{y}_i D_{ij} y_j \\ + \text{terms containing } a_i \text{ or } a_i^* \ + O(\|y\|^3).$$

Therefore, (3.2) equals

$$B + \overline{y}_i D_{ij} y_j + O(\|y\|^3). \qquad (3.4)$$

Hence the matrix $[D_{ij}]$ is positive. □

Note that in cases (1), (3) and (4) the matrix $[D_{ij}]$ does not have to be positive.

References

1. Bardeen, J., Cooper, L. N., Schrieffer, J. R., *Theory of superconductivity*, Phys. Rev. 108 (1957) 1175.
2. Beliaev, S. T., *Effect of pairing correlations on nuclear properties*, Mat. Fys. Medd. Dan. Vid. Selsk. 31, no. 11 (1959).
3. Berezin, F. A., *The Method of Second Quantization*, Academic Press, New York, 1966.
4. Bogoliubov, N. N., J. Phys. (USSR) 9, 23 (1947); J. Phys (USSR) 11, 23 (1947), reprinted in D. Pines, *The Many-Body Problem* (New York, W.A. Benjamin 1962).
5. Cornean, H. D., Dereziński, J., Ziń, P., *On the infimum of the energy-momentum spectrum of a homogeneous Bose gas*, J. Math. Phys. 50, (2009) 062103.
6. Dereziński, J., Meissner, K.A., Napiórkowski, M.: *On the infimum of the energy-momentum spectrum of a homogeneous Fermi gas*, Ann. Henri Poincaré 14, (2013) 1–36.
7. Dereziński, J., Gérard, C., *Mathematics of Quantization and Quantum Fields*, Cambridge University Press, 2013.

8. Fetter, A. L., Walecka, J. D., *Quantum Theory of Many-Particle Systems*, McGraw-Hill Book Company, 1971.
9. Friedrichs, K. O., *Mathematical Aspects of the Quantum Theory of Fields*, Interscience, New York, 1953.
10. Landau, L. D., *On the theory of the Fermi liquid*, Sov. Phys. JETP 8 (1959), 70.
11. Ring, P., Schuck, P., *The Nuclear Many-body Problem*, Springer-Verlag, New York, 1980.
12. Robinson, D. W., *On the ground state energy of the Bose gas*, Commun. Math. Phys. 1, 159 (1965).
13. Ruijsenaars, S. N. M., *On Bogoliubov transformations II. The general case*, Ann. Phys. 116 (1978) 105-132.
14. Thouless, D. J., *Stability conditions and nuclear rotations in the Hartree-Fock theory*, Nucl. Phys. 21 (1960) 225.

VARIATIONAL APPROACH TO ELECTRONIC STRUCTURE CALCULATIONS ON SECOND-ORDER REDUCED DENSITY MATRICES AND THE N-REPRESENTABILITY PROBLEM

Maho Nakata
Advanced Center for Computing and Communication
RIKEN
2-1 Hirosawa, Wako-city, Saitama 351-0198 Japan
maho@riken.jp

Mituhiro Fukuda
Department of Mathematical and Computing Sciences
Tokyo Institute of Technology
2-12-1-W8-41 Ookayama, Meguro-ku, Tokyo 152-8552 Japan
mituhiro@is.titech.ac.jp

Katsuki Fujisawa
Department of Industrial and Systems Engineering
Chuo University
1-13-27 Kasuga, Bunkyo-ku, Tokyo 112-8551 Japan
fujisawa@indsys.chuo-u.ac.jp

The reduced-density-matrix method is a promising next-generation electronic structure calculation method; it is equivalent to solving the Schrödinger equation for the ground state. The number of variables is the same as a four electron system and constant regardless of the number of electrons in the system. Many researchers have been hoping for a simpler method of doing quantum mechanical calculations and this one may be it. In this chapter, we give an overview of the method covering the theory behind it and its methodology. We also give a brief history of the subject and report new computational results. Typically, the results obtained by the reduced-density-matrix method are comparable to those of the CCSD(T), which is a rather sophisticated traditional approach in quantum chemistry.

1. Introduction

Chemistry is an important branch of science which treats changes in matter. It explains, for example, why and how a protein works or the process by which CO_2 releases O_2. The goal is to enable us to predict, understand, and control what happens when we mix substances. To do that, we usually do experiments, which can be explosive, poisonous, expensive, and unstable. This means it is desirable to do chemistry without experiments. Fortunately, the basic equation of chemistry is known, and it is called the Schrödinger equation [13]. It is possible to solve the Schrödinger equation approximately by using various numerical methods and algorithms on computer. The corresponding branch of chemistry that uses these methods and algorithms is called quantum chemistry, and it is our main interest[a].

Determining the exact or an approximate solution to the Schrödinger equation is a fundamental problem in quantum chemistry. The solution is called the *wavefunction*, or is sometimes referred as the *electronic structure*. If we know the electronic structure, we can do chemistry. The methods used to get a solution are called *ab initio* (Latin word which means "from the beginning") or the *first principles* if their approximations are not heuristic and do not employ parameters determined from experiments.

The *ground state energy* calculation of a non-relativistic and time-independent N-electron molecular system under the Born-Oppenheimer approximation is the most important problem [60]. It can be obtained as the lowest eigenvalue E of the electronic Schrödinger equation:

$$H\Psi(\boldsymbol{z}) = E\Psi(\boldsymbol{z}), \tag{1.1}$$

where H is the *Schrödinger operator* or *Hamiltonian* defined by

$$H = -\frac{1}{2}\sum_{i=1}^{N}\nabla_i^2 - \sum_{i=1}^{N}\sum_{A=1}^{M}\frac{Z_A}{r_{iA}} + \sum_{i=1}^{N}\sum_{j>i}^{N}\frac{1}{r_{ij}}, \tag{1.2}$$

in which Z_A is the atomic number of the nucleus A, r_{iA} is the distance between the electron i and nucleus A, and r_{ij} is the distance between two distinct electrons. The solution of (1.1), $\Psi(\boldsymbol{z})$ in $L^2(\mathbb{K}^N)$, $\mathbb{K} = \mathbb{R}^3 \times \{-\frac{1}{2}, \frac{1}{2}\}$ with the inner product $\langle\Psi_1(\boldsymbol{z}), \Psi_2(\boldsymbol{z})\rangle = \int \Psi_1^*(\boldsymbol{z})\Psi_2(\boldsymbol{z})d\boldsymbol{z}$, $\boldsymbol{z} = (\boldsymbol{x}, s) \in \mathbb{K}$ ($\int d\boldsymbol{z}$ includes an integration over spin variables), is the *wavefunction*, and the corresponding eigenvalue E, the *total energy* of the system.

---

[a] Also known as theoretical chemistry or computational chemistry.

Moreover, since electrons are fermions, the Pauli exclusion principle forces the wavefunction itself to be antisymmetric:

$$\Psi(\boldsymbol{z}_1, \ldots, \boldsymbol{z}_i, \ldots, \boldsymbol{z}_j, \ldots, \boldsymbol{z}_N) = -\Psi(\boldsymbol{z}_1, \ldots, \boldsymbol{z}_j, \ldots, \boldsymbol{z}_i, \ldots, \boldsymbol{z}_N).$$

That is, we must solve the Schrödinger equation in the antisymmetric subspace of $L^2(\mathbb{K}^N)$. We denote such a space as $\mathcal{A}L^2(\mathbb{K}^N)$.

Even on computers, treating the N-particle wavefunction is very difficult. To make it tractable, we discretize the Hilbert space $\mathcal{A}L^2(\mathbb{K}^N)$ by taking antisymmetric products of the one-particle Hilbert space $L^2(\mathbb{K})$, whose complete orthonormal system (CONS) is $\{\psi_i\}_{i=1}^{\infty}$. Each ψ_i, called a *single-electron wavefunction* or *spin orbital*, is written as

$$\psi_i : \mathbb{K} \to \mathbb{C} \quad (i = 1, 2, \ldots, \infty). \tag{1.3}$$

We can explicitly construct a CONS of $\mathcal{A}L^2(\mathbb{K}^N)$ by using $\{\psi_i\}_{i=1}^{\infty}$ and *Slater determinants*, defined as follows [60]:

$$\Psi_I(\boldsymbol{z}) = \frac{1}{\sqrt{N!}} \begin{vmatrix} \psi_{i_1}(\boldsymbol{z}_1) & \psi_{i_2}(\boldsymbol{z}_1) & \cdots & \psi_{i_N}(\boldsymbol{z}_1) \\ \psi_{i_1}(\boldsymbol{z}_2) & \psi_{i_2}(\boldsymbol{z}_2) & \cdots & \psi_{i_N}(\boldsymbol{z}_2) \\ \vdots & \vdots & \ddots & \vdots \\ \psi_{i_1}(\boldsymbol{z}_N) & \psi_{i_2}(\boldsymbol{z}_N) & \cdots & \psi_{i_N}(\boldsymbol{z}_N) \end{vmatrix}.$$

Here, we have used an ordered set of indices $I = \{i_1, \ldots, i_j, \ldots, i_k, \ldots, i_N\}$ where $i_j < i_k$, $i_j, i_k \in \mathbb{N}$. It is known that $\{\Psi_I\}_I$ is a CONS for $\mathcal{A}L^2(\mathbb{K}^N)$ [35].

A second approximation used to solve the Schrödinger equation is to choose r functions from a CONS $\{\psi_i\}_{i=1}^{\infty}$ of $L^2(\mathbb{K})$ by carefully using our chemical or physical intuition. Accordingly, we can construct a subspace of $\mathcal{A}L^2(\mathbb{K}^N)$ by taking the Slater determinants considering all possible combinations of N spin orbitals among r possibilities. In this case, solving the Schrödinger equation becomes equivalent to solving the eigenvalue problem of a Hamiltonian matrix, and as such the problem would seem to be calculable on computers. Nevertheless, the dimension of the problem becomes $r!/(N!(r-N)!)$, which is obviously impractical even for small values of r and N. The (approximate) ground state energy obtained by this procedure is considered a reference value, called the *Full Configuration Iteration* (Full CI) (energy). The mainstream approaches in quantum chemistry can be roughly interpreted as linear or nonlinear approximations of this eigenvalue problem, and they include, *e.g.*, Hartree-Fock method, second-order perturbation methods, coupled cluster methods, and truncated CI methods.

Our main motivation is employ the second-order reduced density matrix (2-RDM) as the basic variable with which to construct simpler methods. Since only two-body interactions exist in nature, we can calculate all observables using 2-RDMs. Moreover, the number of variables of the 2-RDM is always four, regardless of the number of electrons in the system (r^4 when discretized), whereas the wavefunction scales as N ($r!/(N!(r-N)!)$ when discretized).

This chapter is organized as follows. In Section 2, we define the first-order and second-order reduced density matrices and introduce the notion of N-representability and its conditions. The reduced-density-matrix method is a viable implementation for approximating the ground state energy of molecular systems. It is formulated as a semidefinite program in Section 3, and numerical results using a parallel optimization code are given for it in Section 5. In Section 4, we present a brief historical note on this approach. In Section 6, we give concluding remarks.

2. The Reduced-Density-Matrix Method

2.1. *Pure states and ensemble states*

Most generally, a quantum system containing N particles is described by the density matrix D, which was introduced independently by von Neumann, Landau, and Bloch. It is an *ensemble average of wavefunctions*,

$$D(\boldsymbol{z}_1, \boldsymbol{z}_2, \ldots, \boldsymbol{z}_N, \boldsymbol{z}'_1, \boldsymbol{z}'_2, \ldots, \boldsymbol{z}'_N) = \sum_b t_b \Psi_b(\boldsymbol{z}_1, \boldsymbol{z}_2, \ldots, \boldsymbol{z}_N) \Psi_b^*(\boldsymbol{z}'_1, \boldsymbol{z}'_2, \ldots, \boldsymbol{z}'_N),$$

where $t_b \geq 0$, $\sum_{b=1}^{\infty} t_b = 1$, and $\{\Psi_b\}_{b=1}^{\infty}$ is a CONS of an N-particle state. For a *pure state* whose system is described by the wavefunction Ψ, D can be written as

$$D(\boldsymbol{z}_1, \boldsymbol{z}_2, \ldots, \boldsymbol{z}_N, \boldsymbol{z}'_1, \boldsymbol{z}'_2, \ldots, \boldsymbol{z}'_N) = \Psi(\boldsymbol{z}_1, \boldsymbol{z}_2, \ldots, \boldsymbol{z}_N)\Psi^*(\boldsymbol{z}'_1, \boldsymbol{z}'_2, \ldots, \boldsymbol{z}'_N).$$

This is equivalent to requiring D to be idempotent; $D^2 = D$. Hereafter, when we refer to a state, it means to be an ensemble if not otherwise specified. We are mainly interested in pure states, but usually, we do not care whether D is pure or an ensemble. This only becomes a problem when the system is degenerate, or we consider its subsystems.

2.2. *The first-order and second-order reduced density matrices*

In this subsection, we define the first- and second-order reduced density matrices in two different representations. One is the coordinate representation, and the other is the second-quantized representation. These two different representations are completely equivalent. The coordinate representation is quite intuitive while the second-quantized representation is handy when manipulating equations. After that, we show some of the conditions which these matrices should satisfy.

2.2.1. *Coordinate representation*

Given an ensemble density matrix $D(\cdot)$, the *first-order Reduced Density Matrix* (1-RDM) [13] is defined by

$$\gamma(\boldsymbol{z}_1, \boldsymbol{z}_1') = N \int D(\boldsymbol{z}_1, \boldsymbol{z}_2, \ldots, \boldsymbol{z}_N, \boldsymbol{z}_1', \boldsymbol{z}_2, \ldots, \boldsymbol{z}_N) d\boldsymbol{z}_2 d\boldsymbol{z}_3 \cdots d\boldsymbol{z}_N.$$

The *second-order Reduced Density Matrix* (2-RDM) [30, 35, 37] is defined by

$$\Gamma(\boldsymbol{z}_1, \boldsymbol{z}_2, \boldsymbol{z}_1', \boldsymbol{z}_2') = \frac{N(N-1)}{2} \int D(\boldsymbol{z}_1, \boldsymbol{z}_2, \ldots, \boldsymbol{z}_N, \boldsymbol{z}_1', \boldsymbol{z}_2', \ldots, \boldsymbol{z}_N) d\boldsymbol{z}_3 \cdots d\boldsymbol{z}_N,$$

and higher-order RDMs are defined in an analogous way. The normalization factor for the p-th order reduced density matrix is then $\frac{N!}{p!(N-p)!}$.

2.2.2. *Second-quantized representation*

Before defining the reduced density matrices in second-quantized representation, we need to introduce some notions such as Fock spaces, occupation number vectors, and creation/annihilation operators.

The Fock space $\mathcal{F}$ is defined by:

$$\mathcal{F} = \bigoplus_{n=1}^{\infty} \{\text{span}\{|0\rangle, |1\rangle\}\}^n$$

where $|0\rangle$ and $|1\rangle$ represent an unoccupied one-particle state and an occupied one-particle state, respectively. Using this notation, we can compactly represent a Fock state where the 2nd, 3rd, 5th, *etc.* states are each occupied by one particle and the other states are unoccupied:

$$|011010\cdots\rangle.$$

Such vectors span the Fock space $\mathcal{F}$, and we call them *occupation number vectors.*

The inner product between an occupation number vector and the adjoint of an occupation vector are defined as follows:

$$\langle 111010\cdots|011010\cdots\rangle = \delta_{111010\cdots,011010\cdots}.$$

In this way, the inner product between $|\Psi\rangle \in \mathcal{F}$ with the adjoint $\langle\Phi| \in \mathcal{F}^\dagger$ of $|\Phi\rangle \in \mathcal{F}$ can also be defined as $\langle\Phi|\Psi\rangle$.

Now let us define the *creation* and the *annihilation operators* $\{a_i, a_i^\dagger\}_{i=1}^\infty$ [20, 60]. The creation operator $a_i^\dagger$ acting on an occupation number vector which has its i-th state unoccupied results in an vector with the corresponding state occupied:

$$a_i^\dagger|\cdots 0\cdots\rangle = (-1)^P|\cdots 1\cdots\rangle;$$

and becomes a null vector when the i-th state is already occupied:

$$a_i^\dagger|\cdots 1\cdots\rangle = 0.$$

On the other hand, the annihilation operator a_i acting on an occupation number vector which has its i-th state occupied results in an vector with its state unoccupied:

$$a_i|\cdots 1\cdots\rangle = (-1)^P|\cdots 0\cdots\rangle;$$

and becomes a null vector when the i-th state is not occupied:

$$a_i|\cdots 0\cdots\rangle = 0.$$

In both cases, P is the phase factor: it takes 1 when the number of occupied states smaller than the i-th occupation state is odd, and 0 when even. These operators satisfy the anti-commutation relation,

$$a_i^\dagger a_j + a_j a_i^\dagger = \delta_{i,j},$$

which indicates that we are treating fermions [20, 60].

It will be convenient hereafter if we define the relationship between the coordinate representation $\Psi(\boldsymbol{z})$ and the Fock space representation $|\Psi\rangle$:

$$\langle \boldsymbol{z}|\Psi\rangle := \Psi(\boldsymbol{z}).$$

Using $\int |\boldsymbol{z}_1\boldsymbol{z}_2\cdots\boldsymbol{z}_N\rangle\langle\boldsymbol{z}_1\boldsymbol{z}_2\cdots\boldsymbol{z}_N|d\boldsymbol{z}_1 d\boldsymbol{z}_2\cdots d\boldsymbol{z}_N$ as the identity operator on the N particle vector of $\mathcal{F}$, the inner product between two states

$|\Phi\rangle, |\Psi\rangle \in \mathcal{F}$ containing the same number of particles becomes

$$\langle \Phi|\Psi\rangle := \int \langle \Phi|\boldsymbol{z}\rangle\langle \boldsymbol{z}|\Psi\rangle d\boldsymbol{z} = \int \Phi^*(\boldsymbol{z})\Psi(\boldsymbol{z})d\boldsymbol{z}.$$

In this way, we can regard the occupation number vectors containing N particles as the Slater determinants:

$$\langle \boldsymbol{z}|\Psi_I\rangle = \Psi_I(\boldsymbol{z}) = \frac{1}{\sqrt{N!}} \begin{vmatrix} \psi_{i_1}(\boldsymbol{z}_1) & \psi_{i_2}(\boldsymbol{z}_1) & \cdots & \psi_{i_N}(\boldsymbol{z}_1) \\ \psi_{i_1}(\boldsymbol{z}_2) & \psi_{i_2}(\boldsymbol{z}_2) & \cdots & \psi_{i_N}(\boldsymbol{z}_2) \\ \vdots & \vdots & \ddots & \vdots \\ \psi_{i_1}(\boldsymbol{z}_N) & \psi_{i_2}(\boldsymbol{z}_N) & \cdots & \psi_{i_N}(\boldsymbol{z}_N) \end{vmatrix}.$$

Finally, the *first-* and *second-order reduced density matrices* in second-quantized form are defined by:

$$\gamma^i_j := \mathrm{tr}(a^\dagger_i a_j D) := \sum_b t_b \langle \Psi_b|a^\dagger_i a_j|\Psi_b\rangle,$$

$$\Gamma^{ij}_{k\ell} := \frac{1}{2}\mathrm{tr}(a^\dagger_i a^\dagger_j a_\ell a_k D) := \frac{1}{2}\sum_b t_b \langle \Psi_b|a^\dagger_i a^\dagger_j a_\ell a_k|\Psi_b\rangle.$$

In this case, the normalization factor for the p-th order reduced density matrix is $1/p!$

2.2.3. *Equivalence between the coordinate and second-quantized representations*

The coordinate representation and the second-quantized representation of the RDMs are of course equivalent. For the particular case of a pure state, using the relation

$$|\Psi\rangle = \sum_I C_I |\Psi_I\rangle,$$

(since $\{\Psi_I\}_I$ is a CONS for $\mathcal{A}L^2(\mathbb{K}^N)$), or equivalently,

$$\Psi(\boldsymbol{z}) = \sum_I C_I \Psi_I(\boldsymbol{z}),$$

the second-quantized representation of 1-RDM γ^i_j can be transformed into a coordinate representation $\gamma(\boldsymbol{z}_1, \boldsymbol{z}'_1)$ such as[b]:

$$
\begin{aligned}
\sum_{ij} \psi_i^*(\boldsymbol{z}'_1)\gamma^i_j\psi_j(\boldsymbol{z}_1) &= \sum_{ij} \psi_i^*(\boldsymbol{z}'_1)\mathrm{tr}(a_i^\dagger a_j D)\psi_j(\boldsymbol{z}_1) \\
&= \sum_{ij} \psi_i^*(\boldsymbol{z}'_1)\langle\Psi|a_i^\dagger a_j|\Psi\rangle\psi_j(\boldsymbol{z}_1) \\
&= \sum_{IJij} \psi_i^*(\boldsymbol{z}'_1)C_I^* C_J\langle\Psi_I|a_i^\dagger a_j|\Psi_J\rangle\psi_j(\boldsymbol{z}_1) \\
&= \sum_{IJij} \psi_i^*(\boldsymbol{z}'_1)(-1)^{P(I\backslash i)+P(J\backslash j)}{C_I}^* C_J\langle\Psi_{I\backslash i}|\Psi_{J\backslash j}\rangle\psi_j(\boldsymbol{z}_1) \\
&= \sum_{IJij} \psi_i^*(\boldsymbol{z}'_1)(-1)^{P(I\backslash i)+P(J\backslash j)}{C_I}^* C_J \int \langle\Psi_{I\backslash i}|\boldsymbol{z}_2\boldsymbol{z}_3\cdots\boldsymbol{z}_N\rangle \\
&\qquad\qquad \langle\boldsymbol{z}_N\ldots\boldsymbol{z}_3\boldsymbol{z}_2|\Psi_{J\backslash j}\rangle\psi_j(\boldsymbol{z}_1)d\boldsymbol{z}_2 d\boldsymbol{z}_3\cdots d\boldsymbol{z}_N \\
&= \left(\frac{1}{\sqrt{N}}\right)^2 \int \sum_{IJij}(-1)^{P(I\backslash i)+P(J\backslash j)}{C_I}^* C_J\psi_i^*(\boldsymbol{z}'_1)\Psi^*_{I\backslash i}(\boldsymbol{z}_2,\ldots,\boldsymbol{z}_N) \\
&\qquad\qquad \psi_j(\boldsymbol{z}_1)\Psi_{J\backslash j}(\boldsymbol{z}_2,\ldots,\boldsymbol{z}_N)d\boldsymbol{z}_2\cdots d\boldsymbol{z}_N \\
&= \frac{1}{N}\int \Psi(\boldsymbol{z}_1,\boldsymbol{z}_2,\ldots,\boldsymbol{z}_N)\Psi^*(\boldsymbol{z}'_1,\boldsymbol{z}_2,\ldots,\boldsymbol{z}_N)d\boldsymbol{z}_2\cdots d\boldsymbol{z}_N \\
&= \gamma(\boldsymbol{z}_1,\boldsymbol{z}'_1).
\end{aligned}
$$

Note that the phase factor does not appear in this case.

Likewise, we can obtain a similar relation for the 2-RDM as well:

$$
\Gamma(\boldsymbol{z}_1,\boldsymbol{z}_2,\boldsymbol{z}'_1,\boldsymbol{z}'_2) = \sum_{ijkl} \psi_i^*(\boldsymbol{z}'_1)\psi_j^*(\boldsymbol{z}'_2)\Gamma^{ij}_{k\ell}\psi_\ell(\boldsymbol{z}_2)\psi_k(\boldsymbol{z}_1).
$$

2.2.4. *Some properties of 1- and 2-RDMs*

The following conditions are inherited by these definitions:

(1) the 1-RDM and 2-RDM are Hermitian,

$$
\gamma^i_j = (\gamma^j_i)^*, \quad \Gamma^{ij}_{k\ell} = (\Gamma^{k\ell}_{ij})^*,
$$

(2) the 2-RDM is antisymmetric,

$$
\Gamma^{ij}_{k\ell} = -\Gamma^{ji}_{k\ell} = -\Gamma^{ij}_{\ell k} = \Gamma^{ji}_{\ell k},
$$

[b]Here, we have assumed that D is equal to $|\Psi\rangle\langle\Psi|$, but this treatment can be trivially extended to ensemble states.

(3) trace conditions are valid,

$$\sum_i \gamma_i^i = N, \quad \sum_{ij} \Gamma_{ij}^{ij} = \frac{N(N-1)}{2},$$

(4) a partial trace condition holds between the 1-RDM and 2-RDM,

$$\frac{N-1}{2}\gamma_j^i = \sum_k \Gamma_{jk}^{ik}.$$

Additionally, we can find more conditions from the symmetry of the system. In particular, the spin symmetry of the (ground state) molecular systems is important:

(5) the total spin S^2; the 2-RDM is the eigenstate of the spin operator,

$$\mathrm{tr}(S^2\Gamma) = S(S+1),$$

where S^2 is defined as follows:

$$\begin{aligned} S^2 &= S_x^2 + S_y^2 + S_z^2 = S_z + S_z^2 + S_- S_+ \\ &= \frac{1}{2}\sum_i \left(a_{i\alpha}^\dagger a_{i\alpha} - a_{i\beta}^\dagger a_{i\beta}\right) + \frac{1}{4}\left(\sum_i a_{i\alpha}^\dagger a_{i\alpha} - a_{i\beta}^\dagger a_{i\beta}\right)^2 \\ &\quad + \sum_{ij} a_{i\beta}^\dagger a_{i\alpha} a_{j\alpha}^\dagger a_{j\beta}. \end{aligned}$$

The indices $i\alpha$, $j\beta$ means that we choose spin eigenfunctions of the z-axis for $\{\psi_i\}_{i=1}^\infty$ and reorder them so that i means the i-th spatial function and α and β denote the eigenfunctions of α-spin and β-spin, *i.e.*, $\{\psi_{i\alpha}, \psi_{i\beta}\}_{i=1}^\infty$, respectively.

(6) The z-component of the spin S_z can be chosen to be integer or half integer,

$$\langle S_z \rangle = \frac{1}{2}\sum_i (\gamma_{i\alpha}^{i\alpha} - \gamma_{i\beta}^{i\beta}).$$

In the subsequent discussion, one will notice that the 1-RDM can be disregarded throughout. However, we shall explicitly use it in order to prioritize the compactness of the notation.

2.3. *Solving the ground state problem using 1- and 2-RDMs*

The Hamiltonian can be written in the most general second-quantization form as

$$H = \sum_{ij} v^i_j a^\dagger_i a_j + \frac{1}{2}\sum_{ijk\ell} w^{ij}_{k\ell} a^\dagger_i a^\dagger_j a_\ell a_k,$$

where v^i_j and $w^{ij}_{k\ell}$ are one- and two-particle terms that can be calculated from the molecular Hamiltonian (1.2) by Slater's rule, as follows:

$$v^i_j = \int \psi^*_i(z_1)\left(-\frac{1}{2}\nabla^2 - \sum_{A=1}^{M}\frac{Z_A}{r_A}\right)\psi_j(z_1)dz_1,$$

$$w^{ij}_{k\ell} = \int \psi^*_i(z_1)\psi^*_j(z_2)\left(\frac{1}{|z_1 - z_2|}\right)\psi_\ell(z_1)\psi_k(z_2)dz_1dz_2,$$

where r_A is the distance between an electron and a nucleus. The derivation of these expressions from (1.2) requires several pages of calculations. The interested reader can find a sketch of the proof on pages 1-19 of [20].

Finally, the ground state energy $E_{\min}$ can be calculated by minimizing the total energy over (1- and) 2-RDMs.

$$\begin{aligned} E_{\min} &= \min \operatorname{tr}(HD) = \min \sum_b t_b\langle\Psi_b|H|\Psi_b\rangle \\ &= \min \sum_b t_b\langle\Psi_b|\sum_{ij} v^i_j a^\dagger_i a_j + \frac{1}{2}\sum_{ijk\ell} w^{ij}_{k\ell} a^\dagger_i a^\dagger_j a_\ell a_k|\Psi_b\rangle \\ &= \min \sum_{ij} v^i_j \operatorname{tr}(a^\dagger_i a_j D) + \frac{1}{2}\sum_{ijk\ell} w^{ij}_{k\ell}\operatorname{tr}(a^\dagger_i a^\dagger_j a_\ell a_k D) \\ &= \min\{\sum_{ij} v^i_j \gamma^i_j + \sum_{ijk\ell} w^{ij}_{k\ell}\Gamma^{ij}_{k\ell}\}. \end{aligned} \tag{2.1}$$

As mentioned above, minimizing over γ^i_j and $\Gamma^{ij}_{k\ell}$ in (2.1) is equivalent to solving the Schrödinger equation for the ground state. Often, such a method is called variational. Furthermore, thanks to the relation (4), (2.1) in fact is a minimization only over 2-RDM elements.

The advantage of using (1- and) 2-RDMs instead of D is that the number of variables is dramatically reduced. However, as we will detail in the next subsections, we need to impose additional conditions on the (1- and) 2-RDMs in order for them to correspond to a true ensemble density matrix D. We only know the formal description of these conditions [28], and their determination would be intractable in a strict sense in terms of computational complexity [34].

2.4. *The N-representability problem and the N-representability conditions*

In the 1950's, researchers armed with the above facts chose the 1- and 2-RDMs as basic variables, and did variational calculations on them. According to Löwdin [36], F. London, J. E. Mayer, A. J. Coleman, P. O. Löwdin, R. McWeeny, N. A. March, C. A. Coulson and others attempted minimizations via (2.1). However, their results were considerably lower than the true energy. The reason is that the trial 2-RDMs were not actually derived from an actual density matrix D. We need to put some more conditions on the trial 2-RDM to ensure that it comes from a *true* D. Such a formalism was first described in 1963 by A. J. Coleman, who named it the *N-representability problem* and the conditions the *N-representability conditions* [10]; Given a trial p-th order RDM $\Gamma^{(p)}$, if there exists some wavefunction or ensemble which reduces to a p-th order RDM $\Gamma^{(p)}$, then this $\Gamma^{(p)}$ is pure or ensemble N-representable.

2.5. *On the complete N-representability conditions*

The 1- and 2-RDM should satisfy relations (1) to (4) of Subsection 2.2.4. The 1- and 2-RDMs for the ground state should additionally satisfy (5) and (6). Therefore, these conditions are necessary ones for the N-representability. Unfortunately, these conditions are not sufficient; consequently, the early attempts obtained very low energies. The necessary and sufficient conditions for the 1-RDM are relatively easy [10, 32]. However, complete (sufficient) N-representability of the 2-RDM is in general very complicated. Garrod and Percus [28] showed that the 2-RDM Γ is ensemble N-representable if and only if

$$\sum_{ijk\ell} (H^{\nu})^{ij}_{k\ell} \Gamma^{ij}_{k\ell} \geq E^{\nu}_{\text{min}},$$

where H^{ν} is every possible Hamiltonian and E^{ν}_{min} is the ground state energy corresponding to H^{ν}. Thus, the ensemble N-representable set $\mathcal{E}^N$ can be defined by:

$$\mathcal{E}^N = \{\Gamma \mid \text{tr}(H^{\nu}\Gamma) \geq E^{\nu}_{\text{min}}, \text{for all possible } H^{\nu} \text{ and } E^{\nu}_{\text{min}}\}.$$

This result is very important, but impractical since if one wants to calculate the *exact* ground state energy of a Hamiltonian, then he/she must know the exact ground state energy of the system beforehand. This is a tautology. Subsequently, many researchers have sought the complete N-

representability condition, but none have succeeded. A meaningful result from complexity theory was obtained by Liu *et al.* [34] in 2007. They showed that the computational complexity of the N-representability problem is QMA-complete, which is the quantum generalization of NP-completeness. Thus, it is almost hopeless to search for an efficient algorithm to decide whether a given 2-RDM is N-representable or not. We can consider a more physical example: the ground state problem of the spin-glass Hamiltonian is known to be very hard; it is equivalent to solving the max-cut problem or the traveling sales person problem, both of which are NP-hard [4]. If the complete N-representability conditions were easier to handle, we could solve such difficult problems in computer science as well. Currently, we do not know how to solve these problems efficiently. We just want to stress: the complete N-representability is really a hard problem.

2.6. *Formulating the variational problem and its geometrical representation*

The problem we want to solve can be formulated using the (1- and) 2-RDMs as basic variables:

$$\begin{aligned} E_{\min} &= \min \operatorname{tr}(HD) \\ &= \min_{\Gamma \in \mathcal{E}^N} \{\sum_{ij} v^i_j \gamma^i_j + \sum_{ijk\ell} w^{ij}_{k\ell} \Gamma^{ij}_{k\ell}\}. \end{aligned}$$

Here, $\mathcal{E}^N$ is known to be a compact and convex set. In addition, all possible Hamiltonians and the corresponding ground state energies serve as a characterization of this convex set. To be precise, a 2-RDM corresponding to the ground state of an N-particle Hamiltonian is a surface point, and any surface point of $\mathcal{E}^N$ corresponds to the ground state of some Hamiltonian [56]. The compact and convex set of the N-representable set is represented as an ellipse in Fig. 1 (although the figure does not show them, there are also cusp points). The Hamiltonians H^1, H^2, H^3, and H^4, and their ground state energies E^1, E^2, E^3, and E^4 serve as their respective N-representability conditions.

2.7. *Some of the known necessary N-representability conditions*

We should not be demotivated by the facts presented above because mathematical theorems do not tell us about chemistry or physics. Rather, it

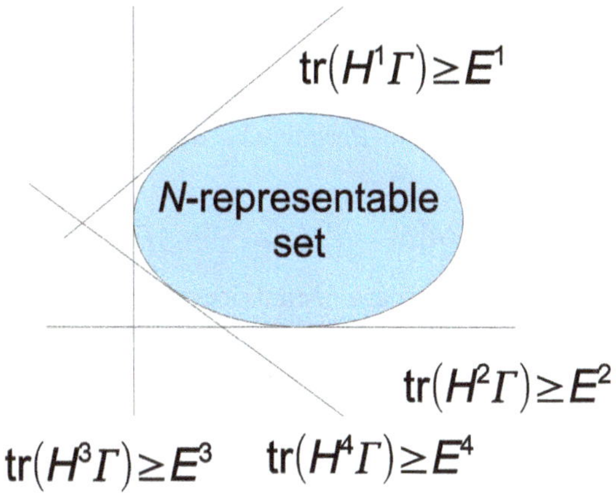

Fig. 1. Schematic representation of an N-representable set; the Hamiltonian and its ground state energy serve as a characterization of the set.

is much more important to gain an understanding of the chemical and/or physical meaning of the necessary N-representability conditions. In practice, establishing N-representability conditions for molecular systems might not be so difficult.

We seek chemically and/or physically meaningful necessary N-representability conditions on the 1- and 2-RDMs. The necessary and sufficient conditions for ensemble N-representability of the 1-RDM are characterized by its eigenvalues lying between 0 and 1 [10, 32]. For a pure state, it is more complicated [1]. Coleman introduced two necessary conditions called the P and Q conditions [10]. These conditions require the P-matrix (Γ), and the Q-matrix to be positive semidefinite:

$$P^{ij}_{k\ell} = \mathrm{tr}(a_i^\dagger a_j^\dagger a_\ell a_k D) \succeq 0, \tag{2.2}$$

$$Q^{ij}_{k\ell} = \mathrm{tr}(a_i a_j a_\ell^\dagger a_k^\dagger D) \succeq 0. \tag{2.3}$$

Here, we shall suppose that the matrices P and Q defined by the elements P_{ij} and Q_{ij} are positive semidefinite [24, 25]. A proof of the non-negativity of P and Q is easy: take an arbitrary operator $A = \sum_{ij} \alpha_{ij} a_i^\dagger a_j^\dagger$; then, $AA^\dagger$ or $A^\dagger A$ is always (a) positive semidefinite (operator). Since α_{ij} is arbitrary, the P and Q-matrices should be non-negative. The conditions listed below can be shown in the same way [6].

Another important necessary condition is called the G condition [28], which also requires a positive semidefiniteness of the G-matrix:

$$G^{ij}_{k\ell} = \mathrm{tr}(a^\dagger_i a_j a^\dagger_\ell a_k D) \succeq 0. \tag{2.4}$$

In the original paper by Garrod-Percus, the definition of the G-matrix is non-linear:

$$G^{ij}_{k\ell} = \mathrm{tr}((a^\dagger_i a_j - \gamma^i_j)(a^\dagger_\ell a_k - \gamma^\ell_k)D) \succeq 0, \tag{2.5}$$

but for a fixed particle state, these G-matrices share the same eigenvalues since γ^i_j can be replaced with $\frac{1}{N}\sum_i a^\dagger_i a_i$ [17]. In Zhao *et al.* [71], we can find explicit formulas for the $T1$ and $T2$ conditions from Erdahl's survey paper [15]:

$$(T1)^{ijk}_{\ell mn} = \mathrm{tr}((a^\dagger_i a^\dagger_j a^\dagger_k a_n a_m a_\ell + a_n a_m a_\ell a^\dagger_i a^\dagger_j a^\dagger_k)D) \succeq 0, \tag{2.6}$$
$$(T2)^{ijk}_{\ell mn} = \mathrm{tr}((a^\dagger_i a^\dagger_j a_k a^\dagger_n a_m a_\ell + a^\dagger_n a_m a_\ell a^\dagger_i a^\dagger_j a_k)D) \succeq 0. \tag{2.7}$$

These conditions are stronger conditions in the sense that they provide better approximations to the ground state energies for atoms/molecules. An important property of these matrices is that Q, G, $T1$, and $T2$-matrices can be *only* expressed in terms of (1- and) 2-RDM elements (with linear dependence), like in the following:

$$\begin{aligned}
Q^{ij}_{k\ell} &= (\delta^i_k\delta^j_\ell - \delta^i_\ell\delta^j_k) - (\delta^i_k\gamma^j_\ell + \delta^j_\ell\gamma^i_k) + (\delta^i_\ell\gamma^j_k + \delta^j_k\gamma^i_\ell) - 2\Gamma^{ij}_{k\ell},\\
G^{ij}_{k\ell} &= \delta^j_\ell\gamma^i_k - 2\Gamma^{i\ell}_{kj},\\
(T1)^{ijk}_{\ell mn} &= \mathcal{A}[i,j,k]\mathcal{A}[\ell,m,n](\frac{1}{6}\delta^i_\ell\delta^j_m\delta^k_n - \frac{1}{2}\delta^i_\ell\delta^j_m\gamma^k_n + \frac{1}{2}\delta^i_\ell\Gamma^{jk}_{mn}),\\
(T2)^{ijk}_{\ell mn} &= \mathcal{A}[j,k]\mathcal{A}[m,n](\frac{1}{2}\delta^j_m\delta^k_n\gamma^i_\ell + \frac{1}{2}\delta^i_\ell\Gamma^{mn}_{jk} - 2\delta^j_m\Gamma^{in}_{\ell k}),
\end{aligned}$$

where $\mathcal{A}$ is the antisymmetrizer operator acting on an arbitrary function $f(i,j,k)$,

$$\mathcal{A}[i,j,k]f(i,j,k) = f(i,j,k) - f(i,k,j) - f(j,i,k) + f(j,k,i) + f(k,i,j) - f(k,j,i). \tag{2.8}$$

In the $T1$ and $T2$'s cases, the 3-RDM terms cancel out. The $T2'$ condition replaces the $T2$ condition and is slightly strengthened by the addition of the one-particle operator [6, 42]. Unfortunately, these N-representability conditions are not exhaustive. We can extend this list [3], which may become a complete N-representability condition, in a similar way to deriving the diagonal inequalities [47].

Other positive semidefinite type representability conditions are known such as the B and C. However, they are implied by the G condition [33].

We can extend these conditions to higher order RDMs. These extensions seem to have been known for a long time. Erdahl and Jin [18] formulated a p-th order approximation to the N-particle density matrix in terms of semidefiniteness conditions on p-th order RDMs, which are generalizations of the P, Q, G, $T1$ and $T2'$ conditions.

2.8. *The reduced-density-matrix method*

The *reduced-density-matrix method* is a variational method having the 2-RDM (and the 1-RDM) as the basic variable(s) and restricted to some approximation $\tilde{\mathcal{E}}^N$ of the N-representability set $\mathcal{E}^N$. It can be formulated as follows:

$$\tilde{E}_{\min} = \min_{\Gamma \in \tilde{\mathcal{E}}^N} \{ \sum_{ij} v^i_j \gamma^i_j + \sum_{ijk\ell} w^{ij}_{k\ell} \Gamma^{ij}_{k\ell} \}. \tag{2.9}$$

Among the possibilities, we usually consider the approximate set obtained by imposing a number of necessary conditions for N-representability. In particular, $\tilde{\mathcal{E}}^N$ should have the following properties.

- it satisfies certain necessary conditions of ensemble N-representability.
- it is compact, so that a linear functional (the Hamiltonian) has a minimum.
- it is a convex set, so that the solution would not become stuck in a local minima.
- it is stringent, so that resulting 2-RDM is physically or chemically meaningful.
- it is computationally feasible and/or efficient.
- it is general: since the form of the Hamiltonian is totally general, it is not only applicable to chemistry but also to physics.
- it should be *ab initio*, *i.e.*, no empirical parameters. Note that many of the very successful methods based on density functional theory employ a lot of empirical parameters.

We can find new N-representability conditions from chemical or physical requirements satisfying the above properties. To do so, we construct a Hamiltonian and obtain an upper bound to the ground state energy; this upper bound then becomes a new condition. Such "cuts" may strengthen $\tilde{\mathcal{E}}^N$.

Trivial N-representability conditions with the P, Q, G, $T1$ and $T2'$ conditions and every possible combination of the P, Q, G, $T1$, $T2$ and $T2'$

conditions satisfy the above criteria. These variational energies have the following property:

- As more necessary conditions are added, the calculated energy usually becomes better and never becomes worse:

$$E_{PQ} \leq E_{PQG} \leq E_{PQGT1} \leq E_{PQGT1T2} \leq E_{PQGT1T2'} \leq E_{\text{fullCI}},$$

where E_{PQ} is the variational energy derived under the P and Q conditions, and $E_{PQGT1T2}$ is the variational energy derived from the P, Q, G, $T1$ and $T2$ conditions, *etc.* (see Fig. 2). This property is totally unlike traditional wavefunction approaches. The variational calculation using the wavefunction gives upper bounds, and the approximation of the total energy becomes closer as the variational space becomes larger.

Note that the obtained 2-RDM may not necessarily be unique, even when the original problem is non-degenerate and the energy is unique.

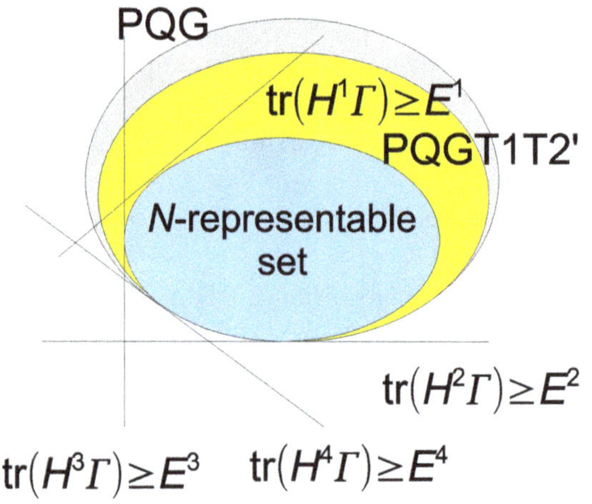

Fig. 2. Schematic representation of approximate N-representable sets; a better set is smaller.

2.9. *Interpreting the conditions*

We usually enforce only the necessary conditions on the trial 2-RDMs. The RDM method therefore gives lower bounds to the exact energy, an N-representable 1-RDM, and a non-physical 2-RDM. Thus, it is important to understand the physical meaning of the necessary conditions employed in the calculations. The following results may be useful for interpreting the results of calculations on molecules and atoms:

(a) If a trial 2-RDM Γ satisfies the P and Q conditions, the original 2-RDM is ensemble N-representable [10].

(b) If a trial 2-RDM Γ satisfies the G condition, then the 1-RDM computed from the original 2-RDM is ensemble N-representable [44].

(c) If the Hamiltonian of a system is time-reversal invariant and the number of particles N is even, the necessary and sufficient condition that an approximate 1-RDM corresponding to a non-degenerate energy eigenstate be N-representable is that its natural spin-orbital occupation numbers are equal in pairs [59]. Moreover Coleman proved that the AGP (anti-symmetrized power) wavefunctions cover all such 1-RDMs [11]. Thus if these conditions apply to the systems, we can always obtain pure representable 1-RDMs satisfying the necessary N-representability conditions.

(d) The G condition is related to the AGP wavefunction, and it gives the correct energy for the Hamiltonian whose ground state can be written in terms of the AGP wavefunction [19]. Moreover, the AGP wavefunction is closely related to the superconductivity [46].

(e) The G condition is exact for the Hubbard model in the high-correlation limit, since the two particle term of the model ($U \sum_i^L a_{i\uparrow}^\dagger a_{i\uparrow} a_{i\downarrow}^\dagger a_{i\downarrow}$, where $U > 0$, L is the number of sites, i is the i-th site, and $\uparrow$ and $\downarrow$ respectively denote the up-spin and down-spin of the electron) is a G-type Hamiltonian ($\sum_{ijk\ell} A_{k\ell}^{ij} a_i^\dagger a_j a_\ell^\dagger a_k$), which is bounded by zero [68].

3. Formulating the RDM Problem as a Semidefinite Program and its Solution Using the Interior-Point Method

3.1. *Semidefinite program*

Semidefinite programs (SDP) comprise an important class of problems in optimization that have been studied since the 1990's. The mathematical theory surrounding them is elegant, and there is an efficient algorithm, called the interior-point method, that can solve one with polynomial-time complexity. The interested reader can refer for instance to [2, 61] for a nice survey about SDPs.

Let $\boldsymbol{C}, \boldsymbol{A}_p$ ($p = 1, 2, \ldots, u$) be block-diagonal real symmetric matrices with prescribed block sizes and $\boldsymbol{b} \in \mathbb{R}^u$ and $\boldsymbol{c}, \boldsymbol{a}_p \in \mathbb{R}^s$ ($p = 1, 2, \ldots, u$) be real vectors. We denote by $\mathbf{Diag}(\boldsymbol{a})$ a diagonal matrix with the elements of the vector $\boldsymbol{a}$ on its diagonal.

An SDP can be defined, for instance, by

$$\begin{cases} \text{maximize } \operatorname{tr}(\boldsymbol{C}\boldsymbol{X}) + \operatorname{tr}(\mathbf{Diag}(\boldsymbol{c})\mathbf{Diag}(\boldsymbol{x})) \\ \text{subject to } \operatorname{tr}(\boldsymbol{A}_p\boldsymbol{X}) + \operatorname{tr}(\mathbf{Diag}(\boldsymbol{a}_p)\mathbf{Diag}(\boldsymbol{x})) = b_p, \quad (p = 1, 2, \ldots, u) \\ \qquad\qquad \boldsymbol{X} \succeq \boldsymbol{O},\ \boldsymbol{x} \in \mathbb{R}^s. \end{cases} \tag{3.1}$$

We shall refer it as the *primal SDP*. The notation $\boldsymbol{X} \succeq \boldsymbol{O}$ means that $\boldsymbol{X}$ is symmetric positive semidefinite. Then, we can define the *dual SDP* as

$$\begin{cases} \text{minimize } \boldsymbol{b}^T\boldsymbol{y} \\ \text{subject to } \boldsymbol{S} = \sum_{p=1}^{u} \boldsymbol{A}_p y_p - \boldsymbol{C} \succeq \boldsymbol{O}, \\ \qquad\qquad \sum_{p=1}^{u} \mathbf{Diag}(\boldsymbol{a}_p) y_p = \mathbf{Diag}(\boldsymbol{c}), \\ \qquad\qquad \boldsymbol{y} \in \mathbb{R}^u. \end{cases} \tag{3.2}$$

The variables for the primal SDP are $(\boldsymbol{X}, \boldsymbol{x})$ while those for the dual SDP are $(\boldsymbol{S}, \boldsymbol{y})$. Under mild assumptions[c] [61], the solution of (3.1)-(3.2) should satisfy

$$\begin{aligned} \operatorname{tr}(\boldsymbol{A}_p\boldsymbol{X}) + \operatorname{tr}(\mathbf{Diag}(\boldsymbol{a}_p)\mathbf{Diag}(\boldsymbol{x})) &= b_p,\ (p = 1, 2, \ldots, u) \\ \boldsymbol{S} = \sum_{p=1}^{u} \boldsymbol{A}_p y_p - \boldsymbol{C} &\succeq \boldsymbol{O}, \\ \sum_{p=1}^{u} \mathbf{Diag}(\boldsymbol{a}_p) y_p &= \mathbf{Diag}(\boldsymbol{c}), \\ \boldsymbol{X}, \boldsymbol{S} &\succeq \boldsymbol{O}, \\ \boldsymbol{b}^T\boldsymbol{y} - \operatorname{tr}(\boldsymbol{C}\boldsymbol{X}) - \operatorname{tr}(\mathbf{Diag}(\boldsymbol{c})\mathbf{Diag}(\boldsymbol{x})) &= 0. \end{aligned} \tag{3.3}$$

These conditions are equivalent to the ones of Erdahl [16] and Bellman and Fan [5].

---

[c] To be precise, we need to eliminate some of the variables by using the equalities $\sum_{p=1}^{u} \mathbf{Diag}(\boldsymbol{a}_p) y_p = \mathbf{Diag}(\boldsymbol{c})$ and assume the Slater's condition, but we shall not do it here to avoid being cumbersome.

The advantage of simultaneously considering the variables in the primal and dual SDPs is that we can check the numerical correctness of the approximate solutions from the above relations.

3.2. *Formulation of the RDM problem as an SDP*

Hereafter, we shall assume that we have chosen r spin orbitals from (1.3), which give a good approximation to the wavefunction we seek. There are plenty of such *bases* in quantum chemistry, and we used them in the numerical experiments described later. Also, notice that all definitions and notions of N-representability and their conditions can be defined accordingly using this finite basis of CONS.

The RDM problem that imposes necessary N-representability conditions such as the P (2.2), Q (2.3), G (2.4), $T1$ (2.6), $T2$ (2.7) or $T2'$ is in fact an SDP. To make its formulation clear, we shall perform some linear transformations on the matrices. In (2.9), the 1-RDM variable γ, and the corresponding one-particle part of the Hamiltonian $\boldsymbol{v}$ have two indexes, which correspond to ordinary matrices in linear algebra. However, the other matrices involved in the calculations have four or even six indices each. To convert from notations convenient for quantum chemists to the notations of elementary linear algebra, we need to map each pair (i, j) or triple (i, j, k) of indices to a composite index on these matrices. For instance, mapping the 2-RDM element $\Gamma^{ij}_{k\ell}$ $(1 \leq i < j \leq r;\ 1 \leq k < \ell \leq r)$ to $\tilde{\Gamma}_{j-i+(2r-i)(i-1)/2,\ell-k+(2r-k)(k-1)/2}$, results in a symmetric matrix of size $r(r-1)/2 \times r(r-1)/2$. We shall assume henceforth that all matrices are transformed into two-index matrices, and we shall keep the same notation as before for simplicity. Furthermore, due to spin symmetry [71], all these matrices will reduce to block-diagonal matrices whose sizes are specified in Table 1 [24, 25][d].

Let us define a linear transformation svec: $\mathbb{S}^d \to \mathbb{R}^{d(d+1)/2}$ from the space of $d \times d$ symmetric matrices $\mathbb{S}^d$. For $\boldsymbol{U} \in \mathbb{S}^d$, define

$$\text{svec}(\boldsymbol{U}) = (U_{11}, \sqrt{2}U_{12}, U_{22}, \sqrt{2}U_{13}, \sqrt{2}U_{23}, U_{33}, \ldots, \sqrt{2}U_{1d}, \ldots, U_{dd})^T.$$

Letting $y = (\text{svec}(\gamma)^T, \text{svev}(\Gamma)^T)^T \in \mathbb{R}^u$, $\boldsymbol{b} = (\text{svec}(\boldsymbol{v})^T, \text{svev}(\boldsymbol{w})^T)^T \in \mathbb{R}^u$, and finding the suitable matrices $\boldsymbol{C}, \boldsymbol{A}_p$ $(p = 1, 2, \ldots, u)$ and vectors $\boldsymbol{c}, \boldsymbol{a}_p$ $(p = 1, 2, \ldots, u)$ for the corresponding necessary N-representability

---

[d]There should be an additional $\binom{\frac{r}{2}+1}{2}$ term in the column of m of Table II [25] that is missing and corresponds to the size of the 1-RDM.

Table 1. Sizes of SDP as a function of the number of spin orbitals r for each necessary N-representability condition.

N-repres. cond.	size of block matrices
$\gamma \succeq \boldsymbol{O}$	$r/2 \times r/2$ (2 blocks)
$\boldsymbol{I} \succeq \gamma$	$r/2 \times r/2$ (2 blocks)
P condition	$(r/2)^2 \times (r/2)^2$ (1 block), $\binom{r/2}{2} \times \binom{r/2}{2}$ (2 blocks)
Q condition	$(r/2)^2 \times (r/2)^2$ (1 block), $\binom{r/2}{2} \times \binom{r/2}{2}$ (2 blocks)
G condition	$2\,(r/2)^2 \times 2\,(r/2)^2$ (1 block), $(r/2)^2 \times (r/2)^2$ (2 blocks)
$T1$ condition	$\frac{r}{2}\binom{r/2}{2} \times \frac{r}{2}\binom{r/2}{2}$ (2 blocks), $\binom{r/2}{3} \times \binom{r/2}{3}$ (2 blocks)
$T2$ condition	$\frac{r}{6}\binom{3r/2}{2} \times \frac{r}{6}\binom{3r/2}{2}$ (2 blocks), $\frac{r}{2}\binom{r/2}{2} \times \frac{r}{2}\binom{r/2}{2}$ (2 blocks)
$T2'$ condition	$\left(\frac{r}{2} + \frac{r}{6}\binom{3r/2}{2}\right) \times \left(\frac{r}{2} + \frac{r}{6}\binom{3r/2}{2}\right)$ (2 blocks), $\frac{r}{2}\binom{r/2}{2} \times \frac{r}{2}\binom{r/2}{2}$ (2 blocks)
u in (3.2)	$\binom{r^2/4+1}{2} + 2\binom{r\,(r/2-1)\,/4+1}{2} + \binom{\frac{r}{2}+1}{2}$
s in (3.2)	$5 + 2\binom{r/2+1}{2}$

here $\binom{a}{b} = \frac{a!}{(b-a)!b!}$, for integers $a \geq b > 0$.

conditions of Table 1, for instance, we can recast problem (2.9) as an SDP (3.2).

Although these transformations may seem a little confusing, they are in fact formulation that Garrod *et al.* arrived at 35 years ago [26]. Nakata *et al.* [49, 50, 51] formulated the problem as a primal SDP (3.1). For a detailed discussion about these transformations and the formulations, see for instance [24, 25, 71].

3.3. *Theoretical computational complexity of the primal-dual interior-point method*

As mentioned previously, SDPs can be solved in polynomial-time using interior-point methods [2, 61]. In particular, employing the parallel code SDPARA [22, 67], which is an implementation of the primal-dual interior-point method, one can theoretically expect that it will take $\mathcal{O}(\sqrt{d_{\max}} \log \varepsilon^{-1})$ iterations with $\mathcal{O}(u^2 f^2/g + u^3/g + u d_{\max}^2 + d_{\max}^3)$ floating-point operations per iteration. Here $d_{\max}$ refers to the size of the largest block matrix in $\boldsymbol{A}_p$ $(p = 1, 2, \ldots, u)$, f is the maximum number of nonzero elements in each of these matrices, g is the total number of available CPU cores in the parallel computer, and ε is the expected accuracy when we replace the rhs of (3.3) "$= 0$" by "$\leq \varepsilon$" (where $\varepsilon > 0$). In our case, $u = \mathcal{O}(r^4)$, $d_{\max} = \mathcal{O}(r^3)$, $f = \mathcal{O}(1)$, and therefore, the total number of theoretical floating-point operations is $\mathcal{O}(r^{13.5} \log \varepsilon^{-1}/g)$ [25].

4. Some Historical Remarks

We shall briefly mention some of the articles related to our work. However, the list is far from complete.

The definition of the RDM was explicitly spelled out by Husimi [30] in 1940. The dependence of the energy on the 2-RDM (and 1-RDM) appeared in Löwdin [35] and Mayer [37] in 1955. The necessary and sufficient conditions for an ensemble 1-RDM, *i.e.*, $\mathbf{0} \preceq \gamma \preceq \boldsymbol{I}$, $\mathrm{tr}(\gamma) = N$, were obtained by Kuhn [32] in 1960 and Coleman [10] in 1963. Coleman's article also gave a precise formulation of the N-representability problem and the P and Q conditions (for the 2-RDM). The following year, Garrod and Percus [28] proposed the G condition.

The restriction of the N-representability problem to only the diagonal elements of the 2-RDM, known as the *diagonal problem*, was investigated by Weinhold and Wilson [66], Davidson [12], McRae-Davidson [43], and Yoseloff [70], beginning in the late 1960's. Progresses on this topic can be found in the survey [3].

The first variational calculations using the 2-RDMs imposing the necessary N-representability conditions were performed by Kijewski and colleagues for the doubly ionized carbon C^{++} ($N = 4$) in the late 1960's and early seventies [31] (see an earlier reference therein). Garrod and co-authors proposed several algorithms, some of which resemble modern optimization algorithms, and reported results for beryllium ($N = 4$) [26, 27, 57]. In particular, Mihailović and Rosina applied one of these algorithms to nuclear physics [45], but obtained large deviations from the full CI calculations on nuclear systems.

In the 1978 survey paper of Erdahl [15], we can find the conditions known as $T1$, $T2$ [71], and $T2'$ conditions [6, 42]. The following year, Erdahl published algorithms based on an exact mathematical characterization of solutions of the lower bound method (RDM method) [16].

This was the golden age of RDM research, but it somehow faded away because it was soon realized that the underlying problem is computationally difficult and poor results were obtained for nuclear systems.

Interest in the 2-RDM approach was revived since 1992 as a result of the work of Valdemoro [62], Nakatsuji and Yasuda [54], and Mazziotti [38]. The approach involves solving the density equation or the contracted Schrödinger equation (CSE), and it has been shown that it is equivalent to solving the Schrödinger equation. Nakatsuji proved that if an N-representable 4-RDM satisfies the CSE, then the original D satisfies the Schrödinger equation and vice versa [53]. Valdemoro, Nakatsuji-Yasuda and Mazziotti considered the 2-RDM as the basic variable. The CSE requires 1- to 4-RDMs; the 3- and 4-RDMs were reconstructed from 1- and 2-RDMs, and the CSE was solved iteratively. The authors assumed the resultant 2-RDM to be nearly N-representable because the reconstruction functional was physically relevant; they did not explicitly impose any N-representability conditions. Their results were quite good, and comparable to single and double CIs for small atoms and molecules such as Be, Ne, and CH_3F. The absolute values of the negative eigenvalues of the P, Q, and G-matrices were small. However, other researchers paid little attention to this approach because it has non-convergence or divergence problems especially when the correlations are strong [14]. Despite improvement, difficulties remained with the systematic refinements of the reconstruction functional [55, 69].

In 2001, Nakata *et al.* [51] were the first to use an optimization software to solve the RDM problem as an SDP. They reported computational results imposing the P, Q, and G conditions on a series of small atoms and molecules. The results were better than the SDCI calculations, and the obtained correlation energies were from 100% to 120%. As mentioned in Section 3.1, these results have a numerical certificate of correctness, which could not be obtained before the advent of interior-point methods. Also, the numerical convergence does not depend on the initial guess, as it does in Hartree-Fock, CCSD methods, *etc.* Moreover, there exists a global minimum, something which is not guaranteed in the CSE approach. The following year, Nakata *et al.* [49] showed results for the dissociation limit of several molecules including triple bonded N_2, demonstrating that the RDM method does not break down in the way that the single reference methods such as CCSD and perturbation methods do. However, it was also shown that the RDM method did have a small size consistency problem.

The inclusion of the Weinhold-Wilson inequalities [66], which are not satisfied if one includes only the P, Q, and G conditions [49], did little to improve the ground state energies [50].

In 2002, Mazziotti reproduced Nakata *et al.*'s results and applied the RDM method to diatomic molecules [39]. The prolific research by Mazziotti and his colleagues in the following years [42] corroborated Nakata *et al.*'s results.

In 2004, Zhao *et al.* made a breakthrough [71]. They included the $T1$ and $T2$ conditions; these became very strong conditions on small molecules and atoms. They noted a "spectacular increase in accuracy", and these results were comparable to CCSD(T); typically, the correlation energies for various atoms and molecules were between 100% to 101% [24, 48, 71].

In the same year, Mazziotti announced the RRSDP method [40]. He reformulates the SDP problem as a nonlinear and nonconvex problem and used a quasi-Newton method to solve it [7]. In 2006, Cancès *et al.* proposed and implemented an algorithm to solve the dual problem of (2.9) [8].

Hammond *et al.* applied the RDM method to the one-dimensional Hubbard model [29]. They calculated the Hubbard models with the P, Q, G, and $T2$ conditions at up to 14 sites. The obtained error per site was -0.0089 for $L = 14$ case under the P, Q, G, and $T2$ conditions for $U = 8$, when its correlation is the strongest. Nakata *et al.* investigated the high correlation limit by using a multiple-precision arithmetic version of the SDP solver, called SDPA-GMP [48]. In the high correlation limit, they reproduced the exact energy and proved that the G condition is also a sufficient N-representability condition in this limit [68].

Size-consistency and size-extensivity are important properties especially when the systems are large. Nakata *et al.* found slight deviations for size-consistency [49], but Van Aggelen *et al.* [63] systematically showed that the RDM method with P, Q and G conditions give an incorrect dissociation limit with fractional charges on the well-separated atoms of diatomic molecules. For instance, in the case of CO, the method gives 5.98 and 8.00 as the Mulliken populations of C and O in the dissociation limit. Even adding the $T1$ and $T2$ conditions did not fix this problem [63, 64]. Moreover, Nakata and Yasuda's numerical investigation calculating 32 non-interacting CH_4 and N_2 showed that size-extensivity is also slightly violated [52]. The inextensive contributions to the energies are 3×10^{-4} Hartrees and 3×10^{-3} Hartrees using the STO-6G basis set. Later, Verstichel *et al.* proposed a method to "cure" this pathological behavior of the RDM method; it is however quite computationally demanding [65]. In fact, all of the current solutions to the size-consistency problem are not practical.

5. Numerical Results for the RDM Method

Here, we describe numerical results for our implementation of the RDM method imposing certain necessary N-representability conditions. Some of the results have not been published elsewhere.

5.1. *New numerical results for larger systems*

Here are numerical results for the largest systems solved so far by our group.

The SDPs obtained by the RDM method imposing the P, Q, G or P, Q, G, $T1$, $T2'$ conditions were solved using the parallel code SDPARA 7.3.2 [22, 23, 67]. The calculations were performed at the Kyoto University's T2K supercomputer using 32 or 128 nodes; each node of this computer has 4

Table 2. Difference in ground state energy from that of full CI of the RDM method imposing the P, Q, G, $T1$, $T2'$ conditions from SDPARA 7.3.2, and those of CCSD(T), SDCI, and Hartree-Fock using Gamess and Gaussian98. The last column shows the full CI energies. The energy units are Hartrees ($= 4.3598 \times 10^{-18}$ J). The correlation energies (relative to 0% for Hartree-Fock and 100% for full CI) are also shown in the second row.

system	state	basis	r	$N(N_\alpha)$	$2S+1$	$\Delta E_{PQGT1T2'}$	$\Delta E_{CCSD(T)}$	ΔE_{SDCI}	ΔE_{HF}	E_{FCI}
NH_2^-	$^1A_1$	double-ζ	28	10 (5)	1	−0.000 6	+0.000 63	+0.008 74	+0.141 98	−55.624 71
						100.4	99.55	93.84	0	100
CH_2	$^1A_1$	double-ζ	28	8 (4)	1	−0.000 4	+0.000 59	+0.005 80	+0.100 67	−38.962 24
						100.4	99.42	94.24	0	100
NH_3	$^1A_1$	valence double-ζ	30	10 (5)	1	−0.000 5	+0.000 49	+0.007 46	+0.128 75	−56.304 89
						100.4	99.62	94.45	0	100
CH_3	$^2A_2''$	valence double-ζ	30	9 (5)	2	−0.000 3	+0.000 31	+0.004 01	+0.094 54	−39.644 14
						100.3	99.67	95.75	0	100
C_2	$^1\Sigma_g^+$	valence double-ζ	36	12 (6)	1	−0.003 5	+0.000 39	+0.055 98	+0.285 66	−75.642 11
						101.2	99.86	80.41	0	100

CPUs (quad-core AMD Opteron 8356 2.3 GHz) and 32 GB of memory, giving a total of 2048 cores. They were also performed on our own custom-made computer cluster with 16 nodes, in which each node has 2 CPUs (quad-core Intel Xeon 5460 3.16 GHz) and 48 GB of memory, giving a total of 128 cores.

Table 2 shows the results for five molecules and $r = 28, 30$ or 36 spin orbitals [23]. The full CI and SDCI (singly and doubly substituted configuration interaction) calculations were performed using the Gamess package [58], while CCSD(T) (coupled cluster singles and doubles with perturbational treatment of triples) and Hartree-Fock calculations were obtained using Gaussian98 [21]. The entries, except the full CI, are the ground state energy differences from the full CI. The RDM method always gives an energy lower than the full CI, while SDCI and Hartree-Fock give higher values. CCSD(T) usually, but not necessarily, results in a higher energy. The energies are in Hartrees. The acceptable accuracy in quantum chemistry is 1 kcal/mol, which corresponds to approximately 0.0016 Hartree. The correlation energy $\varepsilon_{\text{corr}}$ is also an important measure in quantum chemistry. It is defined as a percentage relative to the Hartree-Fock energy (0%) and full CI energy (100%):

$$\varepsilon_{\text{corr}} = \frac{|E - E_{\text{HF}}|}{E_{\text{HF}} - E_{\text{FCI}}} \times 100,$$

where E is the energy calculated by the RDM method, CCSD(T), or SDCI.

From Table 2, we can conclude that the RDM method imposing the $P, Q, G, T1, T2'$ conditions gives equal or better energies than CCSD(T) in terms of the absolute value. The C_2 molecule is unusual. However, it is known to be a "difficult" system to analyze in quantum chemistry.

Table 3. Ground state energies calculated by the RDM method imposing the P, Q, and G conditions from SDPARA 7.3.2, and those obtained by CCSD(T), SDCI, and Hartree-Fock from Gamess and Gaussian98. The energy units are in Hartrees ($= 4.3598 \times 10^{-18}$ J).

system	state	basis	r	$N(N_\alpha)$	$2S+1$	E_{PQG}	$E_{CCSD(T)}$	E_{SDCI}	E_{HF}
O_2^+	$^2\Pi_g$	double-ζ	40	15 (8)	2	−149.450 2	−149.385 95	−149.360 26	−149.091 83

Table 4. Size of SDPs obtained by the RDM method imposing some N-representability conditions shown at Tables 2 and 3, and their computational times when solved at the T2K supercomputer or at the computer cluster (c.c.).

system	r	N-repres. cond.	u	$d_{\max}$	time (s)	system	CPU cores
NH_2^-	28	$P, Q, G, T1, T2'$	27,888	4,032	27,949	T2K	2048
CH_2	28	$P, Q, G, T1, T2'$	27,888	4,032	26,656	T2K	2048
NH_3	30	$P, Q, G, T1, T2'$	36,795	4,965	72,026	T2K	2048
CH_3	30	$P, Q, G, T1, T2'$	36,795	4,965	68,593	T2K	2048
C_2	36	$P, Q, G, T1, T2'$	76,554	8,604	1,554,675	c.c.	128
O_2^+	40	P, Q, G	116,910	800	5,943	T2K	2048

Table 3 shows the same results for the O_2^+ molecule in which $r = 40$ spin orbitals were used. A full CI calculation was not possible for this case due to the limited resources of the computer, and therefore, we show only the ground state energies corresponding to each entry.

Table 4 shows typical sizes of the problem formulated as an SDP (see Section 3.1), and the computational times taken to solve it by SDPARA 7.3.2 on the T2K supercomputer and our custom-made computer cluster.

In the previous study [48], we only could solve the RDM problem with the $P, Q, G, T1, T2'$ conditions for up to $r = 28$ spin orbitals. Here, we give results for up to $r = 36$ spin orbitals. This achievement was possible due to a major update in the parallel code SDPARA 7.3.2 [22, 23, 67], making it faster, and able to take advantage of multi-core (multi-thread) computations in addition to ordinary MPI (message passing interface) computations.

For other physical properties such as the dipole moments, the reader may refer to [48].

5.2. *Summary of the numerical experiments*

Now let us present a graphical summary of the data obtained in our previous study [48] and the ones of Table 2.

Figure 3 plots the differences in ground state energy between the full CI and the RDM method imposing the P, Q, G or $P, Q, G, T1$ conditions and

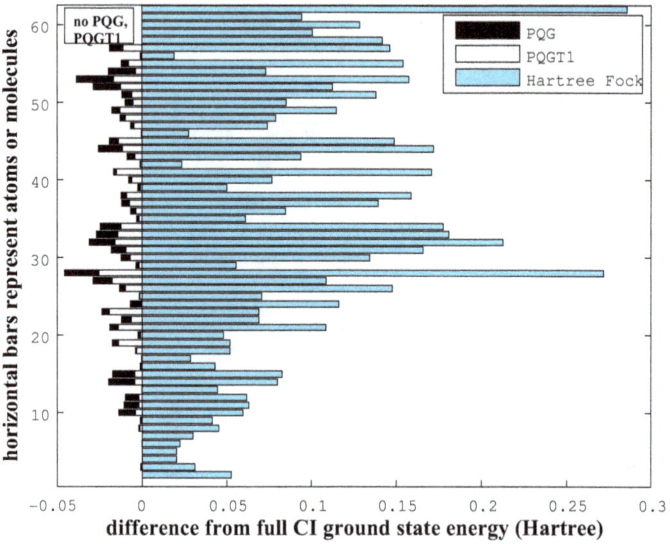

Fig. 3. Difference between the ground state energy of full CI and those of the RDM method imposing the P, Q, G, or $P, Q, G, T1$ conditions, and those obtained by Hartree-Fock for the 57 atomic and molecular systems of [48] and the five systems of Table 2.

between the full CI and Hartree-Fock for 57 atomic or molecular systems [48] and those 5 shown in Table 2. Each horizontal bar corresponds to a system, and the bars are ordered accordingly to the order the system appears in the tables. That is, the bottom entry corresponds to the lithium atom with $r = 10$ spin orbitals [48], while the top five entries correspond to the data of Table 2 (notice that there are no values for P, Q, G and $P, Q, G, T1$ entries for these cases).

Apparently, no correlation exists between the Hartree-Fock and the RDM method's results. However, we clearly see that by imposing the P, Q, G conditions, the RDM methods get much better results than those of Hartree-Fock.

Figure 4 shows the differences in ground state energy between the full CI and the RDM method imposing the $P, Q, G, T1$ (same values as in Figure 3) and $P, Q, G, T1, T2'$ conditions, and between the full CI and CCSD(T) for the same systems as in Figure 3.

Comparing the RDM method with the $P, Q, G, T1, T2'$ conditions and the CCSD(T) values, we can see that they seem equally good. However, in some cases, CCSD(T) fails to converge. There are four cases in [48] which

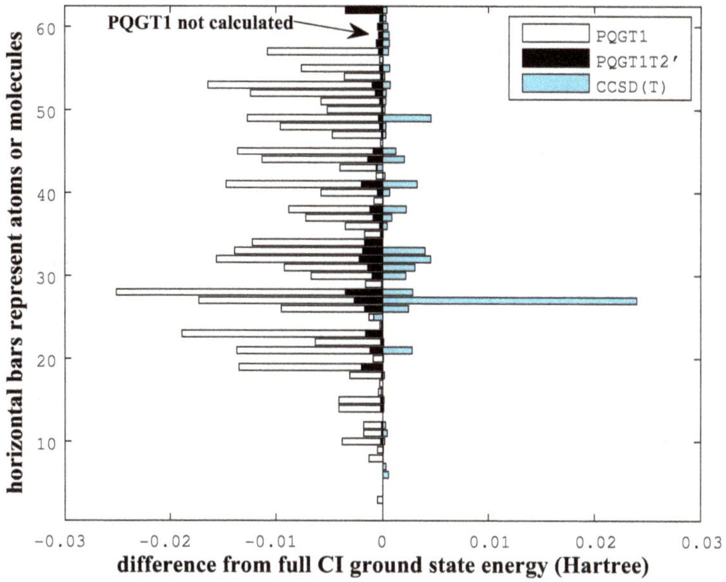

Fig. 4. Difference between the ground state energy of full CI and those of the RDM method imposing the $P, Q, G, T1$, or $P, Q, G, T1, T2'$ conditions, and CCSD(T) for the 57 atomic and molecular systems of [48] and the five systems of Table 2.

are replaced by zero in Figure 4. The largest deviation of CCSD(T) from the full CI value is 0.02397 Hartree for the BN molecule in the $^3\Pi$ state [48].

6. Concluding Remarks

This chapter presented an outline of the reduced-density-matrix method and its applications to atomic and molecular fermionic systems. This method has the following features: (i) it is an *ab initio* method, which is rigorously the same as the Schrödinger equation for the ground state; (ii) the number of variables is always four, regardless of the size of the system; (iii) the existence of a linear scaling method is apparent from the sparsity of the first- and second-order reduced density matrices. The major obstacle of this method is the fundamental difficulty in obtaining complete N-representability conditions for the 2-RDM. However, we know fairly good approximate (necessary) conditions, such as the P, Q, G, $T1$ and $T2'$ conditions, that reproduce ground state energies comparable to those given by CCSD(T), *i.e.*, the gold standard method in quantum chemistry. The method can be cast as a semidefinite programming problem, in which a linear functional is minimized while keeping the eigenvalues of the variable matrices non-negative. The chapter also presented new results, *i.e.*, the energies for NH_2^-, CH_2, NH_3, CH_3, C_2, and O_2^+ computed using a supercomputer with a highly efficient semidefinite programming solver, SDPARA. The semidefinite programming problems for NH_3, CH_3, C_2, and O_2^+ are the largest problems solved so far in the standard formulation. The correlation energies using P, Q, G, $T1$, $T2'$ were 100.4% for NH_2^-, CH_2, NH_3, 100.3% for CH_3, and 101.2% for C_2. The O_2^+ calculation used the double-ζ basis and its correlation energy was 132% because the open-shell systems are difficult and only the P, Q and G conditions were employed in consideration of the space needed.

We would like to close this chapter saying that the RDM method is a promising method for quantum chemistry or condensed matter physics. We believe it will be of fundamental use to researchers in chemistry and physics.

Acknowledgments

The large-scale supercomputer computations in this research were supported by the Collaborative Research Program for Large-Scale Computation of ACCMS and IIMC, Kyoto University. M. N. was supported by the Special Postdoctoral Researchers' Program of RIKEN, and by a Grant-in-Aid for Scientific Research (B) 21300017. M. F. is very thankful for the invitation to the program "Complex Quantum Systems" held at the IMS-NUS, and is especially grateful to the organizers Heinz Siedentop and Matthias Christandl. He also enjoyed the discussions he had with Prof. Robert Erdahl. M. F. is partially supported by a Grant-in-Aid for Young Scientists (B) 21700008.

References

1. M. Altunbulak and A. Klyachko, "The Pauli principle revised", *Commun. Math. Phys.*, **282** (2008), 287–322.
2. M. F. Anjos and J. B. Lasserre (eds.), *Handbook on Semidefinite, Conic and Polynomial Optimization*, Springer, New York, 2012.
3. P. W. Ayers and E. R. Davidson, "Linear inequalities for diagonal elements of density matrices", in [41], 443–483.
4. F. Barahona, M. Grötschel, M. Jünger, and G. Reinelt, "An application of combinatorial optimization to statistical physics and circuit layout design", *Oper. Res.*, **36** (1988), 493–513.
5. R. Bellman and K. Fan, "On systems of linear inequalities in Hermitian ma-

trix variables", in *Convexity*, Vol. **7** of Proc. Sympos. Pure Math., American Mathematical Society, Providence, RI, 1963, 1–11.
6. B. J. Braams, J. K. Percus, and Z. Zhao, "The T1 and T2 representability conditions", in [41], 93–101.
7. S. Burer and R. D. C. Monteiro, "A nonlinear programming algorithm for solving semidefinite programs via low-rank factorization", *Math. Program.*, **95** (2003), 329–357.
8. E. Cancès, G. Stoltz, and M. Lewin, "The electronic ground-state energy problem: A new reduced density matrix approach", *J. Chem. Phys.*, **125** (2006), 064101, 5 pages.
9. J. Cioslowski (ed.), *Many-Electron Densities and Reduced Density Matrices*, Kluwer Academic/Plenum Publishers, New York, 2000.
10. A. J. Coleman, "Structure of fermion density matrices", *Rev. Modern Phys.*, **35** (1963), 668–689.
11. A. J. Coleman and V. I. Yukalov, *Reduced Density Matrices: Coulson's Challenge*, Lecture Notes in Chemistry, Vol. **72**, Springer, New York, 2000.
12. E. R. Davidson, "Linear inequalities for density matrices", *J. Math. Phys.*, **10** (1969), 725–734.
13. P. A. M. Dirac, "Quantum mechanics of many-electron systems", *Proc. Roy. Soc. (London)*, **A 123** (1929), 714–733.
14. M. Ehara, M. Nakata, H. Kou, K. Yasuda, and H. Nakatsuji, "Direct determination of the density matrix using the density equation: Potential energy curves of HF, CH_4, BH_3, NH_3, and H_2O", *Chem. Phys. Lett.*, **305** (1999), 483–488.
15. R. M. Erdahl, "Representability", *Int. J. Quantum Chem.*, **13** (1978), 697–718.
16. R. M. Erdahl, "Two algorithms for the lower bound method of reduced density theory", *Rep. Math. Phys.*, **15** (1979), 147–162.
17. R. M. Erdahl and C. Garrod, "Trace relations, one-particle symmetries and the convex structure of reduced density operators", in R. M. Erdahl (ed.), *Reduced Density Operators with Applications to Physical and Chemical Systems - II*, Queen's Papers in Pure and Applied Mathematics No. **40**, Kingston, Ontario, Canada, 1974, 22–27.
18. R. Erdahl and B. Jin, "On calculating approximate and exact density matrices", in [9], 57–84.
19. R. Erdahl and M. Rosina, "The B-condition is implied by the G-condition", in R. M. Erdahl (ed.), *Reduced Density Operators with Applications to Physical and Chemical Systems - II*, Queen's Papers in Pure and Applied Mathematics No. **40**, Kingston, Ontario, Canada, 1974, 36–43.
20. A. L. Fetter and J. D. Walecka, *Quantum Theory of Many-Particle Systems*, McGraw-Hill, Inc., New York, 1971.
21. M. J. Frisch, G. W. Trucks, H. B. Schlegel *et al.*, *Gaussian 98, Revision A.11.3*, Gaussian, Inc., Pittsburgh, PA, 2002.
22. K. Fujisawa, K. Nakata, M. Yamashita, and M. Fukuda, "SDPA project: Solving large-scale semidefinite programs", J. Oper. Res. Soc. Japan, **50** (2007), 278–298.

23. Fujitsu, press release, http://www.fujitsu.com/global/news/pr/archives/month/2010/20100528-01.html
24. M. Fukuda, B. J. Braams, M. Nakata, M. L. Overton, J. K. Percus, M. Yamashita, and Z. Zhao, "Large-scale semidefinite programs in electronic structure calculation", *Math. Program., Ser. B*, **109** (2007), 553–580.
25. M. Fukuda, M. Nakata, and M. Yamashita, "Semidefinite programming: Formulations and primal-dual interior-point methods", in [41], 103–118.
26. C. Garrod and M. A. Fusco, "A density matrix variational calculation for atomic Be", *Int. J. Quantum Chem.*, **10** (1976), 495–510.
27. C. Garrod, M. V. Mihailović, and M. Rosina, "The variational approach to the two-body density matrix", *J. Math. Phys.*, **16** (1975), 868–874.
28. C. Garrod and J. K. Percus, "Reduction of the N-particle variational problem", *J. Math. Phys.*, **5** (1964), 1756–1776.
29. J. R. Hammond and D. A. Mazziotti, "Variational reduced-density-matrix calculation of the one-dimensional Hubbard model", *Phys. Rev. A*, **73** (2006), 062505, 6 pages.
30. K. Husimi, "Some formal properties of the density matrix", *Proc. Phys. Math. Soc. Jpn.*, **22** (1940), 264–314.
31. L. J. Kijewski, "Strengh of the G-matrix condition in the reduced-density-matrix variational principle", *Phys. Rev. A*, **9** (1974), 2263–2266.
32. H. W. Kuhn, "Linear inequalities and the Pauli principle", *Proc. Symp. Appl. Math.*, **10** (1960), 141–147.
33. H. Kummer, "About the relationship between some necessary conditions for N-representability", *Int. J. Quantum Chem.*, **12** (1977), 1033–1038.
34. Y.-K. Liu, M. Christandl, and F. Verstraete, "Quantum computational complexity of the N-representability problem: QMA complete", *Phys. Rev. Lett.*, **98** (2007), 110503, 4 pages.
35. P.-O. Löwdin, "Quantum theory of many-particle systems. I. Physical interpretations by means of density matrices, natural spin-orbitals, and convergence problems in the method of configurational iteraction", *Phys. Rev.*, **97** (1955), 1474–1489.
36. P.-O. Löwdin, "Some aspects of the development of the theory of reduced density matrices and the representability problem", in R. Erdahl and V. H. Smith, Jr. (eds.), *Density Matrices and Density Functionals*, D. Reidel Publishing Company, Dordrecht, 1987, 21–49.
37. J. E. Mayer, "Electron correlation", *Phys. Rev.*, **100** (1955), 1579–1586.
38. D. A. Mazziotti, "Contracted Schrödinger equation: Determining quantum energies and two-particle density matrices without wave functions", *Phys. Rev. A*, **57** (1998), 4219–4234.
39. D. A. Mazziotti, "Variational minimization of atomic and molecular ground-state energies via the two-particle reduced density matrix", *Phys. Rev. A*, **65** (2002), 062511, 14 pages.
40. D. A. Mazziotti, "Realization of quantum chemistry without wave functions through first-order semidefinite programming", *Phys. Rev. Lett.*, **93** (2004), 213001, 4 pages.
41. D. A. Mazziotti (ed.), *Reduced-Density-Matrix Mechanics: With Applications*

to Many-Electron Atoms and Molecules, Advances in Chemical Physics Vol. **134**, John Wiley & Sons, Inc., Hoboken, NJ, 2007.
42. D. A. Mazziotti, "Variational two-electron reduced-density-matrix theory", in [41], 21–59.
43. W. B. McRae and E. R. Davidson, "Linear inequalities for density matrices II", *J. Math. Phys.*, **13** (1972), 1527–1538.
44. M. V. Mihailović and M. Rosina, "Excitations as ground state variational parameters", *Nuc. Phys.*, **130** (1969), 386–400.
45. M. V. Mihailović and M. Rosina, "The variational approach to the density matrix for light nuclei", *Nucl. Phys.*, **A237** (1975), 221–228.
46. K. Nakamura, "Two-body correlation of interacting fermions", *Progr. Theor. Phys.*, **21** (1959), 713–726.
47. M. Nakata and J. S. M. Anderson, "On the size-consistency of the reduced-density-matrix method and the unitary invariant diagonal N-representability conditions", *AIP Advances*, **2** (2012), 032125, 15 pages.
48. M. Nakata, B. J. Braams, K. Fujisawa, M. Fukuda, J. K. Percus, M. Yamashita, and Z. Zhao, "Variational calculation of second-order reduced density matrices by strong N-representability conditions and an accurate semidefinite programming solver", *J. Chem. Phys.*, **128** (2008), 164113, 14 pages.
49. M. Nakata, M. Ehara, and H. Nakatsuji, "Density matrix variational theory: Application to the potential energy surfaces and strongly correlated systems", *J. Chem. Phys.*, **116** (2002), 5432–5439.
50. M. Nakata, M. Ehara, and H. Nakatsuji, "Density matrix variational theory: Strength of Weinhold-Wilson inequalities", in E. J. Brändas and E. S. Kryachko (eds.), *Fundamental World of Quantum Chemistry: A Tribute to the Memory of Per-Olev Löwdin*, Vol. **I**, Kluwer Academic Publishers, Dordrecht, 2003, 543–557.
51. M. Nakata, H. Nakatsuji, M. Ehara, M. Fukuda, K. Nakata, and K. Fujisawa, "Variational calculations of fermion second-order reduced density matrices by semidefinite programming algorithm", *J. Chem. Phys.*, **114** (2001), 8282–8292.
52. M. Nakata and K. Yasuda, "Size extensivity of the variational reduced-density-matrix method", *Phys. Rev. A*, **80** (2009), 042109, 5 pages.
53. H. Nakatsuji, "Equation for the direct determination of the density matrix", *Phys. Rev. A*, **14** (1976), 41–50.
54. H. Nakatsuji and K. Yasuda, "Direct determination of the quantum-mechanical density matrix using the density equation", *Phys. Rev. Lett.*, **76** (1996), 1039–1042.
55. M. Nooijen, M. Wladyslawski, and A. Hazra, "Cumulant approach to the direct calculation of reduced density matrices: A critical analysis", *J. Chem. Phys.*, **118** (2003), 4832–4848.
56. M. Rosina, "Some theorems on uniqueness and reconstruction of higher-order density matrices", in [9], 19–32.
57. M. Rosina and C. Garrod, "The variational calculation of reduced density matrices", *J. Comput. Phys.*, **18** (1975), 300–310.

58. M. W. Schmidt, K. K. Baldridge, J. A. Boatz, *et al.*, "General atomic and molecular electronic structure system", *J. Comput. Chem.*, **14** (1993), 1347–1363.
59. D. W. Smith, "N-representability problem for fermion density matrices. II. The first-order density matrix with N even", *Phys. Rev.*, **147** (1966), 896–898.
60. A. Szabo and N. S. Ostlund, *Modern Quantum Chemistry: Introduction to Advanced Electronic Structure Theory*, Dover Publications, Inc., Mineola, New York, 1996.
61. M. J. Todd, "Semidefinite optimization", *Acta Numer.*, **10** (2001), 515–560.
62. C. Valdemoro, "Approximating the second-order reduced density matrix in terms of the first-order one", *Phys. Rev. A*, **45** (1992), 4462–4467.
63. H. van Aggelen, P. Bultinck, B. Verstichel, D. van Neck, and P. W. Ayers, "Incorrect diatomic dissociation in variational reduced density matrix theory arises from the flawed description of fractionally charged atoms", *Phys. Chem. Chem. Phys*, **11** (2009), 5558–5560.
64. H. van Aggelen, B. Verstichel, P. Bultinck, D. van Neck, P. W. Ayers, and D. L. Cooper "Chemical verification of variational second-order density matrix based potential energy surfaces for the N_2 isoelectronic series", *J. Chem. Phys*, **132** (2010), 114112, 10 pages.
65. B. Verstichel, H. van Aggelen, D. van Neck, P. W. Ayers, and P. Bultinck, "Subsystem constraints in variational second order reduced density matrix optimization: Curing the dissociative behavior", *J. Chem. Phys.*, **132** (2010), 114113, 6 pages.
66. F. Weinhold and E. B. Wilson Jr., "Reduced density matrices of atoms and molecules. II. On the N-representability problem", *J. Chem. Phys.*, **47** (1967), 2298–2311.
67. M. Yamashita, K. Fujisawa, M. Fukuda, K. Nakata, and M. Nakata, "Parallel solver for semidefinite programming problem having sparse Schur complement matrix", *ACM Transactions on Mathematical Software*, **39** (2012), 6:1–6:22.
68. M. Yamashita, K. Fujisawa, K. Nakata, M. Nakata, M. Fukuda, K. Kobayashi, and K. Goto, "A high-performance software package for semidefinite programs: SDPA 7", *Research Report*, **B-460** Department of Mathematical and Computing Sciences, Tokyo Institute of Technology, 2010.
69. K. Yasuda, "Direct determination of the quantum-mechanical density matrix: Parquet theory", *Phys. Rev. A*, **59** (1999), 4133–4149.
70. M. L. Yoseloff, "A combinatorial approach to the diagonal N-representability problem", *Trans. Amer. Math. Soc.*, **190** (1974), 1–41.
71. Z. Zhao, B. J. Braams, M. Fukuda, M. L. Overton, and J. K. Percus, "The reduced density matrix method for electronic structure calculations and the role of three-index representability conditions", *J. Chem. Phys.*, **120** (2004), 2095–2104.

FERMIONIC QUANTUM MANY-BODY SYSTEMS: A QUANTUM INFORMATION APPROACH

Christina V. Kraus*
Max-Planck Institute for Quantum Optics
Hans-Kopfermann-Str. 1
85748 Garching, Germany
Christina.Kraus@mpq.mpg.de

These lecture notes show how Quantum Information Theory can help us to gain deeper insight into the physics of fermionic quantum many-body systems. In the first part, we start from fermionic correlations as they appear in recent experiments with cold gases and develop a pairing theory for fermionic particles applicable to the current experimental setups. In the second part we introduce and discuss a new class of variational states, the fermionic Projected Entangled Pair States (fPEPS) that allow for an efficient approximation of ground and thermal states of local fermionic Hamiltonians.

1. Introduction

Fermionic particles are of utmost importance in many fields of physics, since they are building blocks of matter and thus central to some of the most fascinating effects in condensed matter physics, like superconductivity, superfluidity or the Quantum Hall effect. In general terms, the plethora of fermionic quantum phases is due to non-trivial quantum correlations which on the other hand turn the theoretical investigation of these systems into a daunting task.

In the case of spin systems, new insights could be gained with the help of Quantum Information Theory (QIT). In brief, this field aims at using quantum mechanical correlations ("entanglement") in systems of distinguishable

*Current addresses: Institute for Quantum Optics and Quantum Information of the Austrian Academy of Sciences, A-6020 Innsbruck, Austria and Institute for Theoretical Physics, Innsbruck University, A-6020 Innsbruck, Austria.

two-level systems ("qubits") for the transmission and processing of information. In this respect, ideas like quantum computation have emerged, and methods to characterize, to detect and to quantify entanglement in systems of qubits have been developed. The theoretical framework of QIT could be successfully applied to spins to gain more insight into physical properties of these systems. Furthermore, new algorithmic techniques were developed.

The goal of this article is to show how methods and tools known in QIT can be applied to learn more about fermionic quantum many-body systems. We address two different topics: In the first part, we aim at characterizing certain fermionic correlations as they appear, for example, in experiments with ultra cold quantum gases, and develop a pairing theory for fermionic particles. In the second part, we show how QIT can lead us to the construction of a new class of variational states, the fPEPS, that allow for an efficient approximation of ground and thermal states of local fermionic Hamiltonians.

2. Pairing in Fermionic Systems: A Quantum Information Perspective

2.1. *Motivation*

The notion of pairing in fermionic systems is at least as old as the seminal work of Bardeen, Cooper and Schrieffer explaining superconductivity [1]. The formation of fermionic pairs with opposite spin and momentum is not only the source for the vanishing resistance in solid state systems, but it can also explain many other interesting phenomena, like superfluidity in helium-3 or inside a neutron star.

Recently, fermionic pairing has gained again a lot of attention, for instance in the field of ultra cold quantum gases [2, 3, 4, 5, 6, 7, 8, 9]. A prominent example is the so-called BEC-BCS crossover, where a transition between a Bose-Einstein condensation of diatomic molecules and loosely bound Cooper pairs can be realized using a Feshbach resonance (see Fig. 1). In both regimes pairs of fermions emerge, but the notion of pairing is not clear and sometimes even controversial [7, 10, 11, 12]. In addition, pairing without superfluidity [13] has been observed in these experiments, raising fundamental questions on quantum correlations in fermionic many-body systems.

Motivated by these exciting experiments and the central role pairing plays in many physical phenomena, and by the lack of accepted criteria to verify the presence of pairing in a quantum state, we propose a clear

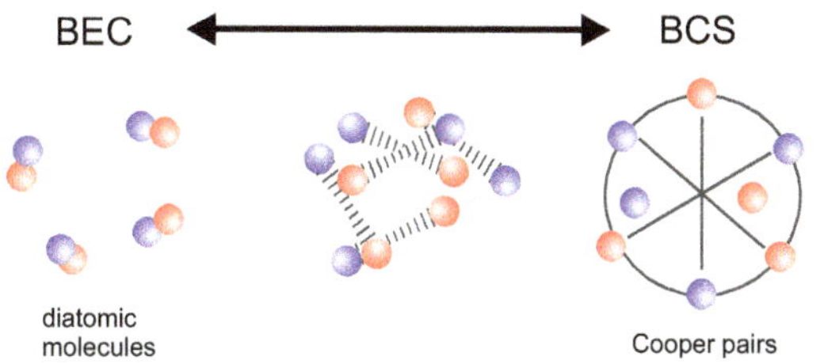

Fig. 1. BEC-BCS crossover of a diatomic fermi gas. A Feshbach resonance can be used to drive a crossover from a Bose-Einstein condensate of diatomic molecules to a BCS-superfluid of Cooper pairs.

and unambiguous definition of pairing intended to capture its two-particle nature and to allow a systematic study of the set of paired states and its properties. We employ methods and tools from QIT to gain a better understanding of the set of paired fermionic states and develop tools for the systematic detection and quantification of pairing which are applicable to current experiments.

2.2. *Pairing theory*

In this section we develop the theoretical framework of our pairing theory. We start with the basic notation.

Basic notation

We consider fermions on an M-dimensional single particle Hilbert space $\mathcal{H} = \mathbb{C}^M$ where all observables are generated by the creation and annihilation operators $a_j^\dagger$ and $a_j, j = 1, \ldots, M$, which satisfy the canonical anti-commutation relations (CAR) $\{a_k, a_l\} = 0$ and $\{a_k, a_l^\dagger\} = \delta_{kl}$. Sometimes a description using the $2M$ hermitian Majorana operators $c_{2j-1} = a_j^\dagger + a_j$, $c_{2j} = (-i)(a_j^\dagger - a_j)$, which satisfy $\{c_k, c_l\} = 2\delta_{kl}$, is more convenient. The Hilbert space of the many-body system, the antisymmetric Fock space over M modes, $\mathcal{A}_M$, is spanned by the orthonormal Fock basis defined by $|n_1, \ldots, n_M\rangle = \left(a_1^\dagger\right)^{n_1} \ldots \left(a_M^\dagger\right)^{n_M} |0\rangle$ where the vacuum state $|0\rangle$ fulfills $a_j|0\rangle = 0\ \forall j$, and the $n_j \in \{0, 1\}$ are the eigenvalues of the mode occupation number operators $n_j = a_j^\dagger a_j$.

Linear transformations of the fermionic operators which preserve the CAR are called canonical transformations. They are of the form $c_k \mapsto c_k' = \sum_i O_{kl} c_l$, where $O \in O(2M)$ is an element of the real orthogonal group.

The subclass of canonical operations which commute with the total particle number $N_{\rm op} = \sum_i n_i$ are called passive transformations. They take a particularly simple form in the complex representation $a_k \mapsto a'_k = \sum_l U_{kl} a_l$, where U is unitary on the single-particle Hilbert space $\mathcal{H}$.

Correlations in fermionic systems

The notion of "pairing" used in the description of superfluids, baryons in nuclei, etc., is always associated with correlated fermions. The subject of quantum correlations in fermionic systems is vast (see [14], and for instance [15, 16]), and has gained a renewed interest from the perspective of QIT in recent years. However, due to the missing notion of locality in these systems, a unique definition of separability (see e.g. [17]) or entanglement for indistinguishable fermions is missing so far. The existing concepts fall into two big classes: Entanglement of modes (see e.g. [18, 19]) and entanglement of particles (see e.g. [20, 21]). We refrain from an exhaustive review of these concepts, and restrict to the following definition:

Definition 2.1: A pure fermionic state $\rho_p^{(N)} = |\Psi_p^{(N)}\rangle\langle\Psi_p^{(N)}| \in \mathcal{S}(\mathcal{A}_M^{(N)})$ is called a *product state*, if there exists a passive transformation $a_k \mapsto a'_k$ such that $|\Psi_p^{(N)}\rangle = \prod_{j=1}^N a_j'^\dagger |0\rangle$. A state ρ_s is called *separable*, if it can be written as the convex combination of product states, i.e.

$$\rho_s = \sum_{p=1}^{K} \lambda_p \rho_p^{(N_p)}, \tag{2.1}$$

where $\sum_{p=1}^K \lambda_p = 1$, $\lambda_p \geq 0$ and all $\rho_p^{(N_p)} \in \mathcal{S}(\mathcal{A}_M^{(N_p)})$ are product states. All other states are said to have "Slater number larger than 1" and are called *entangled* in the sense of [20, 21] ("entanglement of particles"). We denote the set of all separable states by $\mathcal{S}_{sep}$.

Note that the set $\mathcal{S}_{sep}$ of separable states is convex and invariant under passive transformations. Both properties will be useful later on.

Definition of pairing

Separable states have only correlations resulting from their anti-symmetric nature and classical correlations due to mixing. These states will certainly not contain correlations associated with pairing. The simplest system in which we can find pairing consists of two particles and four modes[a]. The prototypical paired state, for example the spin-singlet of two electrons with opposing momenta, is of the form $|\Phi\rangle = \frac{1}{\sqrt{2}}\left(a_1^\dagger a_2^\dagger + a_3^\dagger a_4^\dagger\right)|0\rangle$, and states describing many Cooper pairs in BCS theory are generalizations of $|\Phi\rangle$.

The state $|\Phi\rangle$ describes correlations between the two particles that cannot be reproduced by any uncorrelated state and it can be completely characterized by one- and two-particle expectations (consisting of no more than two creation and annihilation operators each). This is a characteristic of the two-particle property "pairing" that we propose to make the central *defining* property of paired states in the general case of many modes, many particles and mixed states. Since, moreover, we would call the state $|\Phi\rangle$ paired no matter what basis the mode operators a_i refer to and we want it to comprise all BCS states, we are led to the following list of requirements that a sensible definition of pairing should fulfill:

(1) States that have no internal quantum correlation must be unpaired. These are the separable states (2.1).
(2) Pairing must reveal itself by properties related to one- and two-particle expectations only.
(3) Pairing must be a basis-independent property, i.e. it is invariant under passive transformations.
(4) The standard "paired" states appearing in the description of solid state and condensed matter systems, i.e., the so-called BCS-states with wave function

$$|\Psi_{\text{BCS}}^{(N,M)}\rangle = C_N \sum_{k=1}^{M} \left(\alpha_k a_k^\dagger a_{-k}^\dagger\right)^N |0\rangle, \tag{2.2}$$

must be captured by our definition. (C_N is a normalization constant, and $\alpha_k \in \mathbb{R}$.)

Further, it is desirable that there paired states that are a resource for some quantum information application.

Let us define the set of all operators $\{\mathcal{O}_\alpha\}_\alpha$ on $\mathcal{A}_M$ which are the product of at most two creation and two annihilation operators as set of *two-particle operators*. We denote it by A_2. These operators capture all one- and two-particle properties of a state ρ and should therefore contain all information about pairing. We will call a state ρ paired, if it can be distinguished from separable states by looking at observables in A_2 alone. Formally:

[a]For three modes, all pure two-particle states are of product form.

Definition 2.2: A fermionic state ρ is called *paired* if there exists a set of operators $\{\mathcal{O}_\alpha\}_\alpha \subseteq A_2$ such that the expectation values $\{\mathrm{tr}[\rho\mathcal{O}_\alpha]\}_\alpha$ cannot be reproduced by any separable state $\rho_s \in \mathcal{S}_{sep}$. States that are not paired are called *unpaired.*

This definition automatically fulfills our first two requirements. The third, basis independence, clearly holds, since the set of separable states is invariant under passive transformations. We will show later that the last requirement is met, i.e. the BCS-states Eq. (2.2) are all paired. Moreover, we show that there exist paired states that are a resource for quantum metrology.

Pairing and entanglement

Paired states exhibit non-trivial quantum correlations. Since they are inseparable in the sense of [20, 21], we are immediately lead to the following question: Is pairing equivalent to any of the known notions of entanglement?

First, entanglement of modes is basis-dependent. Thus it cannot coincide with our basis-independent definition of pairing. Hence, we have to compare pairing with entanglement of particles. As we have shown in [22], there exist states that are entangled according to Def. 2.2, but which are not paired. An example is the entangled state $|\Psi_4\rangle = \frac{1}{2}(a_1^\dagger a_2^\dagger a_3^\dagger a_4^\dagger + a_5^\dagger a_6^\dagger a_7^\dagger a_8^\dagger)|0\rangle$. Intuitively, this state consists of quadruples of particles, and thus it is not paired. But one sees immediately that the one-and two-particle expectations for $|\Psi_4\rangle$ are the same as for the separable state $\rho_s^{(4)} = \frac{1}{2}|\Phi_1\rangle\langle\Phi_1| + \frac{1}{2}|\Phi_2\rangle\langle\Phi_2|$, where $|\Phi_1\rangle = a_1^\dagger a_2^\dagger a_3^\dagger a_4^\dagger|0\rangle$, $|\Phi_2\rangle = a_5^\dagger a_6^\dagger a_7^\dagger a_8^\dagger|0\rangle$.

Next, one might think that pairing implies entanglement of the two-particle reduced density-matrix. However, we have proven in [22] that there exist certain classes of BCS-states (which are always paired) that have a two-particle reduced density-matrix that is paired but not entangled in four appropriately chosen modes.

In conclusion, the concept of "pairing" according to Def. 2.2 is not equivalent to the known notions of entanglement, but describes a different quantum correlation in systems of indistinguishable fermions.

2.3. *Detection and quantification of pairing*

2.3.1. *Detection of pairing*

Taking Def. 2.2, we would like to find methods that allow to detect pairing of fermionic states. As in the case of separability, this question will be hard

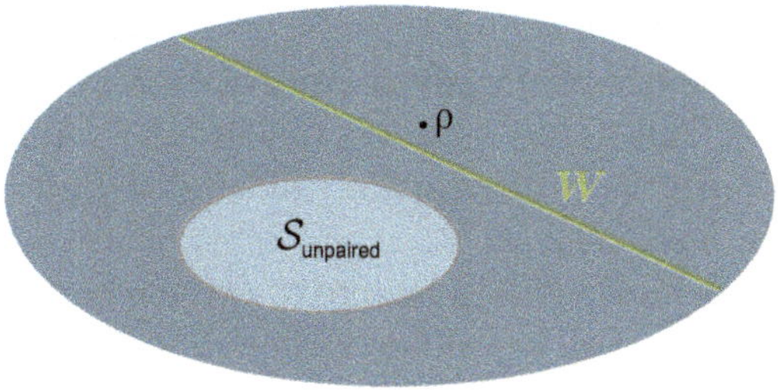

Fig. 2. Pairing witness W detecting the paired state ρ that lies outside the set of unpaired states $\mathcal{S}_{\text{unpaired}}$.

to answer in general. Thus, in the following we will use concepts from QIT to develop methods that allow to give a partial answer to this question.

Pairing witnesses: Following Def. 2.2, it is clear that the set of unpaired states is convex. This suggests the use of the Hahn-Banach separation theorem as a means to certify that a given density operator is not in the set of paired states. In analogy to the entanglement witnesses in quantum information theory [23] we define the following:

Definition 2.3: A *pairing witness* W is a Hermitian operator that fulfills $\text{tr}[W\rho_u] \geq 0$ for all unpaired states ρ_u, and for which there exists a paired state ρ such that $\text{tr}[W\rho] < 0$. We say that W *detects* the paired state ρ.

The witness defines a hyperplane in the space of density operators such that the convex set of unpaired states lies wholly on that side of the plane characterized by $\text{tr}[\rho W] > 0$ (see Fig. 2). For every paired state there exists a witness operator which detects it according to the Hahn-Banach theorem [24].

Since the definition of pairing refers only to expectation values of operators in A_2, it is enough to restrict to witness operators from that set which represents a significant simplification both mathematically and experimentally, since operators involving more than two-body correlations are typically very difficult to measure. However, as in the case of entanglement witnesses we will not be able to construct a complete set of pairing witnesses that detect any paired state. Nonetheless, we will be able to construct witnesses for all BCS-states, some of them being even optimal, in the sense that their distance to the paired state is maximal.

Convex Sets: Whether a state ρ is paired or not can be determined from a finite set of real numbers, namely the expectation values of a hermitian basis $\{O_\alpha\}$ of A_2. This allows us to reformulate the pairing problem as a geometric question on convex sets in finite-dimensional Euclidean space. To this end, consider a set $\{O_\alpha, \alpha = 1, \ldots, K\} \subset A_2$ of hermitian operators in A_2 that are not necessarily a basis. Denote by $\vec{O}$ the vector with components O_α. We define the set of all expectation values of $\vec{O}$ for separable states

$$C_{\vec{O}} = \left\{ \vec{v} = \mathrm{tr}[\vec{O}\rho_s] : \rho_s \in S_{sep} \right\} \subset \mathbb{R}^K. \tag{2.3}$$

For a state ρ let $\vec{v}_\rho \equiv \mathrm{tr}[\vec{O}\rho]$. By definition, ρ is paired if $\vec{v}_\rho \notin C_{\vec{O}}$.

Pairing measure

A theory of pairing should not only answers the question whether a state is paired or not, but also quantify the amount of pairing inherent in a state. For this purpose, we introduce the notion of a pairing measure:

Definition 2.4: Let ρ be an M-mode fermionic state. A pairing measure is a map

$$\mathcal{M} : \rho \mapsto \mathcal{M}(\rho) \in \mathbb{R}_+,$$

which is invariant under passive transformations and fulfills $\mathcal{M}(\rho) = 0$ for every unpaired state ρ.

In the geometric picture of the previous section, a candidate for a pairing measure that immediately comes to mind is the distance of $\vec{v}_\rho$ from the set C. However, the computation of this distance is, in general, very difficult and there is no evident operational meaning to this quantity. In the following sections we will introduce a different measure that can be computed for relevant families of states and allows a physical interpretation in terms of quantifying a resource for precision measurements.

2.4. *Examples: Fermionic Gaussian states and number-conserving states*

In the following section we want to fill our theory with life by applying it to two different classes of states. We will consider fermionic Gaussian States (FGS) and number-conserving states. The latter ones are defined via the property that they commute with the particle number operator, $[\rho, \hat{N}_{op}] = 0$.

2.4.1. *Pairing of Gaussian states*

Fermionic Gaussian states (FGS) are those states that can be represented as an exponential of a quadratic form in the Majorana operators, $\rho \sim \exp[-c^T K c]$, where $K = -K^T \in \mathbb{R}$, and c is a vector of Majorana operators. Thus, FGS are fully characterized by their second moments collected in the real and anti-symmetric covariance matrix (CM) $\Gamma_{kl} = \langle \frac{i}{2}[c_k, c_l] \rangle$ from which all higher correlations can be obtained via Wick's theorem (see e.g. [25]). Every pure FGS is the ground state of a quadratic Hamiltonian $H_Q = i \sum_{kl} h_{kl} c_k c_l$ with real and antisymmetric Hamiltonian matrix h.

Studying pairing within the set of FGS is interesting from a mathematical as well as from a physical perspective. First, since every FGS is completely characterized by its CM we can give a complete solution of the pairing problem with the help of the CM [22]:

Theorem 2.5: *Let ρ be the density operator of a fermionic Gaussian state, and let $Q_{kl} = i\langle a_k a_l \rangle$. Then ρ is paired iff $Q \neq 0$.*

Next, FGS will prove useful to develop methods to detect and quantify pairing for the BCS-states, due to the following reasoning: First, every pure FGS can be brought into a standard form, called Bloch-Messiah reduction [26] via a passive transformation:

$$|\Psi^{(\bar{N})}_{Gauss}\rangle = \prod_k (u_k + v_k a^\dagger_k a^\dagger_{-k})|0\rangle, \tag{2.4}$$

where $u_k, v_k \in \mathbb{C}$, $|u_k|^2 + |v_k|^2 = 1$, $\bar{N} = \sum_k \langle a^\dagger_k a_k \rangle = 2\sum_k |v_k|^2$. We will refer to these states as *Gaussian BCS states.* A relation to the BCS wave function can be established using the series expansion of the exponential function applied to operators:

$$|\Psi^{(\bar{N})}_{Gauss}\rangle = \sum_{N=0}^{2M} \lambda_N |\Psi^{(N)}_{BCS}\rangle, \tag{2.5}$$

where the number-conserving $2N$-particle BCS-state is given by

$$|\Psi^{(N)}_{BCS}\rangle = C_N \left(\sum_{k=1}^{M} \alpha_k P^\dagger_k \right)^N |0\rangle, \tag{2.6}$$

and we have introduced the pair creation operator $P^\dagger_k = a^\dagger_k a^\dagger_{-k}$. The coefficients α_k are related to u_k and v_k via $\alpha_k = v_k/u_k$. Now, consider a number-conserving observable O and denote by $\langle O\rangle_{Gauss}$ and $\langle O\rangle_N$ its expectation value for the Gaussian and $2N$-particle BCS wave function respectively. If the distribution of $|\lambda_N|^2$ is sharply peaked around some average particle number $\bar{N}$ with width Δ, then $\langle O\rangle_{Gauss} \approx \langle O\rangle_N$ for any integer $N \in [\bar{N}-\Delta, \bar{N}+\Delta]$.

Since every pure FGS is the ground state of a quadratic Hamiltonian, we have an easy way to construct an optimal pairing witness for every pure FGS. To be precise, we can show the following [22]:

Theorem 2.6: *Let* $0 < \epsilon < 1$ *and let* $0 \leq |v_k|^2 \leq 1-\epsilon$ *and* $\sum_k |v_k|^2 > 0$. *Then the operator*

$$H = \sum_{k=1}^{M} 2(1-\epsilon-|v_k|^2)(n_k+n_{-k}) - 2v_k u_k^* P_k^\dagger - 2v_k^* u_k P_k \qquad (2.7)$$

is a pairing witness, detecting $|\Psi_{Gauss}^{\bar{N}}\rangle = \prod_k (u_k + v_k P_k^\dagger)|0\rangle$.

Pairing measure for Gaussian states: In the following we propose a pairing measure, inspired by understanding pairing of Gaussian states in terms of an $SU(2)$ angular momentum algebra. Define the $SU(2)$-operators [27, 28] $j_k^{(x)} = \frac{1}{2}\left(P_k^\dagger + P_k\right)$, $j_k^{(y)} = \frac{i}{2}\left(P_k^\dagger - P_k\right)$, and $j_k^{(z)} = \frac{1}{2}(1-n_k-n_{-k})$ that fulfill $\left[j_k^{(a)}, j_k^{(b)}\right] = i\varepsilon_{abc} j_k^{(c)}$, $a,b,c \in \{x,y,z\}$. For pure Gaussian states in the standard form (2.4) we find $\langle j_k^{(x)}\rangle = \mathrm{Re}(u_k\bar{v_k})$, $\langle j_k^{(y)}\rangle = \mathrm{Im}(u_k\bar{v_k})$, and $\langle j_k^{(z)}\rangle = \frac{1}{2}(1-2|v_k|^2)$. As $j^2 = \sum_{i=x,y,z}\langle j_k^{(i)}\rangle^2 = 1/4$ independent of u_k and v_k, the expectation values for every pure Gaussian state lie on the surface of a sphere with radius 1/2. An unpaired state ρ_u fulfills $\langle j_k^{(x)}\rangle_{\rho_u} = \langle j_k^{(y)}\rangle_{\rho_u} = 0$, so that these states are located on the z-axis. The states on the equator have $\langle j_k^{(x)}\rangle^2 + \langle j_k^{(y)}\rangle^2 = 1/4$, i.e. they correspond to $|u_k|^2 = |v_k|^2 = 1/2$. The situation is depicted in Fig. 3.

Referring to the states on the equator as maximally paired is suggested by the fact that they have maximal distance from the set of separable states. This intuitive picture is further borne out by two observations: First, the states on the equator display maximal entanglement between the involved modes [25]. Second, they have the property[b] that they achieve the minimal expectation value of any quadratic witness operator (up to basis change).

[b]Maximally entangled states of two qubits share an analogous property about entanglement witnesses [29].

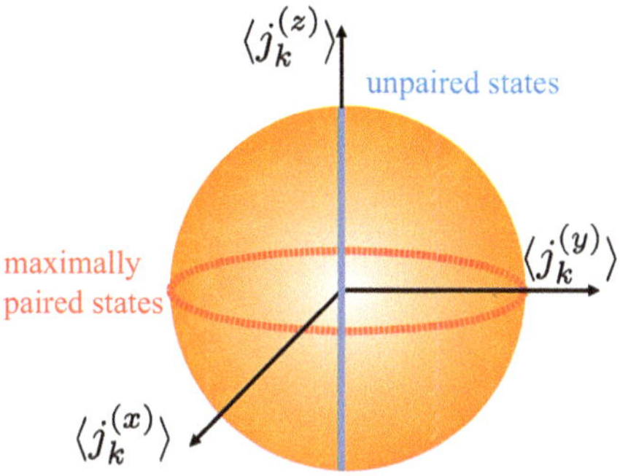

Fig. 3. Bloch sphere representation of the expectation values of $j_k^{(x)}$, $j_k^{(y)}$ and $j_k^{(z)}$ for a variational BCS state. All pure states lie on the surface of the sphere. Unpaired states lie on the z-axis, while the maximally paired states lie on the equator.

This suggests the introduction of a pairing measure via a quantity related to $|\langle j_k^{(x)} \rangle_{\rho G}|^2 + |\langle j_k^{(y)} \rangle_{\rho G}|^2 = |\langle a_k^\dagger a_{-k}^\dagger \rangle_{\rho G}|^2$:

Definition 2.7: Let ρ be a fermionic state, and let $Q_{kl} = i/2\mathrm{tr}(\rho[a_k, a_l])$. Then we define

$$\mathcal{M}_G(\rho) = 2||Q||_2^2 = 2\sum_{kl} |Q_{kl}|^2. \tag{2.8}$$

We have shown in [22] that $\mathcal{M}_G(\rho)$ is indeed a pairing measure fulfilling $\mathcal{M}_G(\rho) \leq M$ for every M-mode Gaussian state. It is easy to show that for every pure Gaussian state with standard form (2.4) the value of the pairing measure is given by $\mathcal{M}_G(\rho) = 4\sum_{k=1}^{M} |u_k|^2|v_k|^2$. Since $|v_k|^2 = 1 - |u_k|^2$ the measure attains its maximum value for $|u_k|^2 = |v_k|^2 = 1/2$, i.e., for the states already identified as maximally paired.

$\mathcal{M}_G(\rho)$ will appear again when we study the use of paired states for metrology applications, linking the pairing measure to the usefulness of a state for quantum phase estimation.

2.4.2. *Pairing of number-conserving states*

We still have to prove pairing of all BCS-states. Remarkably, this can be done by considering only the following vector of operators:

$$\vec{O}_3 = \begin{pmatrix} n_k + n_{-k} + n_l + n_{-l} \\ n_k n_{-k} + n_l n_{-l} \\ a_k^\dagger a_{-k}^\dagger a_{-l} a_l + h.c. \end{pmatrix}. \tag{2.9}$$

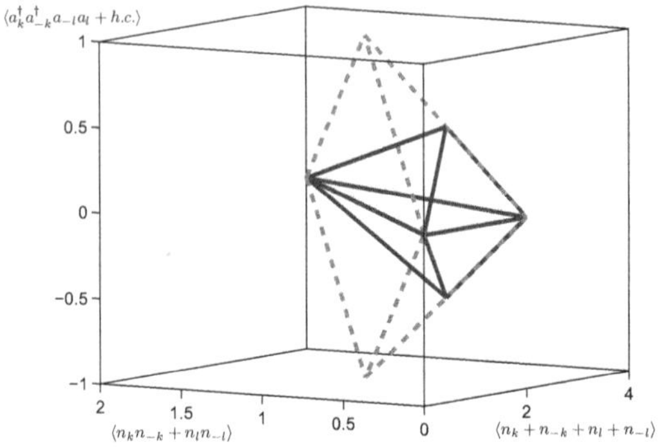

Fig. 4. Expectation values of the vector Eq. (2.9). For all number-conserving states these lie within the convex set $C^{\text{all}}_{\vec{O}_3}$ indicated by the dashed grey lines. The extreme points of the polytope are given by $(0,0,0)$, $(2,0,0)$, $(4,2,0)$ and $(2,1,\pm 1)$. Unpaired states have expectation values in the smaller convex set $C^{\text{unpaired}}_{\vec{O}_3}$ (solid blue) which has extreme points $(0,0,0)$, $(2,0,0)$, $(4,2,0)$ and $(2,1/2,\pm 1/2)$.

We are interested in $C^{\text{unpaired}}_{\vec{O}_3} = \{\text{tr}(\vec{O}_3\rho) : \rho \text{ separable}\}$, the set of all expectation values of $\vec{O}_3$ which correspond to separable states. If for some ρ the vector $\vec{v}_\rho = \text{tr}(\vec{O}_3\rho)$ is found outside of $C^{\text{unpaired}}_{\vec{O}_3}$ then ρ is paired. In Fig. 4 we have depicted the set $C^{\text{unpaired}}_{\vec{O}_3}$ (solid line) as well as the set $C^{\text{all}}_{\vec{O}_3} = \{\text{tr}(\vec{O}_3\rho) : \rho \in \mathcal{S}(\mathcal{A}_M^{(N)})\}$ (dashed line). The edges of $C^{\text{unpaired}}_{\vec{O}_3}$ correspond to the pairing witnesses we are looking for, and one can show that one of them,

$$\begin{aligned} H^{(p)}_{1\pm} &= \frac{1}{2}(n_k + n_{-k} + n_l + n_{-l}) - (n_k n_{-k} + n_l n_{-l}) \\ &\pm (a_k^\dagger a_{-k}^\dagger a_{-l} a_l + h.c.) \end{aligned} \tag{2.10}$$

allows to detect all number-conserving BCS states as paired [22].

While this result proves pairing of all BCS-states, the restriction to a measurement of four modes only is impractical from an experimental perspective. Here, one would desire a global witness operator that has a big negative expectation value for the state one has to detect in order to be robust against noise. We construct now such witness operators for a restricted class of BCS-states, those for which $\alpha_k = \alpha_{k+M}$ for $k = 1, \ldots, M$, and the number of modes $2M$ is much bigger than the number of particles, $2M \gg N$. To this end, we make use of the relation of BCS-states and Gaussian states Eq. (2.5). In Thm. 2.6 we have constructed witnesses H for all Gaussian

BCS states. As these witnesses are optimal, they might lead to an improved witness detecting the corresponding number-conserving BCS state. But H includes terms of the form $P_k^\dagger$ that do not conserve the particle number. Hence, this witness cannot be applied directly to the number-conserving case. However, due to Wick's theorem, $\langle P_k^\dagger P_{k+M}\rangle_{Gauss} = \bar{u}_k v_k \langle P_k^\dagger\rangle_{Gauss}$ holds under our symmetry assumption. This suggests that we replace the non number-conserving operator $\bar{u_k} v_k P_k^\dagger$ by the number-conserving operator $P_k^\dagger P_{k+M}$. We define operators

$$H_k = 2(1-\epsilon-|v_k|^2)N_k - 4(P_k^\dagger P_{k+M} + h.c.), \tag{2.11}$$

$$N_k = n_k + n_{-k} + n_{k+M} + n_{-(k+M)}, \tag{2.12}$$

where $0 \le |v_k|^2 \le 1-\epsilon \;\forall k$ for $\epsilon > 0$. Further, we introduce the notation $\alpha_k = v_k/\sqrt{1-|v_k|^2}$, $\bar{N} = \sum_{k=1}^M |v_k|^2$ and we denote by $\tilde{N}$ the biggest integer fulfilling $\bar{N} - \tilde{N} \ge 0$. Then if $M, \tilde{N} \in \mathbb{N}$ and $1 \ll \tilde{N} < 2M$, and if $1 > \epsilon \ge 18/\sqrt{\pi \tilde{N}}$ the Hamiltonian $H(\{v_k\}) = \sum_{k=1}^M H_k$ is a pairing witness detecting $|\Psi_{BCS,sym}^{(N)}\rangle = C_N \left(\sum_{k=1}^M \alpha_k (P_k^\dagger + P_{k+M}^\dagger)\right)^N |0\rangle$ [22].

Basis-independent criterion: Witness operators prove to be a powerful tool for the detection of pairing, as there exists a witness for every paired state. However, since they depend on the chosen basis, every given witness operator will fail to detect a wide range of states. Thus, we derive a basis independent condition for detecting pairing, providing a sufficient condition for paired states.

Consider the two-particle reduced density matrix $O^{(\rho)}_{(ij)(kl)} = \mathrm{tr}[\rho a_i^\dagger a_j^\dagger a_l a_k]$ which contains all two-particle correlations. A change of basis, $a_i^\dagger \mapsto \sum_k U_{ik} a_k^\dagger$ leaves the spectrum of $O^{(\rho)}$ unchanged and we can prove the following theorem [22]:

Theorem 2.8: *Let ρ be an unpaired state, and let $O^{(\rho)}$ be its two-particle RDM. Then $\lambda_{max}(O^{(\rho)}) \le 2$, where λ_{max} denotes the maximal eigenvalue.*

Pairing measure: In Def. 2.7 we have introduced a pairing measure for Gaussian states. The correspondence with number-conserving BCS states will be a guideline to derive a measure for number conserving states. However, the measure of Def. 2.7 involves expectation values of the form $\langle a_k^\dagger a_{-k}^\dagger\rangle$ that vanish for states with fixed particle number. Yet, Wick's theorem again suggests that a quantity involving expectation values of the form $\langle P_k^\dagger P_l\rangle$

will lead to the pairing measure. This holds indeed true, and we could show that for every pure number-conserving state the following quantity defines a pairing measure [22]:

$$\mathcal{M}(\rho) = \max\left\{ \max_{\{a_i^\dagger\}_i} \sum_{kl=1}^{M} |\langle P_k^\dagger P_l \rangle_\rho| - \frac{1}{2}\sum_k \langle n_k \rangle_\rho, 0 \right\}, \tag{2.13}$$

where $P_k^\dagger = a_k^\dagger a_{-k}^\dagger$ and the maximum is taken over all possible bases of modes $\{a_i^\dagger\}_i$. For mixed states ρ, a measure can be defined via $\mathcal{M}(\rho) = \min \sum_i p_i \mathcal{M}(\rho_i)$, where the minimum is taken over all possible decompositions of $\rho = \sum_i p_i \rho_i$ into pure states ρ_i.

We close the section by calculating the value of the pairing measure for two easy examples. Let $|\Psi_s\rangle = \bigotimes_{k=1}^N \frac{1}{\sqrt{2}} \left(P_k^\dagger + P_{-k}^\dagger\right) |0\rangle$ be the tensor product of N spin-singlet states and $|\Psi_{BCS}^{(N,M)}\rangle = C_N \left(\sum_{k=1}^M P_k^\dagger\right)^N |0\rangle$ BCS state with equal weights. These states have a pairing measure $\mathcal{M}(|\Psi_s\rangle) = N$ resp. $\mathcal{M}(|\Psi_{BCS}^{(N,M)}\rangle = N(M-N)$. Thus, for the spin singlet the pairing measure has in addition the property that it is normalized to 1 and additive, while it is subadditive for $|\Psi_{BCS}^{(N,M)}\rangle$. Further, this example suggests that the pairing of $\mathcal{M}(|\Psi_{BCS}^{(N,M)}\rangle = N(M-N)$ is stronger than for $|\Psi_s\rangle$. We will see indeed in that states of the form $|\Psi_{BCS}^{(N,M)}\rangle$ allow interferometry at the Heisenberg limit.

2.5. *Pairing as a resource*

The goal of quantum phase estimation is to determine an unknown parameter φ of a Hamiltonian $H_\varphi = \varphi H$ with the highest possible accuracy by measuring an observable O on a known input state that has evolved under H_φ. In a region where the function $\varphi \mapsto \langle O(\varphi)\rangle$ is bijective, φ can be inferred by inverting $\langle O(\varphi)\rangle$. In a realistic setup, however, $\langle O(\varphi)\rangle$ cannot be determined, as this would require an infinite number of measurements. Instead, one uses the mean value of the measurement results, o, as an estimate of $\langle O(\varphi)\rangle$. This will result in an error $\delta\varphi$ for the parameter to be estimated. For a given measurement scheme, i.e. for a given input state and a given observable O, the uncertainty in φ can be reduced by using N identical input states and average over the N measurement outcomes. As the preparation of a quantum state is costly, a precision gain which has a strong dependence on N is highly desirable. If these probe states are independent of each other, the precision scales like $1/\sqrt{N}$. This is the so-called

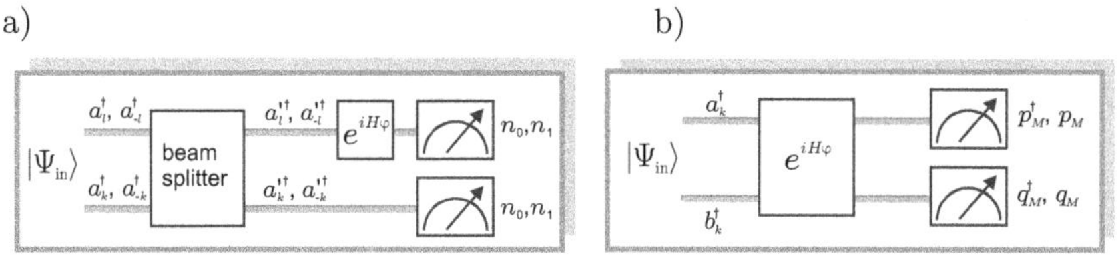

Fig. 5. a) Scheme of the Ramsey interferometer setup. The incoming wave function $|\Psi_{\text{in}}\rangle$ enters a beam splitter. Then particles in the modes $a^\dagger_{\pm l}$ evolve under the Hamiltonian $H_N = \sum_{j=1}^{M}(n_{l_j} + n_{-l_j})$. At the end a particle number measurement is performed on all modes. b) Setup which allows interferometry with paired states at the Heisenberg limit. Particles in modes $a_k^\dagger$ and $b_k^\dagger$ evolve under the complex coupling Hamiltonian $H_F = \sum_{k=1}^{\infty} a_k^\dagger a_{-k}^\dagger b_{-k} b_k + h.c.$. In the end particle numbers are measured.

standard quantum limit (SQL). The main goal of quantum phase estimation is to overcome the standard quantum limit and to reach the so-called Heisenberg limit where the precision scales like $1/N$.

In the following we give two settings where paired states give an improvement over unpaired states. The first setting depicted in Fig. 5a is a standard Ramsey-type interferometer setup. Here, a state in the modes $\{a^\dagger_{k_j}, a^\dagger_{l_j}\}_{j=-M}^{M}$ undergoes mode mixing at a beam splitter, before evolving under the action of the Hamiltonian $H_N = \sum_{j=1}^{M}(n_{l_j} + n_{-l_j})$. Finally, a particle number measurement is performed on the system. We could show in [22] that for any unpaired state of $2N$ particles the phase sensitivity is bounded by $(\delta\phi)^2 \geq 1/(2N)$, while there exist paired states for which this limit can be beaten by a factor of 2, i.e. $(\delta\varphi)^2 = 1/(4N)$. Particularly, in the case of appropriately chosen Gaussian states ρ_G, $(\delta\varphi)^2 \sim 1/(\mathcal{M}_G(\rho_G))$ holds, where $\mathcal{M}_G(\rho_G)$ is the value of the pairing measure defined in Def. 2.7, giving this measure an operational meaning. Due to the relation of the pairing measure for Gaussian states and BCS-states, this meaning can be translated to the pairing measure for number-conserving states as well.

Next, we consider a setup where two fermionic states enter the ports A and B of an interferometer. The particles entering port A can occupy the modes $\{a_k^\dagger\}_{k=-M}^{M}$, while the particles entering through port B can occupy the modes $\{b_k^\dagger\}_{k=-M}^{M}$. Then the two states evolve under the Hamiltonian $H_F = \varphi \sum_{k=1}^{\infty} a_k^\dagger a_{-k}^\dagger b_{-k} b_k + h.c.$, and a particle number measurement is performed at the end. The situation is depicted in Fig. 5b). For every unpaired state we find that $(\delta\varphi)^2 \geq 1/(16N)$. However, when we take the BCS-state $|\Psi_{\text{BCS}}\rangle = |\Psi_a\rangle|\Psi_b\rangle$, where for modes $d = a, b$ the states $|\Psi_d\rangle$ are defined as

$$|\Psi_d\rangle = c_N^{(M)} \left(\frac{1}{\sqrt{M}} \sum_k d_k^\dagger d_{-k}^\dagger \right)^N |0\rangle, \tag{2.14}$$

we can achieve a phase sensitivity $(\delta\varphi)^2 \sim 1/N^2$ for all $M \geq 2N$.

3. Fermionic Projected Entangled Pair State

Quantum lattice models are of central importance in the field of condensed matter physics, since they serve as a powerful framework to describe realistic physical systems. One of the most prominent examples is the fermionic Hubbard model that exhibits a rich phase diagram and is also thought to allow for high temperature superconductivity. However, most of the lattice models appearing in condensed matter physics are too complicated to allow for an analytic solution. Thus, we have to rely on powerful numerical approximation schemes to gain insight into the physical properties of matter. These schemes must be based on variational wave functions that can be efficiently stored on a computer, yet capturing the relevant physical properties of the system.

For spin systems on a lattice with local, i.e., short–range interactions, such families could be found in recent years. In one spatial dimension, Matrix Product States (MPS) [30, 31] provide a good approximation to the ground state of any gapped local Hamiltonian [32, 33]. Projected Entangled Pair States (PEPS) [34, 35], which naturally extend MPS to higher spatial dimensions, approximate spin states at any finite temperature [36, 37], and have been successfully used to simulate spin systems which cannot be dealt with otherwise. In one spatial dimension, it is possible to adapt the methods based on MPS to fermionic systems using the Jordan-Wigner transform [38], which maps fermions into spins while keeping the interactions local. In higher dimensions, however, this is no longer possible: Fermionic operators at different locations anticommute, which effectively induces nonlocal effects when mapping fermions to spins.

In the following we introduce a new family of states, the fermionic Projected Entangled Pair States (fPEPS) [39], which naturally extend the PEPS to fermionic systems. We show that fPEPS can be mapped efficiently to the standard PEPS representation, proving that fPEPS approximate efficiently ground and thermal states of local fermionic Hamiltonians. Further, we show that fPEPS can also arise as the *exact* ground state of local (parent) Hamiltonians and can thus serve as an analytic tool to gain insight into the properties of fermionic systems.

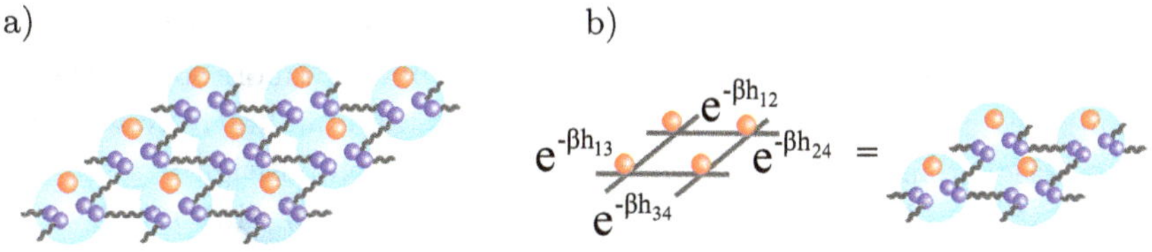

Fig. 6. a) Construction of a PEPS in two dimensions. The balls joined by lines represent pairs of maximally entangled D-dimensional auxiliary spins, which are then mapped to the physical spins (red), as illustrated by the light blue spheres. b) Why PEPS approximate thermal states well: $\exp[-\beta h_{ij}]$ can be implemented using local maps only if an entangled pair is available.

3.1. *A review of the PEPS-construction*

For simplicity, let us consider a $2D$ lattice of $N \equiv N_h \cdot N_v$ spin $1/2$ particles, with states $|0\rangle$ and $|1\rangle$. To each node of coordinates (h, v) we associate four auxiliary spins, with states $|n\rangle$ $(n = 0, \dots, D-1)$, where D is called bond dimension. Each of them is in a maximally entangled state $\sum |n, n\rangle$ with one of its neighbors, as indicated in Fig. 6a. The PEPS $|\Psi\rangle$ is obtained by applying a linear operator ("projector") to each node that maps the auxiliary spins onto the physical ones. This operator can be parametrized as

$$P_{(h,v)} = \sum_{l,r,u,d=0}^{D-1} \sum_{k=0}^{1} (B_{(h,v)})^{[k]}_{l,r,u,d} |k\rangle\langle l, r, u, d|. \tag{3.1}$$

Let us now explain why PEPS are well suited to describe spins in thermal equilibrium in the case of local Hamiltonians, $H = \sum h_\lambda$. For simplicity, we will assume that each h_λ acts on two neighboring spins. We first rewrite the (unnormalized) density operator $e^{-\beta H} = \mathrm{tr}_B[|\chi\rangle\langle\chi|]$, where $|\Psi\rangle = e^{-\beta H/2} \otimes \mathbb{1} |\chi\rangle_{AB}$ is a purification [42] and $|\chi\rangle_{AB}$ a pairwise maximally entangled state of each spin with another one, the latter playing the role of an environment. We will show now that $|\Psi\rangle$ can be expressed as a PEPS. We consider first the simplest case where $[h_\lambda, h_{\lambda'}] = 0$, so that $|\Psi\rangle = \prod_\lambda e^{-\beta h_\lambda/2} \otimes \mathbb{1} |\chi\rangle_{AB}$. The action of each of the terms $e^{-\beta h_\lambda/2}$ on two spins in neighboring nodes can be viewed as follows: we first include two auxiliary spins, one in each node, in a maximally entangled state, and then we apply a local map in each of the nodes which involves the real spin and the auxiliary spin, which ends up in $|0\rangle$. By proceeding in the same way for each term $e^{-\beta h_\lambda/2}$, we end up with the PEPS description (see Fig. 6b).

This is valid for all values of β, in particular for $\beta \to \infty$, i.e., for the ground state. In case the local Hamiltonians do not commute, a more sophisticated proof is required [36]. One can, however, understand qualitatively why the construction remains valid by using a Trotter decomposition to approximate $e^{-\beta H} \approx \prod_{m=1}^{M} \prod_{\lambda} e^{-\beta h_\lambda/2M}$ with $M \gg 1$. Again, this allows for a direct implementation of each $\exp[-\beta h_\lambda/2M]$ using one entangled bond, yielding M bonds for each vertex of the lattice. Since, however, the entanglement induced by each $\exp[-\beta h_\lambda/2M]$ is very small, each of these bonds will only need to be weakly entangled, and the M bonds can thus be well approximated by a maximally entangled state of low dimension. Note that the spins belonging to the purification do not play any special role in this construction, and thus we will omit them in the following.

3.2. *Construction of fPEPS*

We will now extend the above construction to fermionic systems, in such a way that the same arguments apply. We consider fermions on a lattice, and work in second quantization. For a Hamiltonian $H = \sum h_\lambda$, each term h_λ must contain an even number of fermionic operators, in order for the Trotter decomposition to be still possible. Thus, we just have to find out how to express the action of $e^{-\beta h_\lambda}$ in terms of auxiliary systems. This is very simple: one just has to consider that the auxiliary particles are fermions themselves, forming maximally entangled states, and write a general operator which performs the mapping as before. Following this route, we arrive at the definition of fPEPS. More specifically, we define at each node (h, v) four auxiliary fermionic modes, with creation operators $\alpha^\dagger_{(h,v)}, \beta^\dagger_{(h,v)}, \gamma^\dagger_{(h,v)}, \delta^\dagger_{(h,v)}$, respectively. We define $H_{(h,v)} = \frac{1}{\sqrt{2}}(1 + \beta^\dagger_{(h,v)}\alpha^\dagger_{(h+1,v)})$, $V_{(h,v)} = \frac{1}{\sqrt{2}}(1 + \delta^\dagger_{(h,v)}\gamma^\dagger_{(h,v+1)})$ which create maximally entangled states out of the vacuum. We also define the "projectors"

$$Q_{(h,v)} = \sum (A_{(h,v)})^{[k]}_{lrud} a^{\dagger k}_{(h,v)} \alpha^l_{(h,v)} \beta^r_{(h,v)} \gamma^u_{(h,v)} \delta^d_{(h,v)}, \tag{3.2}$$

where $a_{(h,v)}$ is the annihilation operator of the physical fermionic mode, and the sum runs for all the indices from 0 to 1, with the condition that $(u + d + l + r + k) \bmod 2 = c$, were c is fixed for each node[c]. The latter is

[c] In fact one can freely choose c for all but one $Q_{(h,v)}$: Since, e.g., the bond $H_{(h,v)} = \frac{1}{\sqrt{2}}(1 + \beta^\dagger_{(h,v)}\alpha^\dagger_{(h+1,v)})$ is invariant under $(i\beta_{(h,v)} + \beta^\dagger_{(h,v)})(i\alpha_{(h+1,v)} + \alpha^\dagger_{(h+1,v)})$, the corresponding maps (3.2) can be right multiplied with it, switching their parity.

related to the parity of the h_λ and will ensure that the parity of the fPEPS is well defined. The fPEPS is then

$$|\Psi\rangle = \langle \prod_{(h,v)} Q_{(h,v)} \prod_{(h,v)} H_{(v,h)} V_{(v,h)} \rangle_{\text{aux}} \, |\text{vac}\rangle, \tag{3.3}$$

where the expectation value is taken in the vacuum of the auxiliary modes, and $|\text{vac}\rangle$ denotes the vacuum of the physical fermions. Note that the definition of fPEPS straightforwardly extends to systems with both more than one physical mode per site and more than one mode per bond, as well as to open boundaries or higher spatial dimensions.

3.3. *Relation between fPEPS and PEPS*

Next, we will find an efficient description of any fPEPS in terms of standard PEPS. With that, one can readily use the methods introduced for PEPS [34, 35] in order to determine physical observables, as well as to perform simulations of ground or thermal states, and time evolution. We have to identify the Fock space of the fermionic modes with the Hilbert space of spins. For that, we sort the lattice sites according to $M = (v-1)N_h + h$ and associate $a_1^{\dagger k_1} \dots a_N^{\dagger k_N} |\text{vac}\rangle$ to the spin state $|k_1, \dots, k_N\rangle$. Then we write $|\Psi\rangle$ in that basis, and express it as a PEPS in terms of tensors B (3.1). The goal is to find the relation between the tensors B (corresponding to the spin description) and A (fermionic description). In principle, the fPEPS to PEPS transformation can be done straightforwardly by adding extra bonds to the PEPS which take care of the signs which arise from reordering the fermionic operators; however, this would lead to a linear number of bonds per link and thus to a dimension which is exponential in N. Remarkably, it is possible to express every fPEPS as a PEPS by introducing only *one* additional bond per horizontal link as follows: Replace each fermionic bond by a bond of maximally entangled spins, adding one additional horizontal qubit bond everywhere except at the boundaries (see Fig. 7). This means that the tensor B will have now two more indices, say l' and r', which are associated to those new bonds. Then, we find the relation

$$(B_{h,v})^{[k]}_{lrr'ud} = (-1)^{f_{(h,v)}(k,u,d,l,r)} (A_{h,v})^{[k]}_{lrud} (-1)^{(d+l)r'} \tag{3.4}$$

for $h = 1$, while for $h > 1$ we have

$$(B_{h,v})^{[k]}_{ll'rr'ud} = (-1)^{f_{(h,v)}(k,u,d,l,r)} (A_{h,v})^{[k]}_{lrud} (-1)^{dr'} \delta_{l',(r'+u+d) \bmod 2} \tag{3.5}$$

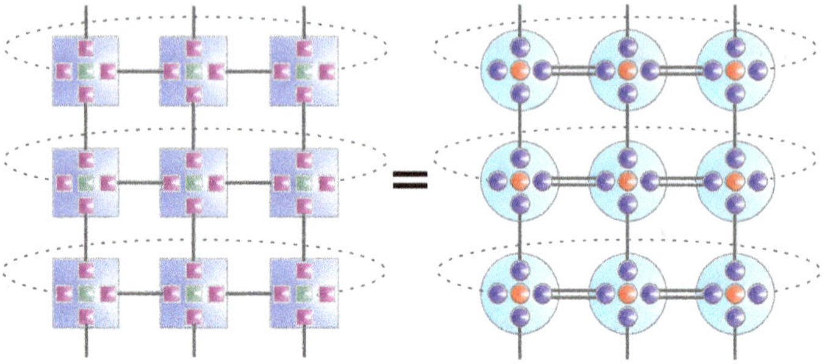

Fig. 7. Every fPEPS can be represented as a PEPS at an extra cost of at most one additional bond per link (shown for a 3×3 PBC lattice).

where $f_{(h,v)}(k, u, d, l, r)$ is a function which only depends on the local indices, and $r' = 0$ for $h = N_h$. The exact form of these functions can be found in [43].

3.4. *Examples*

In the following we will consider the subclass of Gaussian fPEPS. Those are obtained by restricting the map (3.2) to be Gaussian (H and V are already of that form). Using the channel description of Gaussian maps introduced in [25], we could show (see [39] for a proof) that every translationally invariant Gaussian fPEPS is the exact ground state of a *local parent Hamiltonian.*

Parent Hamiltonians have their roots in the theory of valence bond states and MPS/PEPS. Here, a parent Hamiltonian H is a hermitian operator that has an MPS or PEPS as its exact ground state. Further, it is required that H is an operator that is acting on a local set of sites only, as it holds true for lattice models describing interactions appearing in nature. Then, due to its relatively simple structure, the MPS/PEPS representation allows to gain more insight into the physical properties of the system from an analytical perspective. The literature on parent Hamiltonians in the MPS/PEPS setting is large, and we propose [40] as a starting point. As an application, we would like to point out [41], where results on parent Hamiltonians could be used to get insight into topological properties of matter.

In the following we present one particularly interesting example of a Gaussian fPEPS. To this end, we apply the Gaussian fPEPS projector $Q = e^{(i\alpha+\beta)(-\gamma+i\delta)+\alpha\beta+\gamma\delta+a^\dagger(-i\alpha-\beta-\gamma+i\delta)}$ to each site, and obtain an fPEPS which is the exact ground state of the Hamiltonian

$$H_{\text{crit}} = 2i \sum_{(h,v)} a^\dagger_{(h,v)} a^\dagger_{(h,v+1)} - a^\dagger_{(h,v)} a^\dagger_{(h+1,v)} + h.c.$$
$$- \sum_{(h,v)} a^\dagger_{(h,v)} (a_{(h+1,v+1)} + a_{(h+1,v-1)}) + h.c.,$$

and N_h, N_v odd, which will ensure that the ground state is unique. Notably, the ground state possesses correlations that decay as power laws and the Hamiltonian is gapless in the limit $N \to \infty$. In fact, our example provides us with a critical fermionic system obeying the area law, which directly follows from the fact that its ground state is a PEPS with bounded bond dimensions. Note that, although H_{crit} is not number-conserving, it can be converted into a number-conserving one via a simple particle–hole transformation in the B sublattice. This new Hamiltonian possesses a spectrum with a Dirac point separating the modes with positive and negative energies. Thus, the Fermi surface has zero dimension, which explains why our results do not contradict the violation of the area law expected for free Fermionic systems [44].

4. Conclusion and Outlook

In this lecture we have seen how methods and tools known in the field of Quantum Information Theory can be applied to gain a deeper understanding of fermionic quantum many-body systems from two perspectives: In the first part, we have seen how QIT can help to answer fundamental questions on fermionic correlations. In the second part we could construct a new class of variational states that provide a possibility for simulating fermionic many-body systems on a classical computer.

The idea to approach problems in the field of fermionic many-body systems with methods from QIT has not only provided new insights into these systems, but it has also brought up many new interesting questions related to fermionic particles. Concerning fermionic correlations, one could for example try to generalize the pairing theory to higher order correlations, explore their relation to fermionic entanglement and ask for possible applications of these correlations. The field of fPEPS surely demands for a numerical implementations, and benchmark results could already be obtained [45, 46, 47]. Further, when exploring the manifold of parent Hamiltonians the family of fPEPS will help us to learn more about the physics of strongly correlated fermionic many-body systems.

Acknowledgments

These notes are based on a lecture given at IMS of the National University of Singapore in March 2010. The author thanks Profs. Christandl, Siedentop and Winter as well as the IMS for providing such a nice opportunity.

This work was supported by he EU projects QUEVADIS and SCALA, the DFG within Forschergruppe 635 and SFB 631, SFB FOQUS, the excellence cluster Munich Advanced Photonics (MAP), QUANTOP, the Danish Natural Science Research Council (FNU) and the Elite Network of Bavaria programme QCCC.

References

1. J. Bardeen, L. N. Cooper, and J. R. Schrieffer, Phys. Rev. **108**, 1175 (1957).
2. M. Greiner, C. Regal, and D. Jin, Nature **426**, 537 (2003).
3. S. Jochim *et al.*, Science **302**, 2101 (2003).
4. M. W. Zwierlein *et al.*, Phys. Rev. Lett. **91**, 250401 (2003).
5. M. Zwierlein, J. Abo-Shaeer, A. Schirotzek, C. Schunck, and W. Ketterle, Nature **435**, 1047 (2005).
6. M. Zwierlein, C. Schunck, A. Schirotzek, and W. Ketterle, Nature **442**, 154 (2005).
7. G. B. Partridge, K. E. Strecker, R. I. Kamar, M. W. Jack, and R. G. Hulet, Physical Review Letters **95**, 020404 (2005).
8. M. W. Zwierlein *et al.*, Physical Review Letters **92**, 120403 (2004).
9. C. A. Regal, M. Greiner, and D. S. Jin, Physical Review Letters **92**, 040403 (2004).
10. G. B. Partridge, W. Li, R. I. Kamar, Y. Liao, and R. G. Hulet, Science **27**, 311 (2006).
11. M. W. Zwierlein and W. Ketterle, Science **314**, 54a (2006).
12. G. B. Partridge, W. Li, R. I. Kamar, Y. Liao, and R. G. Hulet, Science **314**, 54b (2006).
13. C. Schunck, Y. Shin, A. Schirotzek, M. Zwierlein, and W. Ketterle, Science **316**, 867 (2007).
14. G. D. Mahan, *Many-Particle Physics*, 3rd ed. (Kluwer Academic, 2000).
15. C. Hainzl, E. Hamza, R. Seiringer, and J. P. Solovej, Commun. Math. Phys. **281**, 349 (2008).
16. C. C. Tsuei and J. R. Kirtley, Rev. Mod. Phys. **72**, 969 (2000).
17. M.-C. Banuls, J. I. Cirac, and M. M. Wolf, Physical Review A **76**, 022311 (2007).
18. P. Zanardi, Phys. Rev. A **65**, 042101 (2002).
19. P. Zanardi and X. Wang, J. Phys. A:Math. Gen. **35**, 7947 (2002).
20. J. Schliemann, J. I. Cirac, M. Kuś, M. Lewenstein, and D. Loss, Phys. Rev. A **64**, 022303 (2001).
21. K. Eckert, J. Schliemann, D. Bruss, and M. Lewenstein, Annals of Physics **299**, 88 (2002).

22. C. V. Kraus, M. M. Wolf, J. I. Cirac, and G. Giedke, Physical Review A **79**, 012306 (2009).
23. M. Horodecki, P. Horodecki, and R. Horodecki, Physics Letters A **223**, 1 (1996).
24. W. Rudin, *Functional Analysis*, 2nd ed. (McGraw-Hill, 1991).
25. S. Bravyi, Quant. Inf. Comput. **5**, 216 (2005).
26. C. Bloch and A. Messiah, Nuclear Physics **39**, 95 (1962).
27. P. W. Anderson, Phys. Rev. **112**, 1900 (1958).
28. R. A. Barankov and L. S. Levitov, Phys. Rev. Lett. **93**, 130403 (2004).
29. R. A. Bertlmann, K. Durstberger, B. C. Hiesmayr, and P. Krammer, Physical Review A **72**, 052331 (2005).
30. A. Klümper, A. Schadschneider, and J. Zittartz, J. Phys. A **24**, L955 (1991).
31. A. Klümper, A. Schadschneider, and J. Zittartz, Z. Phys. B **87**, 281 (1992).
32. M. Hastings, J. Stat. Mech., P08024 (2007).
33. F. Verstraete and J. I. Cirac, Phys. Rev. B **73**, 094423 (2006).
34. F. Verstraete and J. I. Cirac, (2004), cond-mat/0407066.
35. F. Verstraete, V. Murg, and J. I. Cirac, Advances in Physics **57**, 143 (2008).
36. M. B. Hastings, Phys. Rev. B **73**, 085115 (2006).
37. M. B. Hastings, Phys. Rev. B **76**, 035114 (2007).
38. P. Jordan and Wigner, Z. Phys. **47**, 631 (1928).
39. C. V. Kraus, N. Schuch, F. Verstraete, and J. I. Cirac, Phys. Rev. A **81**, 052338 (2010).
40. D. Perez-Garcia, F. Verstraete, J. I. Cirac, and M. M. Wolf, Quant. Inf. Comp. **8**, 0650-0663 (2008).
41. N. Schuch, D. Perez-Garcia, J. I. Cirac, Phys. Rev. B 84, 165139 (2011).
42. F. Verstraete, J. J. Garcia-Ripoll, and J. I. Cirac, Phys. Rev. Lett. **93**, 207204 (2004).
43. C. V. Kraus, *A Quantum Information Perspective of Fermionic Quantum Many-Body Systems*, PhD thesis, Technische Universität München (2009).
44. M. M. Wolf, Physical Review Letters **96**, 010404 (2006).
45. P. Corboz, R. Orús, B. Bauer, and G. Vidal, Phys. Rev. B **81**, 165104 (2010).
46. I. Pizorn and F. Verstraete, 1003.2743 (2010).
47. S. Li, Q. Shi, and H. Zhou, arXiv:1001.3343, (2010).

HYDROGEN-LIKE ATOMS IN RELATIVISTIC QED

Martin Könenberg
Fakultät für Mathematik und Informatik, FernUniversität Hagen
Lützowstraße 125, D-58084 Hagen, Germany
Present address:
Fakultät für Physik, Universität Wien
Boltzmanngasse 5, 1090 Vienna, Austria
martin.koenenberg@univie.ac.at

Oliver Matte
Institut für Mathematik, TU Clausthal
Erzstraße 1, D-38678 Clausthal-Zellerfeld, Germany
Present address:
Institut for Matematik, Århus Universitet
Ny Munkegade 118, DK-8000 Århus, Denmark
matte@math.lmu.de

Edgardo Stockmeyer
Mathematisches Institut, Ludwig-Maximilians-Universität
Theresienstraße 39, D-80333 München, Germany
stock@math.lmu.de

In this review we consider two different models of a hydrogenic atom in a quantized electromagnetic field that treat the electron relativistically. The first one is a no-pair model in the free picture, the second one is given by the semi-relativistic Pauli-Fierz Hamiltonian. For both models we discuss the semi-boundedness of the Hamiltonian, the strict positivity of the ionization energy, and the exponential localization in position space of spectral subspaces corresponding to energies below the ionization threshold. Moreover, we prove the existence of degenerate ground state eigenvalues at the bottom of the spectrum of the Hamiltonian in both models. All these results hold true, for arbitrary values of the fine-structure constant, e^2, and the ultra-violet cut-off, and for a general class of electrostatic potentials including the Coulomb potential with nuclear charges less than (sometimes including) the critical charges without radiation field, namely $e^{-2}2/\pi$ for the semi-relativistic Pauli-Fierz operator and $e^{-2}2/(2/\pi+\pi/2)$ for the no-pair operator. Apart from a detailed discussion of diamagnetic inequalities in QED (which are applied to study the semi-boundedness) all results stem from earlier articles written by the authors. While a few proofs are merely sketched, we streamline earlier proofs or present alternative arguments at many places.

Contents

1 Introduction 221
2 Definition of the Models 227
  2.1 Operators in Fock-space 228
  2.2 Interaction term 229
  2.3 The semi-relativistic Pauli-Fierz and no-pair Hamiltonians 230
  2.4 How to deal with the non-local terms 232
3 Self-Adjointness 235
  3.1 Diamagnetic inequalities in QED 235
  3.2 Semi-boundedness 239
4 Bounds on the Ionization Energy 244
5 Exponential Localization 249
  5.1 A general strategy to prove the localization of spectral subspaces 250
  5.2 Choice of the comparison operator Y 252
  5.3 Conjugation of Y with exponential weights 254
6 Existence of Ground States with Mass 258
  6.1 Operators with photon mass 259
  6.2 Discretization of the photon momenta 262
  6.3 Comparison of operators with different coupling functions 263
  6.4 Higher order estimates and their consequences 265
  6.5 Continuity of the ionization thresholds and ground state energies 268
  6.6 Proofs of the existence of ground states with mass 270
7 Infra-Red Bounds 274
  7.1 The gauge transformed operator 275
  7.2 Soft photon bound for the semi-relativistic Pauli-Fierz operator 276
8 Existence of Ground States 277
  8.1 Ground states without photon mass 278
  8.2 Ground state degeneracy 281
9 Commutator Estimates 283
  9.1 Basic estimates 283
  9.2 Commuting projections with the field energy 285
  9.3 Double commutators 286
References 287

1. Introduction

In the late 90's and the past decade the existence of ground states of atoms and molecules interacting with the quantized photon field has been intensively studied by mathematicians in the framework of non-relativistic quantum electrodynamics (QED). The corresponding Hamiltonian is the *non-relativistic Pauli-Fierz operator* which, in the case of a hydrogen-like atom, is given as

$$H_\gamma^{\mathrm{nr}} := \big(\boldsymbol{\sigma}\cdot(-i\nabla_{\mathbf{x}} + \mathbf{A})\big)^2 - \frac{\gamma}{|\mathbf{x}|} + H_{\mathrm{f}}\,. \tag{1.1}$$

Here $\boldsymbol{\sigma} = (\sigma_1, \sigma_2, \sigma_3)$ is a vector containing the Pauli spin matrices, $\mathbf{A}$ is the quantized vector potential in the Coulomb gauge, and H_{f} is the energy of the photon field. The symbol $\mathbf{A}$ includes a prefactor entering into the analysis as a parameter, namely the square-root of the fine structure constant which equals the elementary charge, $e > 0$, in the units chosen in this paper. The Coulomb coupling constant, $\gamma > 0$, is the product of e and the nuclear charge. We shall, however, always consider it as an independent parameter since the interrelationship between e and γ does not play any role in our work. $\mathbf{A}$ additionally depends on some ultra-violet cut-off parameter, $\Lambda > 0$. By now it is well-known that H_γ^{nr} has a self-adjoint realization in the Hilbert space $L^2(\mathbb{R}^3_{\mathbf{x}}, \mathbb{C}^2) \otimes \mathscr{F}_{\mathrm{b}}[\mathscr{K}]$ whose spectrum is bounded below. Here $\mathscr{F}_{\mathrm{b}}[\mathscr{K}]$ denotes the bosonic Fock space modeled over the Hilbert space for a single photon, $\mathscr{K} = L^2(\mathbb{R}^3 \times \mathbb{Z}_2)$. Proving the existence of ground states for H_γ^{nr} means to show that the infimum of the spectrum of H_γ^{nr} is an eigenvalue corresponding to some normalizable ground state eigenvector in $L^2(\mathbb{R}^3_{\mathbf{x}}, \mathbb{C}^2) \otimes \mathscr{F}_{\mathrm{b}}[\mathscr{K}]$. Because of the spin degrees of freedom this ground state eigenvalue will be degenerate. Mathematically, the study of the eigenvalue at the bottom of the spectrum of H_γ^{nr} is very subtle because the spectrum of H_γ^{nr} is continuous up to its minimum and the eigenvalue is, thus, not an isolated one. In particular, many standard methods of spectral theory do not apply and several new mathematical techniques had to be invented in order to overcome this problem.

The first proofs of the existence of ground states for H_γ^{nr} and its molecular analogs have been given in [4, 6], for small values of the involved parameters e and Λ. A few years later the existence of ground states for a molecular non-relativistic Pauli-Fierz Hamiltonian has been established, for arbitrary values of e and Λ, in [17] by means of a certain binding condition which has been verified later on in [33]. Moreover, infra-red finite algorithms and spectral theoretic renormalization group methods have been applied to various models of non-relativistic QED to study their ground state energies and projections [2, 3, 4, 5, 6, 7, 15]. These sophisticated methods yield very precise results as they rely on constructive algorithms rather than on compactness arguments as in [4, 6, 17]. They work, however, only in a regime where e and Λ are sufficiently small.

A question which arises naturally in this context is whether these results still hold true when the electrons are described by a relativistic operator. In this review, which summarizes results from [25, 28, 29, 36, 37, 48], we give a positive answer to this question. We study two different models that seem to be natural candidates for a mathematical analysis: The first one is given by the following no-pair operator,

$$H_\gamma^{\mathrm{np}} := P_{\mathbf{A}}^{+}\left(D_{\mathbf{A}} - \tfrac{\gamma}{|\mathbf{x}|} + H_{\mathrm{f}}\right) P_{\mathbf{A}}^{+}\,, \tag{1.2}$$

or more generally,

$$H_V^{\mathrm{np}} := P_{\mathbf{A}}^{+}\left(D_{\mathbf{A}} + V + H_{\mathrm{f}}\right) P_{\mathbf{A}}^{+}\,, \tag{1.3}$$

for some electrostatic potential V. Here $D_{\mathbf{A}}$ is the free Dirac operator minimally coupled to $\mathbf{A}$ and $P_{\mathbf{A}}^{+}$ denotes the spectral projection onto the positive spectral subspace of $D_{\mathbf{A}}$,

$$P_{\mathbf{A}}^{+} := \mathbb{1}_{(0,\infty)}(D_{\mathbf{A}})\,.$$

The no-pair operator is considered as an operator acting in the projected Hilbert space, $\mathscr{H}_{\mathbf{A}}^{+} := P_{\mathbf{A}}^{+}\,\mathscr{H}$. It is thus acting on a space where the electron and photon degrees of freedom are always linked together. The analog of H_γ^{np} for molecules has been introduced in [32] as a mathematical model to study the stability of matter in relativistic QED. Under certain restrictions on e, Λ, and the nuclear charges it has been shown in [32] that the quadratic form of a molecular no-pair operator is bounded from below by some constant which is proportional to the number of involved electrons and nuclei and uniform in the nuclear positions. Moreover, the (positive) binding energy has been estimated from above in [31]. In fact, there are numerous mathematical contributions on no-pair models where magnetic fields are not taken into account or treated classically; see, e.g., [39] for a list of references. For instance, it is shown in [13] that a no-pair operator with Coulomb potential but without quantized photon field – which is then often called the Brown-Ravenhall operator – has a critical coupling constant, $\gamma_{\mathrm{c}}^{\mathrm{np}} := 2/(2/\pi + \pi/2)$, such that the corresponding quadratic form becomes unbounded below when γ exceeds this value. Moreover, various molecular no-pair models (without quantized fields, however) are widely used in quantum chemistry and in the theoretical and numerical study of highly ionized heavy atoms; see, e.g., [10, 22, 46]. In this context several different choices of the projections determining the model find their applications. For instance, one can include the Coulomb or a Hartree-Fock potential in the projection. These two choices are covered by the results of [39] where two of the present authors provide a spectral analysis of a class of molecular no-pair Hamiltonians with classical magnetic fields. The choice of the projection $P^+_{\mathbf{A}}$ which does not contain any potential terms is referred to as the free picture. We remark that it is essential to include the vector potential in the projection determining the no-pair model. For, if $P^+_{\mathbf{A}}$ is replaced by $P^+_{\mathbf{0}}$, then the analog of (1.2) describing N interacting electrons becomes unstable as soon as $N \geqslant 2$ [18, 32, 34]. Moreover, the operator in (1.2) is formally gauge invariant and this would not hold true anymore with $P^+_{\mathbf{0}}$ in place of $P^+_{\mathbf{A}}$. Gauge invariance plays, however, an important role in the proof of the existence of ground states as it permits to derive bounds on the number of soft photons. In fact, employing a mild infra-red regularization it is possible to prove the existence of ground states for the operator in (1.2) with $P^+_{\mathbf{A}}$ replaced by $P^+_{\mathbf{0}}$ [24, 35]. It seems, however, unlikely that the infra-red regularization can be dropped in this case [24].

The second operator treated in this review, the semi-relativistic Pauli-Fierz operator, is given as

$$\sqrt{(\boldsymbol{\sigma}\cdot(-i\nabla+\mathbf{A}))^2+\mathbb{1}}-\tfrac{\gamma}{|\mathbf{x}|}+H_{\mathrm{f}}\,, \qquad (1.4)$$

where $\boldsymbol{\sigma}$ is a vector containing the Pauli spin matrices. Since $\sqrt{-\Delta}$ and $1/|\mathbf{x}|$ both scale as one over the length there will again be some critical upper bound on all values of $\gamma > 0$ for which (1.4) defines a semi-bounded quadratic form. As we shall see this upper bound is at least as big as (in fact equal to [28]) the critical constant in Kato's inequality, $\gamma_{\mathrm{c}}^{\mathrm{PF}} := 2/\pi$. Again we shall study the semi-relativistic Pauli-Fierz operator also for a more general class of electrostatic potentials V. The latter (straightforward) generalization is relevant in a forthcoming work of the first two authors devoted to the enhanced binding effect [27].

Also the semi-relativistic Pauli-Fierz operator has been investigated earlier in a few mathematical articles. For instance it appears in the mathematical study of Rayleigh scattering [14] where the finite propagation speed of relativistic particles turns out to be an advantageous feature in comparison to models of non-relativistic QED. (The electron spin has, however, been neglected in [14].) For $\gamma = 0$, the fiber decomposition of (1.4) with respect to different values of the total momentum has been studied in [40].

Moreover, there is a remark in [40] relevant for us saying that every (speculative) eigenvalue of the operator in (1.4) is at least doubly degenerate since the Hamiltonian commutes with some anti-linear involution. The existence of the renormalized electron mass in the semi-relativistic Pauli-Fierz model, i.e. twice continuous differentiability of the mass shell in balls about zero, is proved in [26], for small values of e. The last author has shown [48] that, when the speed of light, c, is re-introduced as a parameter and $\gamma \in [0, \gamma_c^{\mathrm{PF}})$, then the operator in (1.4) converges in norm resolvent sense to the non-relativistic Pauli-Fierz operator in (1.1), as c tends to infinity. Finally, there is a contribution [21] on the existence of binding in the semi-relativistic Pauli-Fierz model; see Remark 4.2.

We should also mention that the existence of ground states in a relativistic model describing both the photons and the electrons and positrons by quantized fields has been studied mathematically in [8]. To this end infra-red and ultra-violet cut-offs for the momenta of all involved particles are imposed in the interaction term of the Hamiltonian considered in [8].

In the remaining part of this introduction we explain the organization of this review article and summarize briefly our main results. In Section 2 we recall the definitions of some operators appearing in QED and introduce the no-pair and semi-relativistic Pauli-Fierz Hamiltonians more precisely. Although the general strategy of our whole project relies on the methods developed in [4, 6, 17] the spectral analysis of the operators treated in this article poses a variety of new and non-trivial mathematical obstacles which is mainly caused by their non-locality. In fact, both operators do not act as partial differential operators on the electronic degrees of freedom anymore as it is the case in non-relativistic QED. In this respect the no-pair operator is harder to analyze than the semi-relativistic Pauli-Fierz operator since also the electrostatic potential and the radiation field energy become non-local due to the presence of the spectral projections $P^+_{\mathbf{A}}$. In the last subsection of Section 2 we explain a few mathematical tools used to overcome some of the problems posed by the non-locality thus preparing the reader for the proofs in the succeeding sections.

In Section 3 we provide some basic relative bounds and study the semi-boundedness of the Hamiltonian in both models under consideration. We start with a discussion of various diamagnetic inequalities for quantized vector potentials. They are employed to prove that the quadratic form of the semi-relativistic Pauli-Fierz operator is semi-bounded below on some natural dense domain, for a suitable class of potentials including the Coulomb potential with coupling constants $\gamma \in [0, \gamma_c^{\mathrm{PF}}]$. For the no-pair operator we obtain similar results with Coulomb coupling constants $\gamma \in [0, \gamma_c^{np})$. As a consequence, both operators can be realized as self-adjoint operators in a physically distinguished way by means of a Friedrichs extension. We point out that the results on the semi-boundedness, as well as all further results described below hold true, for arbitrary values of e and Λ.

Section 4 is devoted to the study of binding. For both models treated here we show that the infimum of the spectrum of the Hamiltonian with appropriate non-vanishing potential is strictly less than its ionization threshold which, by definition, is equal to the infimum of the spectrum of the Hamiltonian without electrostatic potential. To this end we employ trial functions which are tensor products of electronic and photonic wave functions and work with unitarily equivalent Hamiltonians in order to separate the electronic and photonic degrees of freedom. The unitary transformation used here represents the free Hamiltonian ($V = 0$) as a direct integral of fiber operators with respect to different values of the total momentum.

Typically, proofs of the existence of ground states in QED require some information on the localization of spectral subspaces corresponding to energies below the ionization threshold (or at least of certain approximate ground state eigenfunctions). Here localization is understood with respect to the electron coordinates. We establish this prerequisite in Section 5 by adapting some ideas from [4, 16]. In this section we present streamlined versions of some of our earlier arguments from [37]. Moreover, we implement later improvements [25] on parts of the results of [37] by providing optimized exponential decay rates in the case of the semi-relativistic Pauli-Fierz operator that reduce to the typical relativistic decay rates known for the electronic Dirac and square-root operators, when the radiation field is turned off. The class of potentials allowed for in Section 5 covers Coulomb potentials with $\gamma \in [0, \gamma_c^{PF}]$ in the case of the semi-relativistic Pauli-Fierz model and with $\gamma \in [0, \gamma_c^{np})$ in the no-pair model. It is, however, possible to prove the exponential localization for the no-pair operator with Coulomb potential also in the critical case $\gamma = \gamma_c^{np}$ by another modification of the arguments [25].

The main results of our joint project are the proofs of the existence of ground states for the semi-relativistic Pauli-Fierz and no-pair operators. As already stressed above our proofs work, for arbitrary values of e and Λ and for a class of potentials including the Coulomb potential with $\gamma \in (0, \gamma_c^{PF})$ and $\gamma \in (0, \gamma_c^{np})$, respectively. Starting from these results it is actually possible to prove the existence of ground states also in the critical cases $\gamma = \gamma_c^{PF}$ and $\gamma = \gamma_c^{np}$, respectively, by means of an additional approximation argument. We refrain from explaining any details of the latter in the present article and refer the interested reader to [25] instead.

The proofs of the existence of ground states given here are divided into two steps:

First, one introduces a photon mass, $m > 0$, and shows that the resulting Hamiltonians possess normalized ground state eigenfunctions, ϕ_m [4, 6, 17]. In this first step, which is presented in Section 6, we employ a discretization of the photon momenta as in [6]. Roughly speaking, by discretizing the photon momenta one may replace the Fock space $\mathscr{F}_{\mathrm{b}}[\mathscr{K}]$ by a Fock space modeled over some ℓ^2 space. As a consequence the spectrum of the radiation field energy becomes discrete and one can in fact argue that the total Hamiltonian has discrete eigenvalues at the bottom of its spectrum when all small photon momenta are discarded. At this point we add a new observation based on the localization estimates to the arguments of [6] which allows to carry through the proof, for all values of e and Λ. (In [4, 6] these parameters were assumed to be sufficiently small.) Another technical tool turns out to be very helpful in order to compare discretized and non-discretized Hamiltonians (or those with and without photon mass), namely, certain higher order estimates allowing to control higher powers of the radiation field energy by corresponding powers of the resolvent of the total Hamiltonian. For the semi-relativistic Pauli-Fierz operator such estimates have been established in [14]. In [36] one of the present authors re-proves the higher order estimates from [14] by means of a different and more model-independent method which also permits to derive higher order estimates for a (molecular) no-pair operator for the first time. We discuss these higher order estimates in Subsection 6.4 but refrain from repeating their proofs. We remark that many of the arguments presented in Section 6 are alternatives to those used in [28, 29].

The second step in the proof of the existence of ground states comprises of a compactness argument showing that every sequence $\{\phi_{m_j}\}$ with $m_j \searrow 0$ contains a strongly convergent subsequence. In fact, one readily verifies that the limit of such a subsequence is a ground state eigenfunction of the original Hamiltonian with massless photons. This step is performed in Section 8, in parts by means of arguments alternative to those in [28, 29]. The compactness argument requires, however, a number of non-trivial ingredients. First, we need two infra-red bounds, namely a bound on the number of photons with low energy in the eigenfunctions ϕ_m (soft photon bound) [6, 17] and a certain bound on the weak derivatives of ϕ_m with respect to the photon momenta (photon derivative bound) [17]. To derive the infra-red bounds one can either adapt a procedure proposed in [17] (this is carried through in earlier preprint versions of [28, 29] available on the arXiv) or establish a formula for $a(k)\,\phi_m$ by means of a virial type argument and infer the bounds from that representation. We outline the proof of the latter formula and of the soft photon bound for the semi-relativistic Pauli-Fierz operator in Section 7. The photon derivative bound and the infra-red bounds for the no-pair operator are derived by very similar procedures and we refer the interested reader to our original articles [28, 29] for the rather dull technical details. The arguments presented in Section 7 are also intended to emphasize the role of the gauge invariance of the models treated here. In fact, one first applies a unitary operator-valued gauge transformation (Pauli-Fierz transformation) and the infra-red bounds are then derived in the new gauge. Without the gauge transformation one would encounter infra-red divergent integrals.

As soon as the infra-red estimates are established, the soft photon bound and the exponential localization estimates show that the eigenvectors ϕ_m are localized uniformly in m with respect to the electron and photon coordinates and that their components in all but finitely many Fock space sectors are negligible. Moreover, the photon derivative bound implies that their weak derivatives with respect to the photon momenta are uniformly bounded in a suitable L^p-space and since their energies are uniformly bounded we also know that the vectors have uniformly bounded half-derivatives with respect to the electron coordinates in L^2. It is an idea of [17] to exploit such information by applying compact embedding theorems for Sobolev-type spaces to single out subsequences that converge strongly in L^2. In the semi-relativistic setting considered in Section 8 some classical embedding theorems by Nikol′skiĭ turn out to be useful substitutes for the Rellich-Kondrachov theorem employed in [17]. At the end of Section 8 we also show that the ground state energies of both Hamiltonians are degenerate eigenvalues.

At last, in Section 9, we present the proofs of some technical results we have referred to in earlier sections so that most parts of this review become essentially self-contained.

2. Definition of the Models

In order to introduce the models treated in this article more precisely we first fix our notation and recall some standard facts.

2.1. *Operators in Fock-space*

The state space of the quantized photon field is the bosonic Fock space,

$$\mathscr{F}_{\mathrm{b}}[\mathscr{K}] := \bigoplus_{n=0}^{\infty} \mathscr{F}_{\mathrm{b}}^{(n)}[\mathscr{K}] \ni \psi = (\psi^{(0)}, \psi^{(1)}, \psi^{(2)}, \dots).$$

It is modeled over the one photon Hilbert space

$$\mathscr{K} := L^2(\mathbb{R}^3 \times \mathbb{Z}_2, dk), \qquad \int dk := \sum_{\lambda \in \mathbb{Z}_2} \int_{\mathbb{R}^3} d^3\mathbf{k}.$$

$k = (\mathbf{k}, \lambda)$ denotes a tuple consisting of a photon wave vector, $\mathbf{k} \in \mathbb{R}^3$, and a polarization label, $\lambda \in \mathbb{Z}_2$. Moreover, $\mathscr{F}_{\mathrm{b}}^{(0)}[\mathscr{K}] := \mathbb{C}$ and $\mathscr{F}_{\mathrm{b}}^{(n)}[\mathscr{K}]$ is the subspace of all complex-valued, square integrable functions on $(\mathbb{R}^3 \times \mathbb{Z}_2)^n$ that remain invariant under permutations of the $n \in \mathbb{N}$ wave vector/polarization tuples. The subspace

$$\mathscr{C}_0 := \mathbb{C} \oplus \bigoplus_{n \in \mathbb{N}} C_0((\mathbb{R}^3 \times \mathbb{Z}_2)^n) \cap \mathscr{F}_{\mathrm{b}}^{(n)}[\mathscr{K}] \quad \text{(Algebraic direct sum)}$$

is dense in $\mathscr{F}_{\mathrm{b}}[\mathscr{K}]$. The energy of the photon field, $H_{\mathrm{f}} := d\Gamma(\omega)$, is given as the second quantization of the dispersion relation $\omega(k) := |\mathbf{k}|$, $k = (\mathbf{k}, \lambda) \in \mathbb{R}^3 \times \mathbb{Z}_2$. We recall that the second quantization of some real-valued Borel measurable function ϖ is given by $(d\Gamma(\varpi)\,\psi)^{(0)} = 0$ and

$$(d\Gamma(\varpi)\,\psi)^{(n)}(k_1, \dots, k_n) = \sum_{j=1}^{n} \varpi(k_j)\,\psi^{(n)}(k_1, \dots, k_n), \qquad n \in \mathbb{N},$$

for all $\psi = (\psi^{(n)})_{n=0}^{\infty} \in \mathscr{F}_{\mathrm{b}}[\mathscr{K}]$ such that $([d\Gamma(\varpi)\,\psi]^{(n)})_{n=0}^{\infty} \in \mathscr{F}_{\mathrm{b}}[\mathscr{K}]$. We further recall that the creation and the annihilation operators of a photon state $f \in \mathscr{K}$ are given, for $n \in \mathbb{N}$, by

$$(a^\dagger(f)\,\psi)^{(n)}(k_1, \dots, k_n) = n^{-1/2} \sum_{j=1}^{n} f(k_j)\,\psi^{(n-1)}(\dots, k_{j-1}, k_{j+1}, \dots),$$

$$(a(f)\,\psi)^{(n-1)}(k_1, \dots, k_{n-1}) = n^{1/2} \int \overline{f}(k)\,\psi^{(n)}(k, k_1, \dots, k_{n-1})\,dk,$$

and $(a^\dagger(f)\,\psi)^{(0)} = 0$, $a(f)\,\Omega = 0$, where $\Omega := (1, 0, 0, \dots) \in \mathscr{F}_{\mathrm{b}}[\mathscr{K}]$ is the vacuum vector. We define $a^\dagger(f)$ and $a(f)$ on their maximal domains. For $f, g \in \mathscr{K}$, the following canonical commutation relations hold true on $\mathscr{C}_0$,

$$[a(f)\,,\,a(g)] = [a^\dagger(f)\,,\,a^\dagger(g)] = 0, \qquad [a(f)\,,\,a^\dagger(g)] = \langle\, f \,|\, g \,\rangle\,\mathbb{1}.$$

For a three-vector of functions $\mathbf{f} = (f^{(1)}, f^{(2)}, f^{(3)}) \in \mathscr{K}^3$, the symbol $a^\sharp(\mathbf{f})$ denotes the triple of operators $a^\sharp(\mathbf{f}) := \big(a^\sharp(f^{(1)}), a^\sharp(f^{(2)}), a^\sharp(f^{(3)})\big)$, where $a^\sharp$ is always either a or $a^\dagger$.

2.2. *Interaction term*

Next, we describe the interaction between four-spinors and the photon field. The full Hilbert space underlying our models is

$$\mathscr{H} := L^2(\mathbb{R}^3_{\mathbf{x}}, \mathbb{C}^4) \otimes \mathscr{F}_{\mathrm{b}}[\mathscr{K}] .$$

It contains the dense subspace

$$\mathscr{D} := C_0^\infty(\mathbb{R}^3_{\mathbf{x}}, \mathbb{C}^4) \otimes \mathscr{C}_0 . \quad \text{(Algebraic tensor product)}$$

We introduce the self-adjoint Dirac matrices $\alpha_1, \alpha_2, \alpha_3$, and β that act on the four spinor components of an element from $\mathscr{H}$, that is, on the second tensor factor in $\mathscr{H} \cong L^2(\mathbb{R}^3_{\mathbf{x}}) \otimes \mathbb{C}^4 \otimes \mathscr{F}_{\mathrm{b}}[\mathscr{K}]$. They are given by

$$\alpha_j := \begin{pmatrix} 0 & \sigma_j \\ \sigma_j & 0 \end{pmatrix}, \quad j \in \{1,2,3\}, \qquad \beta := \alpha_0 := \begin{pmatrix} \mathbb{1} & 0 \\ 0 & -\mathbb{1} \end{pmatrix},$$

where $\sigma_1, \sigma_2, \sigma_3$ denote the standard Pauli matrices, and fulfill the Clifford algebra relations

$$\alpha_i \, \alpha_j + \alpha_j \, \alpha_i = 2 \, \delta_{ij} \, \mathbb{1} , \qquad i, j \in \{0,1,2,3\} . \tag{2.1}$$

The interaction between the electron/positron and photon degrees of freedom in the Coulomb gauge is given as $\boldsymbol{\alpha} \cdot \mathbf{A} := \alpha_1 \, A^{(1)} + \alpha_2 \, A^{(2)} + \alpha_3 \, A^{(3)}$, where

$$\mathbf{A} := (A^{(1)}, A^{(2)}, A^{(3)}) := a^\dagger(\mathbf{G}) + a(\mathbf{G}) , \quad a^\sharp(\mathbf{G}) := \int_{\mathbb{R}^3}^{\oplus} \mathbb{1}_{\mathbb{C}^4} \otimes a^\sharp(\mathbf{G}_{\mathbf{x}}) \, d^3\mathbf{x} .$$

The physical choice of the coupling function $\mathbf{G}_{\mathbf{x}} = (G^{(1)}_{\mathbf{x}}, G^{(2)}_{\mathbf{x}}, G^{(3)}_{\mathbf{x}})$ is

$$\mathbf{G}_{\mathbf{x}}(k) := -e \, \frac{\mathbb{1}_{\{|\mathbf{k}| \leqslant \Lambda\}}}{2\pi \sqrt{|\mathbf{k}|}} \, e^{-i\mathbf{k}\cdot\mathbf{x}} \, \boldsymbol{\varepsilon}(k), \tag{2.2}$$

for $\mathbf{x} \in \mathbb{R}^3$ and almost every $k = (\mathbf{k}, \lambda) \in \mathbb{R}^3 \times \mathbb{Z}_2$. The parameter $\Lambda > 0$ is an ultraviolet cut-off and $e \in \mathbb{R}$. (In nature $e^2 \approx 1/137$ is Sommerfeld's fine-structure constant which equals the square of the elementary charge in our units[a].) The values of e and Λ can be chosen arbitrarily in the whole article. Writing

$$\mathbf{k}_\perp := (k^{(2)} , -k^{(1)} , 0) , \qquad \mathbf{k} = (k^{(1)}, k^{(2)}, k^{(3)}) \in \mathbb{R}^3 , \tag{2.3}$$

---

[a] Energies are measured in units of mc^2, m denoting the rest mass of an electron and c the speed of light. Length, i.e. $\mathbf{x}$, are measured in units of $\hbar/(mc)$, which is the Compton wave length divided by 2π. $\hbar$ is Planck's constant divided by 2π. The photon wave vectors $\mathbf{k}$ are measured in units of 2π times the inverse Compton wavelength, $mc/\hbar$.

the polarization vectors are given as

$$\varepsilon(\mathbf{k},0) = \frac{\mathbf{k}_\perp}{|\mathbf{k}_\perp|}, \qquad \varepsilon(\mathbf{k},1) = \frac{\mathbf{k}}{|\mathbf{k}|} \wedge \varepsilon(\mathbf{k},0), \tag{2.4}$$

for $\mathbf{k} \in \mathbb{R}^3 \setminus \{0\}$ with $\mathbf{k}_\perp \neq 0$. It is sufficient to determine ε almost everywhere. Many of our results and estimates do not depend on the special choice of $\mathbf{G}_\mathbf{x}$. If we consider a larger class of coupling functions at a certain point in this review we shall explain the required properties of $\mathbf{G}_\mathbf{x}$ at the beginning of the corresponding (sub)section.

2.3. *The semi-relativistic Pauli-Fierz and no-pair Hamiltonians*

In order to define the no-pair and semi-relativistic Pauli-Fierz operators we recall that the free Dirac operator minimally coupled to $\mathbf{A}$ is given as

$$D_\mathbf{A} := \boldsymbol{\alpha}\cdot(-i\nabla+\mathbf{A})+\beta := \sum_{j=1}^{3} \alpha_j \big(-i\partial_{x_j} + a^\dagger(G_\mathbf{x}^{(j)}) + a(G_\mathbf{x}^{(j)})\big) + \beta\,. \tag{2.5}$$

A straightforward application of Nelson's commutator theorem shows that $D_\mathbf{A}$ is essentially self-adjoint on $\mathscr{D}$; see, e.g., [32, 40]. We denote its closure starting from $\mathscr{D}$ again by the same symbol. As a consequence of (2.1) we further have

$$D_\mathbf{A}^2 = \mathcal{T}_\mathbf{A} \oplus \mathcal{T}_\mathbf{A}\,, \qquad \mathcal{T}_\mathbf{A} := (\boldsymbol{\sigma}\cdot(-i\nabla+\mathbf{A}))^2 + \mathbb{1}\,, \tag{2.6}$$

on $\mathscr{D}$. In particular, $D_\mathbf{A}^2 \geqslant 1$ on $\mathscr{D}$, and since $D_\mathbf{A}$ is essentially self-adjoint on $\mathscr{D}$ we see that $\|(D_\mathbf{A} - z)\,\psi\| \geqslant (1-|z|)\,\|\psi\|$, $\psi \in \mathcal{D}(D_\mathbf{A})$, $z \in \mathbb{C}$, whence

$$\sigma(D_\mathbf{A}) \subset (-\infty, -1] \cup [1, \infty)\,.$$

Contrary to the usual convention used also in the introduction we define the *semi-relativistic Pauli-Fierz operator* as an operator acting in $\mathscr{H}$ a priori by

$$H_\gamma^{\mathrm{PF}}\,\varphi := \big(|D_\mathbf{A}| - \tfrac{\gamma}{|\mathbf{x}|} + H_\mathrm{f}\big)\,\varphi\,, \qquad \varphi \in \mathscr{D}\,, \tag{2.7}$$

$$H_V^{\mathrm{PF}}\,\varphi := \big(|D_\mathbf{A}| + V + H_\mathrm{f}\big)\,\varphi\,, \qquad \varphi \in \mathscr{D}\,. \tag{2.8}$$

We shall impose appropriate conditions on the general potential $V \in L^2_{\mathrm{loc}}(\mathbb{R}^3, \mathbb{R})$ later on. In fact, the operator defined in (2.7) is a two-fold copy of the one given in (1.4) since $|D_\mathbf{A}| = \mathcal{T}_\mathbf{A}^{1/2} \oplus \mathcal{T}_\mathbf{A}^{1/2}$ by (2.6). We prefer to consider the operator defined by (2.7) to have a unified notation throughout the following sections. The *no-pair operator* acts in a projected Hilbert space, $\mathscr{H}_{\mathbf{A}}^+$, given by

$$\mathscr{H}_{\mathbf{A}}^+ := P_{\mathbf{A}}^+ \mathscr{H}, \qquad P_{\mathbf{A}}^+ := \mathbb{1}_{[0,\infty)}(D_{\mathbf{A}}), \qquad P_{\mathbf{A}}^- := \mathbb{1} - P_{\mathbf{A}}^+. \tag{2.9}$$

A priori we define it on the dense domain $P_{\mathbf{A}}^+ \mathscr{D} \subset \mathscr{H}_{\mathbf{A}}^+$,

$$H_\gamma^{\mathrm{np}} \varphi^+ := P_{\mathbf{A}}^+ \left(D_{\mathbf{A}} - \tfrac{\gamma}{|\mathbf{x}|} + H_{\mathrm{f}}\right) \varphi^+, \qquad \varphi^+ \in P_{\mathbf{A}}^+ \mathscr{D}, \tag{2.10}$$

$$H_V^{\mathrm{np}} \varphi^+ := P_{\mathbf{A}}^+ \left(D_{\mathbf{A}} + V + H_{\mathrm{f}}\right) \varphi^+, \qquad \varphi^+ \in P_{\mathbf{A}}^+ \mathscr{D}, \tag{2.11}$$

where $V \in L^2_{\mathrm{loc}}(\mathbb{R}^3, \mathbb{R})$ satisfies $\mathcal{D}(D_{\mathbf{0}}) \subset \mathcal{D}(V)$. In the above definitions we have to take care that the right hand sides are actually well-defined as it is, for instance, not obvious that $\frac{\gamma}{|\mathbf{x}|} P_{\mathbf{A}}^+ \mathscr{D} \subset \mathscr{H}$. It follows, however, from the proof of Lemma 3.4(ii) in [37] that $P_{\mathbf{A}}^+$ maps $\mathscr{D}$ into $\mathcal{D}(D_{\mathbf{0}}) \cap \mathcal{D}(H_{\mathrm{f}}^\nu)$, for every $\nu > 0$, so that the definitions (2.10) and (2.11) make sense.

As soon as we have shown in Section 3 that the quadratic forms of H_V^{PF} and H_V^{np} are semi-bounded below on the dense domains $\mathscr{D}$ and $P_{\mathbf{A}}^+ \mathscr{D}$, respectively, we may extend them to self-adjoint operators by means of a Friedrichs extension. As already mentioned in the introduction there will, however, be critical values for γ above which the quadratic forms are unbounded below in the case of the Coulomb potential. We prove in Section 3 that, for H_γ^{PF}, this critical value is not less than

$$\gamma_{\mathrm{c}}^{\mathrm{PF}} := 2/\pi,$$

which is the critical constant in Kato's inequality, $(2/\pi)|\mathbf{x}|^{-1} \leqslant \sqrt{-\Delta}$. In the case of the no-pair operator we prove the semi-boundedness of the quadratic form of H_γ^{np}, for all $\gamma \in [0, \gamma_{\mathrm{c}}^{\mathrm{np}})$, where

$$\gamma_{\mathrm{c}}^{\mathrm{np}} := 2/(2/\pi + \pi/2). \tag{2.12}$$

The instability of both models above the respective critical values for γ is shown in [28, 29] by means of suitable test functions that drive the energy to minus infinity. For the definition of H_γ^{np} in the case $\gamma = \gamma_{\mathrm{c}}^{\mathrm{np}}$ see [29].

It has been shown in [13] that the quadratic form associated with the (electronic) Brown-Ravenhall operator,

$$B_\gamma^{\mathrm{el}} := P_{\mathbf{0}}^+ \left(D_{\mathbf{0}} - \tfrac{\gamma}{|\mathbf{x}|}\right) P_{\mathbf{0}}^+, \tag{2.13}$$

is semi-bounded on $P_{\mathbf{0}}^+ C_0^\infty(\mathbb{R}^3, \mathbb{C}^4)$ if and only if $\gamma \leqslant \gamma_{\mathrm{c}}^{\mathrm{np}}$. Thus, for $\gamma \leqslant \gamma_{\mathrm{c}}^{\mathrm{np}}$, it has a self-adjoint Friedrichs extension which we again denote by B_γ^{el} and which actually satisfies Tix' inequality, $B_\gamma^{\mathrm{el}} \geqslant (1-\gamma)\, P_{\mathbf{0}}^+$, $\gamma \in [0, \gamma_{\mathrm{c}}^{\mathrm{np}}]$;

see [49]. We exploit the semi-boundedness of B_γ^{el} in the proof of Theorem 3.6 below.

Finally, we introduce a convention used throughout this review: We will frequently use the symbols $H_V^\sharp$, $H_\gamma^\sharp$, $\gamma_c^\sharp$, etc. when we treat both the relativistic Pauli-Fierz and no-pair operators at the same time; that is, $\sharp$ is PF or np.

2.4. *How to deal with the non-local terms*

Although general strategies to prove the existence of ground states have been developed in the framework of non-relativistic QED [4, 6, 17] the application of these ideas to the models discussed in this review poses a variety of new mathematical problems. This is mainly due to the non-locality of the operators $|D_{\mathbf{A}}|$ and $P_{\mathbf{A}}^+$ appearing in $H_V^\sharp$. In this respect the no-pair operator is considerably more difficult to analyze than the semi-relativistic Pauli-Fierz operator since also the projected potential and radiation field energies become non-local. As a consequence a variety of commutator estimates involving $|D_{\mathbf{A}}|$, $P_{\mathbf{A}}^+$, H_{f}, cut-off functions etc. is required for a spectral analysis of $H_V^\sharp$. Most of these commutator estimates are based on the observations and facts we collect in this subsection. We shall only present one proof in the present subsection in order to illustrate some simple ideas and defer other technical arguments to Section 9. We introduce a general hypothesis on the coupling function which is sometimes used in the sequel:

Hypothesis 2.1: The map $\mathbb{R}^3 \times (\mathbb{R}^3 \times \mathbb{Z}_2) \ni (\mathbf{x}, k) \mapsto \mathbf{G}_{\mathbf{x}}(k)$ is measurable such that $\mathbf{x} \mapsto \mathbf{G}_{\mathbf{x}}(k)$ is continuously differentiable, for almost every k, and

$$\mathbf{G}_{\mathbf{x}}(-\mathbf{k}, \lambda) = \epsilon(\lambda)\,\overline{\mathbf{G}}_{\mathbf{x}}(\mathbf{k}, \lambda)\,, \qquad \mathbf{x} \in \mathbb{R}^3,\ \text{a.e. } \mathbf{k}\,,\ \lambda \in \mathbb{Z}_2\,, \tag{2.14}$$

where $\epsilon(\lambda) \in \{-1, 1\}$, $\lambda \in \mathbb{Z}_2$. There exist $d_{-1}, d_0, d_1, \ldots \in (0, \infty)$ satisfying

$$2\int \omega(k)^\ell\, \|\mathbf{G}(k)\|_\infty^2\, dk \leqslant d_\ell^2\,, \quad 2\int \frac{\|\nabla_{\mathbf{x}} \wedge \mathbf{G}(k)\|_\infty^2}{\omega(k)}\, dk \leqslant d_1^2\,, \tag{2.15}$$

where $\|\mathbf{G}(k)\|_\infty := \sup_{\mathbf{x}} |\mathbf{G}_{\mathbf{x}}(k)|$, etc.

We remark that, if (2.14) is fulfilled, then $[A^{(j)}(\mathbf{x}), A^{(k)}(\mathbf{y})] = 0$, for all $j, k \in \{1, 2, 3\}$, $\mathbf{x}, \mathbf{y} \in \mathbb{R}^3$. For later reference we also recall the following well-known relative bounds, valid for every $\psi \in \mathcal{D}(H_{\mathrm{f}}^{1/2})$,

$$\|\boldsymbol{\alpha} \cdot a(\mathbf{G})\,\psi\|^2 \leqslant d_{-1}^2\, \|H_{\mathrm{f}}^{1/2}\psi\|^2, \tag{2.16}$$

$$\|\boldsymbol{\alpha} \cdot a^\dagger(\mathbf{G})\,\psi\|^2 \leqslant d_{-1}^2\, \|H_{\mathrm{f}}^{1/2}\psi\|^2 + d_0^2\, \|\psi\|^2. \tag{2.17}$$

In order to cope with the non-locality of $P^+_{\mathbf{A}}$ we write

$$R_{\mathbf{A}}(iy) := (D_{\mathbf{A}} - iy)^{-1}, \qquad y \in \mathbb{R},$$

and use the following representation of the sign function of $D_{\mathbf{A}}$ as a strongly convergent principal value (see Lemma VI.5.6 in [23]),

$$S_{\mathbf{A}}\,\varphi := D_{\mathbf{A}}\,|D_{\mathbf{A}}|^{-1}\,\varphi = \lim_{\tau\to\infty}\int_{-\tau}^{\tau} R_{\mathbf{A}}(iy)\,\varphi\,\frac{dy}{\pi}, \qquad \varphi \in \mathscr{H}. \tag{2.18}$$

In addition we observe that

$$|D_{\mathbf{A}}| = S_{\mathbf{A}}\,D_{\mathbf{A}}, \qquad P^+_{\mathbf{A}} = \frac{1}{2}\,\mathbb{1} + \frac{1}{2}\,S_{\mathbf{A}}. \tag{2.19}$$

These formulas reduce computations involving $|D_{\mathbf{A}}|$ or $P^+_{\mathbf{A}}$ to computations involving $D_{\mathbf{A}}$ and integrals over its resolvent. To study the exponential localization it is hence useful to recall that, for all $y \in \mathbb{R}$, $a \in [0,1)$, and $F \in C^\infty(\mathbb{R}^3_{\mathbf{x}}, \mathbb{R})$ having a fixed sign and satisfying $|\nabla F| \leqslant a$, we have $iy \in \varrho(D_{\mathbf{A}} + i\boldsymbol{\alpha}\cdot\nabla F)$,

$$R^F_{\mathbf{A}}(iy) := e^F\,R_{\mathbf{A}}(iy)\,e^{-F} = (D_{\mathbf{A}} + i\boldsymbol{\alpha}\cdot\nabla F - iy)^{-1}\restriction_{\mathcal{D}(e^{-F})}, \tag{2.20}$$

and

$$\|R^F_{\mathbf{A}}(iy)\| \leqslant \frac{\sqrt{6}}{\sqrt{1+y^2}}\cdot\frac{1}{1-a^2}. \tag{2.21}$$

For classical vector potentials this essentially follows from a computation we learned from [9]; see also [38] where (2.20) and (2.21) are proved in the form stated above. It is, however, clear that the arguments in [38] work for a quantized vector potential, too. Moreover, it is easy to verify that

$$[R_{\mathbf{A}}(iy)\,,\,\chi\,e^F]\,e^{-F} = R_{\mathbf{A}}(iy)\,i\boldsymbol{\alpha}\cdot(\nabla\chi + \chi\,\nabla F)\,R^F_{\mathbf{A}}(iy), \tag{2.22}$$

where $\chi \in C^\infty(\mathbb{R}^3_{\mathbf{x}}, [0,1])$ is some smooth function of the electron coordinates and F is as above. Finally, we note that

$$\|i\boldsymbol{\alpha}\cdot(\nabla\chi + \chi\,\nabla F)\| \leqslant \|\nabla\chi\|_\infty + a, \tag{2.23}$$

since $\|\boldsymbol{\alpha}\cdot\mathbf{v}\| = |\mathbf{v}|$, $\mathbf{v} \in \mathbb{R}^3$, by the Clifford algebra relations, and $|\nabla F| \leqslant a$. As an example we treat some commutator estimates whose proofs make use of these remarks and a few further useful observations.

Lemma 2.2: *Assume that $\mathbf{G}_{\mathbf{x}}$ fulfills Hypothesis 2.1. Let χ and F be as above, assume additionally that F is bounded, and set $\check{H}_{\mathrm{f}} := H_{\mathrm{f}} + E$, for some sufficiently large $E \geqslant 1$ (depending on d_1). Let $V \in L^1_{\mathrm{loc}}(\mathbb{R}^3, \mathbb{R})$ be* *relatively form-bounded with respect to* $\sqrt{-\Delta}$. *Then, for all* $a_0, \kappa \in [0,1)$, $\nu \geqslant 0$, *and* $a \in [0, a_0]$,

$$\| \, |D_{\mathbf{A}}|^{\kappa} \, [P^{+}_{\mathbf{A}} \, , \, \chi \, e^{F}] \, e^{-F} \, \| \leqslant \mathrm{const}(a_0, \kappa) \cdot (a + \|\nabla\chi\|_{\infty}) \, , \tag{2.24}$$

$$\| \, \check{H}_{\mathrm{f}}^{\nu} \, [P^{+}_{\mathbf{A}} \, , \, \chi \, e^{F}] \, e^{-F} \, \check{H}_{\mathrm{f}}^{-\nu} \, \| \leqslant \mathrm{const}(a_0, \nu) \cdot (a + \|\nabla\chi\|_{\infty}) \, , \tag{2.25}$$

$$\| \, |V|^{1/2} \, [P^{+}_{\mathbf{A}} \, , \, \chi \, e^{F}] \, e^{-F} \, \check{H}_{\mathrm{f}}^{-1/2} \, \| \leqslant \mathrm{const}(a_0, V) \cdot (a + \|\nabla\chi\|_{\infty}) \, . \tag{2.26}$$

Notice that the a_0-dependence of the constants originates from the singularity at $a = 1$ in (2.21). Notice also that we may choose $V = |\mathbf{x}|^{-1}$ in (2.26) in view of Kato's inequality.

Before the proof we further remark that all operators appearing in the norms in (2.24)–(2.26) and in similar estimates below are always well-defined a priori on $\mathscr{D}$ and have unique bounded extensions to the whole Hilbert space. In fact, $P^{+}_{\mathbf{A}} \, \mathscr{D} \subset \mathcal{D}(D_{\mathbf{0}}) \cap \bigcap_{\nu > 0} \mathcal{D}(H_{\mathrm{f}}^{\nu})$ as we have recalled from [37] above already. To simplify the presentation we shall not comment on this anymore from now on.

Proof: We use the fact that an operator, T, acting in some Hilbert space is bounded if and only if $\sup_{\|\varphi\|, \|\psi\| = 1} |\langle \, \varphi \, | \, T \, \psi \, \rangle|$ is bounded in which case it is equal to $\|T\|$. Here it is sufficient to take the supremum over all normalized φ and ψ from a dense set which is a core for T. Combining (2.18), (2.19), and (2.22) we find, for all normalized $\varphi, \psi \in \mathscr{D}$,

$$\begin{aligned} &\langle \, |D_{\mathbf{A}}|^{\kappa} \, \varphi \, | \, [P^{+}_{\mathbf{A}} \, , \, \chi \, e^{F}] \, e^{-F} \, \psi \, \rangle \\ &= \lim_{\tau \to \infty} \int_{-\tau}^{\tau} \langle \, |D_{\mathbf{A}}|^{\kappa} \, \varphi \, | \, R_{\mathbf{A}}(iy) \, i\boldsymbol{\alpha} \cdot (\nabla\chi + \chi \, \nabla F) \, R^{F}_{\mathbf{A}}(iy) \, \psi \, \rangle \, \frac{dy}{2\pi} \, . \end{aligned}$$

On account of $\| \, |D_{\mathbf{A}}|^{\kappa} \, R_{\mathbf{A}}(iy) \| \leqslant \mathrm{const}(\kappa)(1 + y^2)^{-1/2 + \kappa/2}$, (2.21), and (2.23) we see that the scalar product under the integral sign defines some Lebesgue integrable function of y and

$$|\langle \, |D_{\mathbf{A}}|^{\kappa} \, \varphi \, | \, [P^{+}_{\mathbf{A}} \, , \, \chi \, e^{F}] \, e^{-F} \, \psi \, \rangle| \leqslant \mathrm{const}(\kappa) \, (\|\nabla\chi\|_{\infty} + a) \int_{\mathbb{R}} \frac{dy}{(1 + y^2)^{1 - \kappa/2}} \, ,$$

where the last integral is finite. Therefore, $[P^{+}_{\mathbf{A}} \, , \, \chi \, e^{F}] \, e^{-F} \, \psi$ belongs to the domain of $(|D_{\mathbf{A}}|^{\kappa})^{*} = |D_{\mathbf{A}}|^{\kappa}$ and the first bound (2.24) follows.

In order to prove the second bound (2.25) we introduce another little tool which turns out to be useful in our whole analysis. Namely, if $E \geqslant 1$ is sufficiently large depending on d_1, ν we can construct $\Upsilon^{F}_{\nu}(iy) \in \mathscr{L}(\mathscr{H})$ such that $R^{F}_{\mathbf{A}}(iy) \, \check{H}_{\mathrm{f}}^{-\nu} = \check{H}_{\mathrm{f}}^{-\nu} \, R^{F}_{\mathbf{A}}(iy) \, \Upsilon^{F}_{\nu}(iy)$, for every $y \in \mathbb{R}$, and such that the norm of $\Upsilon^{F}_{\nu}(iy)$ is uniformly bounded with respect to $y \in \mathbb{R}$;

see Corollary 9.2 below. (In particular, $R^F_{\mathbf{A}}(iy)$ maps $\mathcal{D}(H_\mathrm{f}^\nu)$ into itself.) Therefore,

$$\begin{aligned}&\big|\langle\, H_\mathrm{f}^\nu\varphi\,|\,R_{\mathbf{A}}(iy)\,i\boldsymbol{\alpha}\cdot(\nabla\chi+\chi\,\nabla F)\,R^F_{\mathbf{A}}(iy)\,\check{H}_\mathrm{f}^{-\nu}\,\psi\,\rangle\big|\\ &\quad=\big|\langle\,\varphi\,|\,R_{\mathbf{A}}(iy)\,\Upsilon^0_\nu(iy)\,i\boldsymbol{\alpha}\cdot(\nabla\chi+\chi\,\nabla F)\,R^F_{\mathbf{A}}(iy)\,\Upsilon^F_\nu(iy)\,\psi\,\rangle\big|\\ &\quad\leqslant C\,(\|\nabla\chi\|_\infty+a)(1+y^2)^{-1}\,,\end{aligned}$$

and it is clear from the argument above how to derive (2.25).

The last bound (2.26) follows from the first two and the inequality $|V|/C\leqslant|D_{\mathbf{A}}|+H_\mathrm{f}+E$ proved later on in Theorem 3.4. □

3. Self-Adjointness

As it is obvious from the definitions in the preceding section the operators $H_0^\sharp$ are positive. In this section we present some basic relative bounds that allow to define the perturbed operators $H_V^\sharp$ as self-adjoint Friedrichs extensions. As a rule we denote the self-adjoint extensions of $H_V^\sharp$ or $H_\gamma^\sharp$ – which are only defined on $\mathscr{D}$ and $P^+_{\mathbf{A}}\,\mathscr{D}$ so far – again by the same symbols. For suitable V, we are also able to characterize the quadratic form domains of $H_V^\sharp$ which turn out to be the spaces of all vectors with finite kinetic and radiation field energy; see Theorems 3.4 and 3.6 below.

Before we present the afore-mentioned results we discuss various (essentially well-known) diamagnetic inequalities in QED; see Theorem 3.2 below. Since these estimates are of independent interest we decided to present one way to derive them (adapted from [47]) in detail which has not been worked out in the literature before, as it seems to us.

3.1. *Diamagnetic inequalities in QED*

In this subsection it is sufficient to assume that

$$\mathbf{A}=\int_{\mathbb{R}^3}^{\oplus}\mathbf{A}(\mathbf{x})\,d^3\mathbf{x}\,,\qquad\mathbf{A}(\mathbf{x}):=\mathbb{1}_{\mathbb{C}^4}\otimes\big(a^\dagger(\mathbf{G}_\mathbf{x})+a(\mathbf{G}_\mathbf{x})\big)\,,$$

where $\mathbb{R}^3\times(\mathbb{R}^3\times\mathbb{Z}_2)\ni(\mathbf{x},k)\mapsto\mathbf{G}_\mathbf{x}(k)$ is measurable such that $\mathbf{x}\mapsto\mathbf{G}_\mathbf{x}(k)$ is continuously differentiable, for almost every k, and $\int(\omega(k)^{-1}+\omega(k)^2)\sup_\mathbf{x}|\mathbf{G}_\mathbf{x}(k)|^2dk<\infty$ as well as $\int(1+\omega(k)^{-1})\sup_\mathbf{x}|\nabla_\mathbf{x}\mathbf{G}_\mathbf{x}(k)|^2dk<\infty$. The following result is probably well-known but the argument sketched in its proof might be new.

Lemma 3.1: *Let $\lambda\geqslant 0$. Under the above condition on $\mathbf{G}_\mathbf{x}$ the operator $(-i\nabla+\mathbf{A})^2+\lambda\,H_\mathrm{f}$ is essentially self-adjoint on $\mathscr{D}$.*

Proof: It is a standard exercise to show that $\{-i\nabla, \mathbf{A}\} + \mathbf{A}^2$ is a small operator perturbation of $-\Delta + c\,H_{\mathrm{f}}$, provided that $c > 0$ is chosen sufficiently large depending on $\mathbf{G}_{\mathbf{x}}$. In particular, $\mathscr{N} := (-i\nabla + \mathbf{A})^2 + c\,H_{\mathrm{f}} + 1$ is essentially self-adjoint on any core of $-\Delta + c\,H_{\mathrm{f}}$ and, in particular, on $\mathscr{D}$. In the next step we apply Nelson's commutator theorem with the closure of $\mathscr{N}$ starting from $\mathscr{D}$ as test operator to conclude. □

We denote the closure of $(-i\nabla + \mathbf{A})^2$ starting from $\mathscr{D}$ by $\tau_{\mathbf{A}}$. For every $\phi, \psi \in \mathscr{H} = L^2(\mathbb{R}^3_{\mathbf{x}}, \mathbb{C}^4 \otimes \mathscr{F}_{\mathrm{b}})$, we write $(\,\phi\,|\,\psi\,)$ for the (partial) scalar product on $\mathbb{C}^4 \otimes \mathscr{F}_{\mathrm{b}}$ and denote $[\![\varphi]\!](\mathbf{x}) := (\,\varphi(\mathbf{x})\,|\,\varphi(\mathbf{x})\,)^{1/2}$. Furthermore, we set $S_\phi(\mathbf{x}) := \frac{1}{[\![\phi]\!](\mathbf{x})}\,\phi(\mathbf{x})$, for $\phi(\mathbf{x}) \neq 0$, and $S_\phi(\mathbf{x}) = 0$, for $\phi(\mathbf{x}) = 0$.

Theorem 3.2: *(i) Let $\phi \in \mathcal{D}(\tau_{\mathbf{A}})$. Then $[\![\phi]\!] \in H^1(\mathbb{R}^3)$, and*

$$\langle\,\eta\,|\,-\Delta\,[\![\phi]\!]\,\rangle_{L^2(\mathbb{R}^3)} \leqslant \mathrm{Re}\int_{\mathbb{R}^3} \eta(\mathbf{x})\,\big(\,S_\phi(\mathbf{x})\,\big|\,(\tau_{\mathbf{A}}\,\phi)(\mathbf{x})\big)\big)\,d^3\mathbf{x}\,, \tag{3.1}$$

for all $\eta \in H^1(\mathbb{R}^3)$, $\eta \geqslant 0$. In particular, for $\eta = [\![\phi]\!]$,

$$\langle\,[\![\phi]\!]\,|\,-\Delta\,[\![\phi]\!]\,\rangle_{L^2(\mathbb{R}^3)} \leqslant \langle\,\phi\,|\,\tau_{\mathbf{A}}\,\phi\,\rangle\,. \tag{3.2}$$

(ii) Let $\phi \in \mathcal{D}(\tau_{\mathbf{A}}^{1/2})$. Then $[\![\phi]\!] \in H^{1/2}(\mathbb{R}^3)$, and

$$\langle\,\eta\,|\,\sqrt{-\Delta}\,[\![\phi]\!]\,\rangle_{L^2(\mathbb{R}^3)} \leqslant \mathrm{Re}\int_{\mathbb{R}^3} \eta(\mathbf{x})\,\big(\,S_\phi(\mathbf{x})\,\big|\,(\tau_{\mathbf{A}}^{1/2}\phi)(\mathbf{x})\,\big)\,d^3\mathbf{x}\,, \tag{3.3}$$

for all $\eta \in H^{1/2}(\mathbb{R}^3)$, $\eta \geqslant 0$. In particular, for $\eta = [\![\phi]\!]$,

$$\langle\,[\![\phi]\!]\,|\,\sqrt{-\Delta}\,[\![\phi]\!]\,\rangle_{L^2(\mathbb{R}^3)} \leqslant \langle\,\phi\,|\,\tau_{\mathbf{A}}^{1/2}\,\phi\,\rangle\,. \tag{3.4}$$

(iii) For all $\psi \in \mathscr{H}$ and $t \in [0, \infty)$, we have, almost everywhere on $\mathbb{R}^3$,

$$[\![\,e^{-t\tau_{\mathbf{A}}}\,\psi\,]\!] \leqslant e^{-t(-\Delta)}\,[\![\,\psi\,]\!]\,, \tag{3.5}$$

$$[\![\,e^{-t\tau_{\mathbf{A}}^{1/2}}\,\psi\,]\!] \leqslant e^{-t\sqrt{-\Delta}}\,[\![\,\psi\,]\!]\,. \tag{3.6}$$

Remark 3.3: (1) Arguing as in Theorem 7.21 of [30] with the corresponding changes as in the proof below one can easily verify that, for $\phi \in \mathscr{D}$, we have $[\![\phi]\!] \in H^1(\mathbb{R}^3)$ and $|\nabla\,[\![\,\phi\,]\!](\mathbf{x})| \leqslant [\![(-i\nabla + \mathbf{A})\phi\,]\!](\mathbf{x})$, for a.e. $\mathbf{x} \in \mathbb{R}^3$.

(2) In [19] diamagnetic inequalities for infra-red regularized vector potentials have been proved by means of dressing transformations. For an alternative proof using functional integrals see [20]. If all components of the vector potential commute, $[A^{(j)}(\mathbf{x}), A^{(k)}(\mathbf{y})] = 0$, then one can also reduce the diamagnetic inequalities to classical ones by diagonalizing all components $A^{(j)}(\mathbf{x})$ simultaneously; this argument due to J. Fröhlich is mentioned in [1]. The proofs given here are variants of the ones presented in [43, 47].

Proof: Let $\varepsilon > 0$. First, we assume that $\phi \in \mathscr{D}$ and set $u_\varepsilon := \sqrt{[\![\phi]\!]^2 + \varepsilon^2} \in C^\infty(\mathbb{R}^3, \mathbb{R})$. Since $A^{(j)}(\mathbf{x})$ is symmetric on $\mathscr{C}_0$, for every $\mathbf{x} \in \mathbb{R}^3$, we have $\mathrm{Re}\,(\,\phi(\mathbf{x})\,|\,iA^{(j)}(\mathbf{x})\,\phi(\mathbf{x})\,) = 0$, thus

$$u_\varepsilon \, \nabla \, u_\varepsilon \,=\, \frac{1}{2}\,\nabla\, u_\varepsilon^2 \,=\, \mathrm{Re}\,(\,\phi\,|\,\nabla\,\phi\,) \,=\, \mathrm{Re}\,(\,\phi\,|\,(\nabla + i\mathbf{A})\,\phi\,)\,. \tag{3.7}$$

In particular,

$$|\nabla\, u_\varepsilon| \,\leqslant\, \frac{[\![\phi]\!]}{u_\varepsilon}\,[\![(\nabla + i\mathbf{A})\,\phi]\!] \,\leqslant\, [\![(\nabla + i\mathbf{A})\,\phi]\!]\,. \tag{3.8}$$

Taking the divergence of (3.7) we obtain

$$\begin{aligned} |\nabla\, u_\varepsilon|^2 + u_\varepsilon\,\Delta\, u_\varepsilon \,&=\, \mathrm{Re}\,(\,\nabla\phi\,|\,(\nabla + i\mathbf{A})\,\phi\,) + \mathrm{Re}\,(\,\phi\,|\,\nabla\,(\nabla + i\mathbf{A})\,\phi\,) \\ &=\, [\![(\nabla + i\mathbf{A})\,\phi]\!]^2 - \mathrm{Re}\,(\,\phi\,|\,\tau_{\mathbf{A}}\,\phi\,)\,. \end{aligned} \tag{3.9}$$

Combining this identity with (3.8) we arrive at

$$-\Delta\, u_\varepsilon \,\leqslant\, \mathrm{Re}\,(\,u_\varepsilon^{-1}\,\phi\,|\,\tau_{\mathbf{A}}\,\phi\,)\,. \tag{3.10}$$

Now, assume that $\phi \in \mathcal{D}(\tau_{\mathbf{A}})$. Since $\tau_{\mathbf{A}}$ is essentially self-adjoint on $\mathscr{D}$ we find $\phi_n \in \mathscr{D}$, $n \in \mathbb{N}$, such that $\phi_n \to \phi$ and $\tau_{\mathbf{A}}\phi_n \to \tau_{\mathbf{A}}\phi$ in $\mathscr{H}$. On account of (3.10) we have

$$\int_{\mathbb{R}^3} (-\Delta\,\eta)(\mathbf{x})\, u_\varepsilon^{(n)}(\mathbf{x})\, d^3\mathbf{x} \,\leqslant\, \mathrm{Re}\,\langle\,\eta\,(u_\varepsilon^{(n)})^{-1}\,\phi_n\,|\,\tau_{\mathbf{A}}\,\phi_n\,\rangle\,, \tag{3.11}$$

for all Schwartz functions $\eta \in \mathscr{S}(\mathbb{R}^3)$, $\eta \geqslant 0$, where $u_\varepsilon^{(n)} := \sqrt{[\![\phi_n]\!]^2 + \varepsilon^2}$, $n \in \mathbb{N}$. Passing to appropriate subsequences if necessary we may assume that $[\![\phi_n]\!] \to [\![\phi]\!]$ and, hence, $u_\varepsilon^{(n)} \to u_\varepsilon$ almost everywhere. Using that $u_\varepsilon^{-1}, (u_\varepsilon^{(n)})^{-1} \leqslant 1/\varepsilon$, it is easy to see that $\eta\,(u_\varepsilon^{(n)})^{-1}\,\phi_n \to \eta\, u_\varepsilon^{-1}\,\phi$ in $\mathscr{H}$. By virtue of the Riesz-Fischer theorem we further find a square-integrable majorant for the sequence $([\![\phi_n]\!])$. We can thus pass to the limit $n \to \infty$ in (3.11) to get, for all $\eta \in \mathscr{S}(\mathbb{R}^3)$, $\eta \geqslant 0$, and $\phi \in \mathcal{D}(\tau_{\mathbf{A}})$,

$$\int_{\mathbb{R}^3} (-\Delta\eta)(\mathbf{x})\, u_\varepsilon(\mathbf{x})\, d^3\mathbf{x} \leqslant \mathrm{Re} \int_{\mathbb{R}^3} \eta(\mathbf{x})\,(\,u_\varepsilon^{-1}(\mathbf{x})\phi(\mathbf{x})\,|\,(\tau_{\mathbf{A}}\phi)(\mathbf{x})\,)\, d^3\mathbf{x}. \tag{3.12}$$

Here we may take the limit $\varepsilon \to 0$ by means of the dominated convergence theorem (with the majorant $\eta\,[\![\tau_{\mathbf{A}}\phi]\!]$ on the right hand side) to obtain, for all $\eta \in \mathscr{S}(\mathbb{R}^3)$, $\eta \geqslant 0$, and $\phi \in \mathcal{D}(\tau_{\mathbf{A}})$,

$$\int_{\mathbb{R}^3} (-\Delta\,\eta)(\mathbf{x})\,[\![\,\phi\,]\!](\mathbf{x})\, d^3\mathbf{x} \,\leqslant\, \mathrm{Re} \int_{\mathbb{R}^3} \eta(\mathbf{x})\,(\,S_\phi(\mathbf{x})\,|\,(\tau_{\mathbf{A}}\,\phi)(\mathbf{x})\,)\, d^3\mathbf{x}\,. \tag{3.13}$$

Adding $\int \eta\lambda[\![\phi]\!]$ with $\lambda > 0$ to both sides we obtain

$$\int_{\mathbb{R}^3} [(-\Delta+\lambda)\eta](\mathbf{x})\,[\![\,\phi\,]\!](\mathbf{x})\,d^3\mathbf{x} \leqslant \mathrm{Re}\int_{\mathbb{R}^3} \eta(\mathbf{x})\,(\,S_\phi\,|\,(\tau_{\mathbf{A}}+\lambda)\phi\,)(\mathbf{x})\,d^3\mathbf{x}$$
$$\leqslant \int_{\mathbb{R}^3} \eta(\mathbf{x})\,[\![(\tau_{\mathbf{A}}+\lambda)\,\phi]\!](\mathbf{x})\,d^3\mathbf{x}\,.$$

Let $0 \leqslant \chi \in C_0^\infty(\mathbb{R}^3)$ and $\psi \in \mathscr{H}$. Since $(-\Delta+\lambda)^{-1}$, $\lambda > 0$, is positivity preserving, we may then choose $\eta := (-\Delta+\lambda)^{-1}\chi \in \mathscr{S}(\mathbb{R}^3)$ and $\phi := (\tau_{\mathbf{A}}+\lambda)^{-1}\psi \in \mathcal{D}(\tau_{\mathbf{A}})$ and arrive at

$$\int_{\mathbb{R}^3} \chi(\mathbf{x})\,[\![\,(\tau_{\mathbf{A}}+\lambda)^{-1}\psi\,]\!](\mathbf{x})\,d^3\mathbf{x} \leqslant \int_{\mathbb{R}^3} \chi(\mathbf{x})\,[\,(-\Delta+\lambda)^{-1}[\![\,\psi\,]\!]\,](\mathbf{x})\,d^3\mathbf{x}\,.$$

Since $\chi \in C_0^\infty(\mathbb{R}^3)$ is arbitrary we find $[\![\,(\tau_{\mathbf{A}}+\lambda)^{-1}\psi\,]\!] \leqslant (-\Delta+\lambda)^{-1}[\![\,\psi\,]\!]$, almost everywhere on $\mathbb{R}^3$, and by induction (see [47], for the same argument) we get, for all $n \in \mathbb{N}$ and $t > 0$,

$$[\![\,(\tfrac{n}{t})^n\,(\tau_{\mathbf{A}}+\tfrac{n}{t})^{-n}\,\psi\,]\!] \leqslant (\tfrac{n}{t})^n(-\Delta+\tfrac{n}{t})^{-n}[\![\,\psi\,]\!]\,.$$

Both sides converge almost everywhere along some subsequence to $[\![e^{-t\tau_{\mathbf{A}}}\psi]\!]$ and $e^{-t(-\Delta)}[\![\psi]\!]$ respectively, and (3.5) follows. Equation (3.6) follows from (3.5), the spectral calculus, the subordination identity

$$e^{-t\lambda^{1/2}} = \int_0^\infty e^{-s-t^2\lambda/(4s)}\,\frac{ds}{\sqrt{\pi s}}\,, \qquad t,\lambda \geqslant 0\,,$$

and the properties of the Bochner-Lebesgue integral. In the remaining part of the proof we derive (following again [47]) (3.1) and (3.3) at the same time. To this end let $\nu \in \{1/2, 1\}$ and $\phi \in \mathcal{D}(\tau_{\mathbf{A}}^\nu)$.

Since $\langle\,\phi\,|\,e^{-t\tau_{\mathbf{A}}^\nu}\phi\,\rangle \leqslant \int [\![\phi]\!]\,[\![e^{-t\tau_{\mathbf{A}}^\nu}\phi]\!]$ Equations (3.5) and (3.6) imply $\langle\,\phi\,|\,e^{-t\tau_{\mathbf{A}}^\nu}\phi\,\rangle \leqslant \int [\![\phi]\!] e^{-t(-\Delta)^\nu}[\![\phi]\!]$, thus

$$\langle\,\phi\,|\,t^{-1}(1-e^{-t\tau_{\mathbf{A}}^\nu})\phi\,\rangle \geqslant \int_{\mathbb{R}^3} t^{-1}(1-e^{-t|\xi|^{2\nu}})|[\![\phi]\!]\hat{\ }|^2(\xi)\,d^3\xi\,.$$

Since $\phi \in \mathcal{D}(\tau_{\mathbf{A}}^\nu)$ the limit $t \to 0$ exists on the left hand side of the previous inequality. By the monotone convergence theorem we conclude that the limit $t \to 0$ of the right hand side exists, too, and

$$\langle\,\phi\,|\,\tau_{\mathbf{A}}^\nu\phi\,\rangle \geqslant \int_{\mathbb{R}^3} |\xi|^{2\nu}|[\![\phi]\!]\hat{\ }|^2(\xi)\,d^3\xi\,.$$

Hence $[\![\phi]\!] \in H^\nu(\mathbb{R}^3)$ and (3.2) and (3.4) hold true. Using this we may take the derivatives at $t = 0$ on the left and right sides of the following consequence of (3.5) and (3.6),

$$\begin{aligned}\mathrm{Re}\int_{\mathbb{R}^3}\eta(\mathbf{x})(\,S_\phi\,|\,e^{-t\tau_{\mathbf{A}}^\nu}\,\phi\,)(\mathbf{x})\,d^3\mathbf{x} &\leqslant \int_{\mathbb{R}^3}\eta(\mathbf{x})[\![\,e^{-t\tau_{\mathbf{A}}^\nu}\,\phi\,]\!](\mathbf{x})\,d^3\mathbf{x}\\ &\leqslant \int_{\mathbb{R}^3}\eta(\mathbf{x})\,e^{-t(-\Delta)^\nu}\,[\![\,\phi\,]\!](\mathbf{x})\,d^3\mathbf{x}\,,\end{aligned}\tag{3.14}$$

to get $-\mathrm{Re}\,\langle\,\eta S_\phi\,|\,\tau_{\mathbf{A}}^\nu\,\phi\,\rangle \leqslant -\langle\,\eta\,|\,(-\Delta)^\nu[\![\,\phi\,]\!]\,\rangle_{L^2(\mathbb{R}^3)}$, for all $\eta\in H^\nu(\mathbb{R}^3), \eta\geqslant 0$. Here we have also used that all expressions in (3.14) are equal to $\int\eta[\![\,\phi\,]\!]$ at $t=0$. □

3.2. *Semi-boundedness*

The following theorem is a slight generalization of a result from [37]. Its proof is based on two basic steps: The first one follows immediately from the diamagnetic inequalities by means of which the form bounds the potential satisfies by assumption can be turned into form bounds with respect to the *scalar* operators $\tau_{\mathbf{A}}^{1/2}$ or $\tau_{\mathbf{A}}$. After that we use relative bounds on the magnetic field to include spin, $\tau_{\mathbf{A}}\leqslant D_{\mathbf{A}}^2+H_{\mathrm{f}}^2+\text{const}$. To complete the square on the right hand side of the previous bound we employ the inequality (3.20) below. After completing the square we take square roots on both sides to obtain a bound on $\tau_{\mathbf{A}}^{1/2}$.

Theorem 3.4: *For $\nu\in\{1/2,1\}$, let $V_\nu\in L^1_{\mathrm{loc}}(\mathbb{R}^3,\mathbb{R})$ and assume that there is some $c_\nu\geqslant 0$ such that, for every $\varphi\in H^\nu(\mathbb{R}^3)$,*

$$\langle\,\varphi\,|\,V_\nu\,\varphi\,\rangle_{L^2(\mathbb{R}^3)} \leqslant \langle\,\varphi\,|\,(-\Delta)^\nu\,\varphi\,\rangle_{L^2(\mathbb{R}^3)} + c_\nu\,\|\,\varphi\,\|^2_{L^2(\mathbb{R}^3)}. \tag{3.15}$$

Then there is some $C\in(0,\infty)$ such that, for all $\mathbf{G}_{\mathbf{x}}$ fulfilling Hypothesis 2.1, $\phi\in\mathscr{D}$, and $\delta>0$,

$$\big\langle\,\phi\,\big|\,V_\nu\phi\,\big\rangle \leqslant \big\langle\,\phi\,\big|\,\big(|D_{\mathbf{A}}|+\delta\,H_{\mathrm{f}}+(\delta^{-1}+C\,\delta)\,d_1^2\big)^{2\nu}\,\phi\,\big\rangle + c_\nu\,\|\phi\|^2. \tag{3.16}$$

In particular,

$$\frac{1}{4}\,\big\|\,|\mathbf{x}|^{-1}\,\phi\,\big\|^2 \leqslant \big\|\,\big(|D_{\mathbf{A}}|+\delta\,H_{\mathrm{f}}+(\delta^{-1}+C\,\delta)\,d_1^2\big)\,\phi\,\big\|^2\,, \tag{3.17}$$

for all $\phi\in\mathscr{D}$, and

$$\frac{2}{\pi}\frac{1}{|\mathbf{x}|} \leqslant |D_{\mathbf{A}}|+\delta\,H_{\mathrm{f}}+(\delta^{-1}+C\,\delta)\,d_1^2\,, \tag{3.18}$$

in the sense of quadratic forms on $\mathscr{D}$. Therefore, $H^{\mathrm{PF}}_{V_{1/2}}$ and H^{PF}_γ, $\gamma\in[0,\gamma_{\mathrm{c}}^{\mathrm{PF}}]$, have self-adjoint Friedrichs extensions – henceforth again denoted by the same symbols – and $\mathscr{D}$ is a form core for these extensions. Moreover, *for $a \in [0,1)$ and $\gamma \in [0,\gamma_{\mathrm{c}}^{\mathrm{PF}})$, we know that* $\mathscr{Q}(H_{aV_{1/2}}^{\mathrm{PF}}) = \mathscr{Q}(H_{\gamma}^{\mathrm{PF}}) = \mathscr{Q}(H_0^{\mathrm{PF}}) = \mathscr{Q}(|D_{\mathbf{0}}|) \cap \mathscr{Q}(H_{\mathrm{f}})$.

Proof: First, we show that $\mathscr{Q}(H_0^{\mathrm{PF}}) = \mathscr{Q}(|D_{\mathbf{0}}|) \cap \mathscr{Q}(H_{\mathrm{f}})$; see [28, 48]. The remaining statements on form domains will then be a consequence of (3.16) and (3.18). In fact, this follows from the bounds [37]

$$\big\|\,|D_{\mathbf{0}}|^{1/2}(S_{\mathbf{A}} - S_{\mathbf{0}})\,\check{H}_{\mathrm{f}}^{-1/2}\big\| \leqslant C\,, \qquad \big\|\,|D_{\mathbf{A}}|^{1/2}(S_{\mathbf{A}} - S_{\mathbf{0}})\,\check{H}_{\mathrm{f}}^{-1/2}\big\| \leqslant C\,,$$

where $\check{H}_{\mathrm{f}} := H_{\mathrm{f}} + E$ with $E \geqslant 1 + (4d_1)^2$. (These bounds are derived exactly as in Lemma 6.3 below.) Together with $|D_{\mathbf{A}}| - |D_{\mathbf{0}}| = D_{\mathbf{0}}\,(S_{\mathbf{A}} - S_{\mathbf{0}}) + \boldsymbol{\alpha}\cdot\mathbf{A}\,S_{\mathbf{A}}$ and (2.16) and (2.17) the first bound implies

$$\begin{aligned} |\langle\,\varphi\,|\,(|D_{\mathbf{A}}| - |D_{\mathbf{0}}|)\,\varphi\,\rangle| &\leqslant C'\,\big\|\,|D_{\mathbf{0}}|^{1/2}\varphi\big\|\,\big\|\check{H}_{\mathrm{f}}^{1/2}\varphi\big\| \\ &\leqslant C''\langle\,\varphi\,|\,(|D_{\mathbf{0}}| + H_{\mathrm{f}})\,\varphi\,\rangle\,, \end{aligned} \tag{3.19}$$

for all $\varphi \in \mathscr{D}$. Analogously, the second bound implies (3.19) with $|D_{\mathbf{0}}|$ replaced by $|D_{\mathbf{A}}|$ on the right hand side. Consequently, the form norms of $|D_{\mathbf{A}}| + H_{\mathrm{f}}$ and $|D_{\mathbf{0}}| + H_{\mathrm{f}}$ are equivalent on $\mathscr{D}$ which implies $\mathscr{Q}(H_0^{\mathrm{PF}}) = \mathscr{Q}(|D_{\mathbf{0}}|) \cap \mathscr{Q}(H_{\mathrm{f}})$.

All details missing in the proof of (3.16)–(3.18) sketched below can be found in [37]. We set $\check{H}_{\mathrm{f}} := H_{\mathrm{f}} + E$, for $E > 0$. Besides some standard arguments the main ingredient in this proof is the following bound proven in [37] : We find some constant, $C > 0$, such that, for all $E > C\,d_1^2$ and $\phi \in \mathscr{D}$,

$$\mathrm{Re}\,\big\langle\,|D_{\mathbf{A}}|\,\phi\,\big|\,\check{H}_{\mathrm{f}}\,\phi\,\big\rangle \geqslant 0\,. \tag{3.20}$$

This estimate follows from the following identity $\mathrm{Re}\,(|D_{\mathbf{A}}|\,\check{H}_{\mathrm{f}}) = \check{H}_{\mathrm{f}}^{1/2}(|D_{\mathbf{A}}| - \mathcal{T})\check{H}_{\mathrm{f}}^{1/2}$ on $\mathscr{D}$, where $\mathcal{T} := \mathrm{Re}\,\{[|D_{\mathbf{A}}|, \check{H}_{\mathrm{f}}^{-1/2}]\check{H}_{\mathrm{f}}^{1/2}\} \leqslant \varepsilon\,|D_{\mathbf{A}}| + \varepsilon^{-1}\,\mathrm{const}\,d_1/E^{1/2}$, for $\varepsilon \in (0,1]$ and $E \geqslant (4d_1)^2$, as we shall see at the end of Subsection 9.2. To make use of the bound (3.20) we recall that, since $[A^{(j)}(\mathbf{x}), A^{(k)}(\mathbf{y})] = 0$, we have

$$D_{\mathbf{A}}^2\,\phi = \tau_{\mathbf{A}}\,\phi + \mathbf{S}\cdot\mathbf{B}\,\phi + \phi\,, \qquad \phi \in \mathscr{D}\,, \tag{3.21}$$

where the entries of the formal three-vector $\mathbf{S}$ are $S_j = \sigma_j \otimes \mathbb{1}_2$ and $\mathbf{B}$ is the magnetic field, i.e. $\mathbf{S}\cdot\mathbf{B} = \mathbf{S}\cdot a^{\dagger}(\nabla_{\mathbf{x}} \wedge \mathbf{G}) + \mathbf{S}\cdot a(\nabla_{\mathbf{x}} \wedge \mathbf{G})$. By (2.16) with $(\nabla_{\mathbf{x}} \wedge \mathbf{G}, d_1)$ instead of $(\mathbf{G}, d_{-1})$ we have, for all $\delta > 0$ and $\phi \in \mathscr{D}$,

$$\big|\langle\,\phi\,|\,\mathbf{S}\cdot\mathbf{B}\,\phi\,\rangle\big| \leqslant 2\,d_1\,\|\phi\|\,\big\|\,H_{\mathrm{f}}^{1/2}\,\phi\,\big\| \leqslant \delta\,\langle\,\phi\,|\,(H_{\mathrm{f}} + \delta^{-2}\,d_1^2)\,\phi\,\rangle\,. \tag{3.22}$$

Choosing $E = (\delta^{-2} + C)\, d_1^2$ we infer from (3.20)–(3.22), for all $\phi \in \mathscr{D}$,

$$\begin{aligned} \langle\, \phi \,|\, \tau_{\mathbf{A}}\, \phi \,\rangle &\leqslant \langle\, D_{\mathbf{A}}\, \phi \,|\, D_{\mathbf{A}}\, \phi \,\rangle + \delta \,\langle\, \phi \,|\, \check{H}_{\mathrm{f}}\, \phi \,\rangle - \|\phi\|^2 \qquad (3.23) \\ &\leqslant \langle\, D_{\mathbf{A}}\, \phi \,|\, D_{\mathbf{A}}\, \phi \,\rangle + \langle\, \phi \,|\, \delta^2\, \check{H}_{\mathrm{f}}^2\, \phi \,\rangle + 2\mathrm{Re}\, \langle\, |D_{\mathbf{A}}|\, \phi \,|\, \delta\, \check{H}_{\mathrm{f}}\, \phi \,\rangle \\ &= \big\|\, (|D_{\mathbf{A}}| + \delta\, \check{H}_{\mathrm{f}})\, \phi \,\big\|^2 . \end{aligned}$$

Furthermore, since the square root is operator monotone it follows from (3.23) that $\langle\, \phi \,|\, \tau_{\mathbf{A}}^{1/2}\, \phi \,\rangle \leqslant \langle\, \phi \,|\, (|D_{\mathbf{A}}| + \delta\, \check{H}_{\mathrm{f}})\, \phi \,\rangle$. Using the diamagnetic inequalities (3.2) and (3.4) we further find, for $\nu \in \{1/2, 1\}$,

$$\begin{aligned} \langle\, \phi \,|\, V_\nu\, \phi \,\rangle &= \langle\, [\![\phi]\!] \,|\, V_\nu\, [\![\phi]\!] \,\rangle_{L^2(\mathbb{R}^3)} \leqslant \langle\, [\![\phi]\!] \,|\, (-\Delta)^\nu\, [\![\phi]\!] \,\rangle_{L^2(\mathbb{R}^3)} + c_\nu\, \|\phi\|^2 \\ &\leqslant \langle\, \phi \,|\, \tau_{\mathbf{A}}^{2\nu}\, \phi \,\rangle + c_\nu \|\phi\|^2, \qquad (3.24) \end{aligned}$$

and we conclude that (3.16) holds true. Inequalities (3.17) and (3.18) follow from (3.24) together with Hardy's and Kato's inequality, respectively. □

Theorem 3.4 has a straightforward extension to the case of N electrons [36]. We discuss this extension in the next corollary mainly since its proof gives the opportunity to introduce some identities and estimates which are used later on. Let $\mathscr{H}_N$ and $\mathscr{D}_N$, $N \in \mathbb{N}$, be defined in the same way as $\mathscr{H}$ and $\mathscr{D}$ but with the $L^2(\mathbb{R}^3, \mathbb{C}^4) = L^2(\mathbb{R}^3 \times \mathbb{Z}_4)$ replaced by $L^2((\mathbb{R}^3 \times \mathbb{Z}_4)^N)$. The spatial coordinates of the i-th electron are denoted by $\mathbf{x}_i \in \mathbb{R}^3$ and we designate an operator acting only on $\mathbf{x}_i$, the i-th spinor components, and on the photon field by a superscript (i).

Corollary 3.5: *Assume that $N, K \in \mathbb{N}$, $e > 0$, $\gamma_1, \ldots, \gamma_K \in (0, 2/\pi]$, and $\{\mathbf{R}_1, \ldots, \mathbf{R}_K\} \subset \mathbb{R}^3$. Then*

$$\sum_{i=1}^N |D_{\mathbf{A}}^{(i)}| - \sum_{i=1}^N \sum_{k=1}^K \frac{\gamma_k}{|\mathbf{x}_i - \mathbf{R}_k|} + \sum_{i<j} \frac{e^2}{|\mathbf{x}_i - \mathbf{x}_j|} + \delta\, H_{\mathrm{f}} > -\infty\,, \qquad (3.25)$$

for every $\delta > 0$, in the sense of quadratic forms on $\mathscr{D}_N$.

Proof: In view of (3.18) we only have to explain how to localize the non-local kinetic energy terms. To begin with we note the following special cases of (2.24) and (9.10), respectively: For every $\chi \in C^\infty(\mathbb{R}^3_{\mathbf{x}}, [0,1])$,

$$\|\, [\chi, S_{\mathbf{A}}] \,\| \leqslant \|\nabla\chi\|_\infty\,, \qquad \big\|D_{\mathbf{A}}\, \big[\chi\,,\, [\chi, S_{\mathbf{A}}]\big]\big\| \leqslant 2\, \|\nabla\chi\|_\infty^2\,. \qquad (3.26)$$

Let $\mathcal{B}_r(\mathbf{z})$ denote the open ball in $\mathbb{R}^3$ of radius $r > 0$ centered at $\mathbf{z} \in \mathbb{R}^3$. We set $\varrho := \min\{|\mathbf{R}_k - \mathbf{R}_\ell| : k \neq \ell\}/2$ and pick a smooth partition of unity on $\mathbb{R}^3$, $\{\chi_k\}_{k=0}^K$, such that $\chi_k \equiv 1$ on $\mathcal{B}_{\varrho/2}(\mathbf{R}_k)$ and $\mathrm{supp}(\chi_k) \subset \mathcal{B}_\varrho(\mathbf{R}_k)$, for

$k = 1, \ldots, K$, and such that $\sum_{k=0}^{K} \chi_k^2 = 1$. Combining the following IMS type localization formula,

$$|D_{\mathbf{A}}| = \sum_{k=0}^{K} \Big\{ \chi_k \, |D_{\mathbf{A}}| \, \chi_k + \frac{1}{2} \big[\chi_k \, , \, [\chi_k, |D_{\mathbf{A}}|] \big] \Big\} \quad \text{on } \mathscr{D} , \tag{3.27}$$

and the identities

$$\big[\chi_k \, , \, [\chi_k, |D_{\mathbf{A}}|] \big] = 2 \, i \boldsymbol{\alpha} \cdot (\nabla \chi_k) \, [\chi_k, S_{\mathbf{A}}] + D_{\mathbf{A}} \big[\chi_k \, , \, [\chi_k, S_{\mathbf{A}}] \big] \tag{3.28}$$

and $\| \boldsymbol{\alpha} \cdot \nabla \chi_k \| = |\nabla \chi_k|$ with (3.26), we obtain

$$\big\| \big[\chi_k \, , \, [\chi_k, |D_{\mathbf{A}}|] \big] \big\| \leqslant 4 \, \| \nabla \chi_k \|_\infty^2 \, , \tag{3.29}$$

for all $k \in \{0, \ldots, K\}$. Since we are able to localize the kinetic energy terms and since, by the choice of the partition of unity, the functions $\mathbb{R}^3 \ni \mathbf{x} \mapsto |\mathbf{x} - \mathbf{R}_k|^{-1} \, \chi_\ell^2(\mathbf{x})$ are bounded, for $k \in \{1, \ldots, K\}$, $\ell \in \{0, \ldots, K\}$, $k \neq \ell$, the bound (3.25) is now an immediate consequence of (3.18). □

Next, we discuss the semi-boundedness of the no-pair operator. The idea is to reduce the stability of the no-pair operator to the one of the purely electronic Brown-Ravenhall operator.

Theorem 3.6: *Assume that $\mathbf{G}_{\mathbf{x}}$ fulfills Hypothesis 2.1. Let $\delta \in (0,1]$ and let $V \in L^2_{\mathrm{loc}}(\mathbb{R}^3, \mathbb{R})$ be form bounded with respect to $\sqrt{-\Delta}$ and satisfy*

$$\langle \, P_{\mathbf{0}}^+ \, \varphi \, | \, V \, P_{\mathbf{0}}^+ \, \varphi \, \rangle \leqslant a \, \langle \, \varphi \, | \, D_{\mathbf{0}} \, P_{\mathbf{0}}^+ \, \varphi \, \rangle + b \, \| P_{\mathbf{0}}^+ \, \varphi \|^2, \quad \varphi \in C_0^\infty(\mathbb{R}^3) \, , \tag{3.30}$$

for some $a \in (0,1)$ and $b \geqslant 0$. Then there exist constants $c_V, C \in (0, \infty)$, $C \equiv C(\delta, V, d_{-1}, d_1)$, such that, for all $\varphi^+ \in P_{\mathbf{A}}^+ \, \mathscr{D}$, $\| \varphi^+ \| = 1$,

$$\langle \, \varphi^+ \, | \, (D_{\mathbf{A}} + V + \delta H_{\mathrm{f}}) \, \varphi^+ \, \rangle \geqslant c_V \langle \, \varphi^+ \, | \, |D_{\mathbf{0}}| \, \varphi^+ \, \rangle - C \, , \tag{3.31}$$

and in particular, for every $\gamma \in [0, \gamma_{\mathrm{c}}^{\mathrm{np}})$,

$$\langle \, \varphi^+ \, | \, (D_{\mathbf{A}} - \tfrac{\gamma}{|\mathbf{x}|} + \delta H_{\mathrm{f}}) \, \varphi^+ \, \rangle \geqslant c_\gamma \langle \, \varphi^+ \, | \, |D_{\mathbf{0}}| \, \varphi^+ \, \rangle - C(\delta, \gamma, d_{-1}, d_1) \, . \tag{3.32}$$

Therefore, H_V^{np} and H_γ^{np} have self-adjoint Friedrichs extensions – henceforth again denoted by the same symbols – and $P_{\mathbf{A}}^+ \, \mathscr{D}$ is a form core for these extensions. Furthermore, $\mathcal{Q}(H_V^{\mathrm{np}}) = \mathcal{Q}(H_\gamma^{\mathrm{np}}) = \mathcal{Q}(H_0^{\mathrm{np}}) = \mathcal{Q}(|D_{\mathbf{0}}|) \cap \mathcal{Q}(H_{\mathrm{f}}) \cap \mathrm{Ran} P_{\mathbf{A}}^+$.

Proof: The statement on the form domains follows from Theorem 3.2 of [29]. (See also Section 3.4 of the *first* preprint version of [29] available on the arXiv for an alternative proof.) The estimate (3.32) is derived in [37] and we shall outline its proof in what follows.

We pick some $\rho > 1$ with $\rho a < 1$ and write, for $\varphi^+ \in P^+_{\mathbf{A}} \mathscr{D}$,

$$\begin{aligned}\langle \varphi^+ \,|\, (D_{\mathbf{A}} + V)\,\varphi^+ \rangle = \rho^{-1}&\langle \varphi^+ \,|\, P^+_{\mathbf{0}}\,(D_{\mathbf{0}} + \rho V)\, P^+_{\mathbf{0}}\,\varphi^+ \rangle \\ &+ (1-\rho^{-1})\,\langle \varphi^+ \,|\, P^+_{\mathbf{0}}\, D_{\mathbf{0}}\,\varphi^+ \rangle \\ &+ \langle \varphi^+ \,|\, \boldsymbol{\alpha}\cdot\mathbf{A}\,\varphi^+ \rangle \\ &+ \langle \varphi^+ \,|\, P^-_{\mathbf{0}}\,(D_{\mathbf{0}} + V)\,P^-_{\mathbf{0}}\,\varphi^+ \rangle \\ &+ 2\,\mathrm{Re}\,\langle \varphi^+ \,|\, P^+_{\mathbf{0}}\, V\, P^-_{\mathbf{0}}\,\varphi^+ \rangle\,.\end{aligned} \tag{3.33}$$

The estimate (3.32) is based on the identity above and the following bound on the difference between the spectral projections with and without field (see Lemma 6.3 for similar statements): For $E \geqslant 1 + (4d_1)^2$, there is a constant, $C \equiv C(d_{-1}, d_0) > 0$, such that

$$\big\|\, |D_{\mathbf{0}}|^{3/4} (P^\pm_{\mathbf{0}} - P^\pm_{\mathbf{A}}) \check{H}_{\mathrm{f}}^{-1/2} \big\| \leqslant C\,, \tag{3.34}$$

where $\check{H}_{\mathrm{f}} = H_{\mathrm{f}} + E$. On account of $P^-_{\mathbf{0}}\,\varphi^+ = (P^-_{\mathbf{0}} - P^-_{\mathbf{A}})\,\varphi^+$, for $\varphi^+ \in P^+_{\mathbf{A}} \mathscr{D}$, and (3.34) we have, for every $\varepsilon > 0$,

$$\begin{aligned}\big\|\, |D_{\mathbf{0}}|^{1/2}\, P^-_{\mathbf{0}}\,\varphi^+ \big\|^2 &\leqslant \big\|\, |D_{\mathbf{0}}|^{1/4}\, P^-_{\mathbf{0}}\,\varphi^+ \big\|\, \big\|\, |D_{\mathbf{0}}|^{3/4}\,(P^-_{\mathbf{0}} - P^-_{\mathbf{A}})\,\varphi^+ \big\| \\ &\leqslant \frac{1}{2}\big\|\, |D_{\mathbf{0}}|^{1/2}\, P^-_{\mathbf{0}}\,\varphi^+ \big\|^2 + \frac{C(\varepsilon, d_{-1}, d_0)}{2}\big\| P^-_{\mathbf{0}}\,\varphi^+ \big\|^2 + \frac{\varepsilon}{2}\big\| \check{H}_{\mathrm{f}}^{1/2}\,\varphi^+ \big\|^2,\end{aligned}$$

that is,

$$\big\|\, |D_{\mathbf{0}}|^{1/2}\, P^-_{\mathbf{0}}\,\varphi^+ \big\|^2 \leqslant \varepsilon \big\| \check{H}_{\mathrm{f}}^{1/2}\,\varphi^+ \big\|^2 + C(\varepsilon, d_{-1}, d_0)\,\big\| P^-_{\mathbf{0}}\,\varphi^+ \big\|^2. \tag{3.35}$$

By virtue of $|V| \leqslant C\,|D_{\mathbf{0}}|$, the previous estimate further implies, for every $\tau > 0$,

$$\begin{aligned}&\big|\langle P^+_{\mathbf{0}}\,\varphi^+ \,|\, V\, P^-_{\mathbf{0}}\,\varphi^+ \rangle\big| \\ &\leqslant \tau\, \big\|\, |D_{\mathbf{0}}|^{1/2}\, P^+_{\mathbf{0}}\,\varphi^+ \big\|^2 + \varepsilon\, \big\| \check{H}_{\mathrm{f}}^{1/2}\,\varphi^+ \big\|^2 + C_{\varepsilon,\tau}\, \big\| P^-_{\mathbf{0}}\,\varphi^+ \big\|^2.\end{aligned} \tag{3.36}$$

Here the second term on the RHS of (3.33) can be used to control the first term on the RHS of (3.36). Recalling the definition (2.13) and applying (2.16), (3.35), (3.36), and (3.30) to the various terms in (3.33) we thus find, for every $\delta \in (0,1]$, some constant, $C \equiv C(\delta, \rho, d_{-1}, d_1) \in (0, \infty)$, such that

$$\langle \varphi^+ \,|\, (D_{\mathbf{A}} + V + \delta H_{\mathrm{f}})\,\varphi^+ \rangle \geqslant c_{a,\rho} \langle \varphi^+ \,|\, D_{\mathbf{0}}\, P^+_{\mathbf{0}}\,\varphi^+ \rangle - C\,\|\varphi^+\|^2\,.$$

Using (3.35) once more to replace $D_{\mathbf{0}}\, P^+_{\mathbf{0}}$ by $|D_{\mathbf{0}}|$ on the right hand side, we arrive at the first asserted estimate. According to the remarks made below (2.13) the first estimate applies in particular to the Coulomb potential, as long as $\gamma \in [0, \gamma_{\mathrm{c}}^{\mathrm{np}})$. □

From the previous theorem and our commutator estimates one can also infer the semi-boundedness of a no-pair operator for $N \in \mathbb{N}$ electrons and $K \in \mathbb{N}$ static nuclei, analogously to Corollary 3.5, as long as all Coulomb coupling constants $\gamma_1, \ldots, \gamma_K$ are less than γ_c^{np}; see Proposition A.2 of [36].

Since we are addressing the question of finding distinguished self-adjoint realizations of $H_\gamma^\sharp$ it is also natural to state the following theorem whose proof can be found in Corollary 3.4 of [29].

Theorem 3.7: *Let $\gamma \in [0, 1/2)$ and assume that $\mathbf{G_x}$ fulfills Hypothesis 2.1. Then H_γ^{PF} and H_γ^{np} are essentially self-adjoint on $\mathscr{D}$ and $P_{\mathbf{A}}^+ \mathscr{D}$, respectively.*

For sufficiently small values of $|e|$ and/or Λ, the essential self-adjointness of H_0^{PF} has been shown earlier in [40].

4. Bounds on the Ionization Energy

As a first step towards the proof of the existence of ground states we need to show that binding occurs in the atomic system defined by $H_V^\sharp$ in the sense that $\inf \sigma[H_V^\sharp] < \inf \sigma[H_0^\sharp]$. This information will be exploited mathematically when we apply a bound on the spatial localization of low-lying spectral subspaces of $H_V^\sharp$ from [37]. The localization estimate in turn enters into the proof of the existence of ground states at various places, for instance, into the derivation of the infra-red estimates and into the compactness argument given in Subsection 8.1. Theorem 4.1 below is the main result of this section. In its statement we abbreviate ($\sharp \in \{\mathrm{PF}, \mathrm{np}\}$)

$$E_V^\sharp := \inf \sigma[H_V^\sharp], \quad E_\gamma^\sharp := \inf \sigma[H_\gamma^\sharp], \ \gamma \in (0, \gamma_c^\sharp), \quad \Sigma^\sharp := \inf \sigma[H_0^\sharp],$$

where V satisfies the conditions under which $H_V^\sharp$ has been defined in the previous section. To simplify the exposition we only consider the physical choice of the coupling function $\mathbf{G_x}$ given in (2.2), as always for arbitrary values of e and Λ. Our proofs work, however, equally well for other coupling functions, for instance, for the infra-red cut-off and discretized coupling functions introduced in Section 6, and we obtain uniform bounds on the binding energies in these cases. If we consider coupling functions other than (2.2) then the unitary transformation U introduced below has to be changed accordingly; see [28, 29].

Theorem 4.1: *(i) Let $V \in L^2_{\mathrm{loc}}(\mathbb{R}^3, \mathbb{R})$ be form bounded with respect to $\sqrt{-\Delta}$ with form bound less than or equal to one. (So V fulfills* (3.15) *with $\nu = 1/2$.) Define the self-adjoint operator $h_V := \sqrt{1 - \Delta} + V$ by means of* *a Friedrichs extension starting from* $C_0^\infty(\mathbb{R}^3)$ *and assume that* $\inf \sigma[h_V]$ *is an eigenvalue. Then*

$$\Sigma^{\mathrm{PF}} - E_V^{\mathrm{PF}} \geqslant 1 - \inf \sigma[h_V]\,. \tag{4.1}$$

In particular, for $\gamma \in [0, \gamma_c^{\mathrm{PF}}]$,

$$\Sigma^{\mathrm{PF}} - E_\gamma^{\mathrm{PF}} \geqslant 1 - \inf \sigma[\sqrt{1-\Delta} - \tfrac{\gamma}{|\mathbf{x}|}]\,. \tag{4.2}$$

(ii) Let $V \leqslant 0$ *be relatively form bounded with respect to* $\sqrt{-\Delta}$ *and assume that* V *satisfies* (3.30) *with* $a \in (0,1)$. *Additionally, assume there exist* $r \geqslant 1$, $c > 0$, *and* $\theta \in (0,2)$ *such that*

$$V(\mathbf{x}) \leqslant -c\,|\mathbf{x}|^{\theta-2}, \qquad |\mathbf{x}| \geqslant r\,.$$

Then

$$\Sigma^{\mathrm{np}} - E_V^{\mathrm{np}} > 0\,, \tag{4.3}$$

and in particular, for $\gamma \in (0, \gamma_c^{\mathrm{np}})$,

$$\Sigma^{\mathrm{np}} - E_\gamma^{\mathrm{np}} > 0\,.$$

Remark 4.2: (i) The bound (4.1) has been obtained first in [21] (under the assumption that H_V^{PF} be essentially self-adjoint which, in the case $V = -\frac{\gamma}{|\mathbf{x}|}$, is true, at least for all $\gamma < 1/2$). The result of [21] improved a lower bound on the binding energy in an earlier preprint version of [28]. The latter was given in terms of the *non-relativistic* ground state energy of an electronic Schrödinger operator.

(ii) In a recent paper by the first two authors [27] is it shown that the inequalities (4.1) and (4.2) are actually *strict*, for all $e, \Lambda > 0$. Moreover, there is a certain class of short-range potentials V such that $\Sigma^{\mathrm{PF}} - E_V^{\mathrm{PF}} > 0$ and in particular – according to the present article – E_V^{PF} is an eigenvalue of H_V^{PF} although $\inf \sigma[h_V] = 1$ and 1 is not an eigenvalue of h_V. This effect is called *enhanced binding* due to the quantized radiation field and we are again able to prove its occurrence, for arbitrary large values of e and Λ. There are numerous results on enhanced binding in non-relativistic QED; up to now complete proofs were, however, available only for small e. The proofs in [27] extend the ideas and methods underlying the proof of Theorem 4.1 given below.

Because of lack of space we shall only describe the proof of Theorem 4.1 for the semi-relativistic Pauli-Fierz operator [28] in detail. The proof of (4.3) follows similar lines and can be found in [29]; see also Remark 4.3 below.

Our proof of (4.1) and (4.3) is based on a direct fiber decomposition of $\mathscr{H}$ with respect to fixed values of the total momentum $\mathbf{p}\otimes\mathbb{1}+\mathbb{1}\otimes\mathbf{p}_{\mathrm{f}}$, where $\mathbf{p}:=-i\nabla_{\mathbf{x}}$ and

$$\mathbf{p}_{\mathrm{f}} := d\Gamma(\mathbf{k}) := \big(d\Gamma(k^{(1)})\,,\,d\Gamma(k^{(2)})\,,\,d\Gamma(k^{(3)})\big) \tag{4.4}$$

is the photon momentum operator. In fact, a conjugation of the Dirac operator with the unitary operator $e^{i\mathbf{p}_{\mathrm{f}}\cdot\mathbf{x}}$ – which is simply a multiplication with the phase $(\mathbb{R}^3)^n \ni (\mathbf{k}_1,\dots,\mathbf{k}_n)\mapsto e^{i(\mathbf{k}_1+\dots+\mathbf{k}_n)\cdot\mathbf{x}}$ in each Fock space sector $\mathscr{F}_{\mathrm{b}}^{(n)}[\mathscr{K}]$ – yields

$$e^{i\mathbf{p}_{\mathrm{f}}\cdot\mathbf{x}}\,D_{\mathbf{A}}\,e^{-i\mathbf{p}_{\mathrm{f}}\cdot\mathbf{x}} \;=\; \boldsymbol{\alpha}\cdot(\mathbf{p}-\mathbf{p}_{\mathrm{f}}+\mathbf{A}(\mathbf{0}))+\beta\,,$$

and a further conjugation with the Fourier transform, $\mathcal{F}:L^2(\mathbb{R}^3_{\mathbf{x}})\to L^2(\mathbb{R}^3_{\mathbf{P}})$, turns the latter expressions into

$$(\mathcal{F}\otimes\mathbb{1})\,e^{i\mathbf{p}_{\mathrm{f}}\cdot\mathbf{x}}\,D_{\mathbf{A}}\,e^{-i\mathbf{p}_{\mathrm{f}}\cdot\mathbf{x}}\,(\mathcal{F}^{-1}\otimes\mathbb{1}) \;=\; \int_{\mathbb{R}^3}^{\oplus}\widehat{D}(\mathbf{P})\,d^3\mathbf{P}\,. \tag{4.5}$$

Here the operators

$$\widehat{D}(\mathbf{P}) := \boldsymbol{\alpha}\cdot(\mathbf{P}-\mathbf{p}_{\mathrm{f}}+\mathbf{A}(\mathbf{0}))+\beta\,,\qquad \mathbf{P}\in\mathbb{R}^3\,,$$

acting in $\mathbb{C}^4\otimes\mathscr{F}_{\mathrm{b}}[\mathscr{K}]$, are fiber Hamiltonians of the transformed Dirac operator in (4.5) with respect to the isomorphism

$$\mathscr{H} \;=\; L^2(\mathbb{R}^3_{\mathbf{P}},\mathbb{C}^4)\otimes\mathscr{F}_{\mathrm{b}}[\mathscr{K}] \;\cong\; \int_{\mathbb{R}^3}^{\oplus}\mathbb{C}^4\otimes\mathscr{F}_{\mathrm{b}}[\mathscr{K}]\,d^3\mathbf{P}\,. \tag{4.6}$$

(In particular, the transformed Dirac operator in (4.5) again acts in $\mathscr{H}$, where the variable in the first tensor factor, $\mathbf{P}$, is now interpreted as the total momentum of the combined electron-photon system.) Accordingly, we have the direct integral representation (compare, e.g., Theorem XIII.85 in [44])

$$(\mathcal{F}\otimes\mathbb{1})\,e^{i\mathbf{p}_{\mathrm{f}}\cdot\mathbf{x}}\,H_0^{\mathrm{PF}}\,e^{-i\mathbf{p}_{\mathrm{f}}\cdot\mathbf{x}}\,(\mathcal{F}^{-1}\otimes\mathbb{1}) \;=\; \int_{\mathbb{R}^3}^{\oplus}H_0^{\mathrm{PF}}(\mathbf{P})\,d^3\mathbf{P}\,, \tag{4.7}$$

where

$$H_0^{\mathrm{PF}}(\mathbf{P}) := |\widehat{D}(\mathbf{P})|+H_{\mathrm{f}}\,.$$

Let $\varepsilon>0$. Then we know that the Lebesgue measure of the set of all $\mathbf{P}\in\mathbb{R}^3$ satisfying $\sigma[H_0^{\mathrm{PF}}(\mathbf{P})]\cap(\Sigma^{\mathrm{PF}}-\varepsilon,\Sigma^{\mathrm{PF}}+\varepsilon)\neq\varnothing$ is strictly positive. In particular, we find some $\mathbf{P}_\star\in\mathbb{R}^3$ and some normalized $\varphi_\star\in\mathcal{Q}(H_0^{\mathrm{PF}}(\mathbf{P}_\star))$ such that

$$\big\langle\,\varphi_\star\,\big|\,H_0^{\mathrm{PF}}(\mathbf{P}_\star)\,\varphi_\star\,\big\rangle_{\mathbb{C}^4\otimes\mathscr{F}_{\mathrm{b}}[\mathscr{K}]} \;<\; \Sigma^{\mathrm{PF}}+\varepsilon\,. \tag{4.8}$$

We define the unitary transformation

$$U := e^{i(\mathbf{p}_\mathrm{f}-\mathbf{P}_\star)\cdot\mathbf{x}} \tag{4.9}$$

and observe as above that

$$U\,D_\mathbf{A}\,U^* = \widehat{D}_\mathbf{p}(\mathbf{P}_\star) := \boldsymbol{\alpha}\cdot(\mathbf{p}+\mathbf{P}_\star-\mathbf{p}_\mathrm{f}+\mathbf{A}(\mathbf{0}))+\beta\,.$$

It is sufficient to prove the bound (4.1) for the unitarily equivalent operator

$$UH_V^{\mathrm{PF}}U^* = |\widehat{D}_\mathbf{p}(\mathbf{P}_\star)|+V+H_\mathrm{f}\,. \tag{4.10}$$

Proof of Theorem 4.1: (The Semi-Relativistic Pauli-Fierz Case) Let $\varepsilon>0$ and $\mathbf{P}_\star$ be as in the preceding paragraphs. We abbreviate $\mathbf{t}_\star := \mathbf{P}_\star-\mathbf{p}_\mathrm{f}+\mathbf{A}(\mathbf{0})$ and, for $\eta\geqslant 0$,

$$R_1(\eta) := \big(\mathbf{p}^2+(\boldsymbol{\alpha}\cdot\mathbf{t}_\star)^2+\eta+1\big)^{-1},\quad R_2(\eta) := \big((\boldsymbol{\alpha}\cdot(\mathbf{p}+\mathbf{t}_\star))^2+\eta+1\big)^{-1}.$$

Since the anti-commutator of $\boldsymbol{\alpha}\cdot\mathbf{p}$ and $\boldsymbol{\alpha}\cdot\mathbf{t}_\star$ is equal to $2\,\mathbf{p}\cdot\mathbf{t}_\star$ it holds $(\boldsymbol{\alpha}\cdot(\mathbf{p}+\mathbf{t}_\star))^2=(\boldsymbol{\alpha}\cdot\mathbf{t}_\star)^2+2\,\mathbf{p}\cdot\mathbf{t}_\star+\mathbf{p}^2$. We deduce that, for any $\varphi\in\mathscr{D}$,

$$\begin{aligned}-R_2(\eta)\varphi &= -R_2(\eta)\,[\,\mathbf{p}^2+(\boldsymbol{\alpha}\cdot\mathbf{t}_\star)^2+1+\eta\,]\,R_1(\eta)\varphi\\ &= R_2(\eta)\,[2\,\mathbf{p}\cdot\mathbf{t}_\star]\,R_1(\eta)\varphi-R_1(\eta)\varphi\,.\end{aligned} \tag{4.11}$$

We use the following formula, for a self-adjoint operator $T>0$,

$$\sqrt{T}\,\varphi = \int_0^\infty\Big(1-\frac{\eta}{T+\eta}\Big)\,\varphi\,\frac{d\eta}{\pi\sqrt{\eta}}\,,\qquad \varphi\in\mathcal{D}(T)\,, \tag{4.12}$$

and the resolvent identity (4.11) to obtain, for any $\varphi\in\mathscr{D}$,

$$\begin{aligned}&\langle\,\varphi\,|\,\big(\sqrt{(\boldsymbol{\alpha}\cdot(\mathbf{p}+\mathbf{t}_\star))^2+1}-\sqrt{\mathbf{p}^2+(\boldsymbol{\alpha}\cdot\mathbf{t}_\star)^2+1}\,\big)\,\varphi\,\rangle\\ &= \int_0^\infty\langle\,\varphi\,|\,(R_1(\eta)-R_2(\eta))\,\varphi\,\rangle\sqrt{\eta}\,\frac{d\eta}{\pi}\\ &= \int_0^\infty\langle\,R_2(\eta)\,\varphi\,|\,[2\,\mathbf{p}\cdot\mathbf{t}_\star]\,R_1(\eta)\,\varphi\,\rangle\sqrt{\eta}\,\frac{d\eta}{\pi}\\ &= \int_0^\infty\langle\,\varphi\,|\,R_1(\eta)\,[2\,\mathbf{p}\cdot\mathbf{t}_\star]\,R_1(\eta)\,\varphi\,\rangle\sqrt{\eta}\,\frac{d\eta}{\pi}\\ &\quad-\int_0^\infty\langle\,\varphi\,|\,R_1(\eta)\,[2\,\mathbf{p}\cdot\mathbf{t}_\star]\,R_2(\eta)\,[2\,\mathbf{p}\cdot\mathbf{t}_\star]\,R_1(\eta)\,\varphi\,\rangle\sqrt{\eta}\,\frac{d\eta}{\pi}\\ &\leqslant \int_0^\infty\langle\,\varphi\,|\,R_1(\eta)\,[2\,\mathbf{p}\cdot\mathbf{t}_\star]\,R_1(\eta)\,\varphi\,\rangle\sqrt{\eta}\,\frac{d\eta}{\pi}\,.\end{aligned} \tag{4.13}$$

In the last step we used the positivity of $R_2(\eta)$. We consider now $\varphi := \varphi_1 \otimes \varphi_2$ where $\varphi_1 \in C_0^\infty(\mathbb{R}^3, \mathbb{R})$ and $\varphi_2 \in \mathbb{C}^4 \otimes \mathscr{C}_0$ with $\|\varphi_j\| = 1$, $j = 1, 2$. Writing $\Phi_2(\boldsymbol{\xi}, \eta) := (\boldsymbol{\xi}^2 + (\boldsymbol{\alpha} \cdot \mathbf{t}_\star)^2 + 1 + \eta)^{-1} \varphi_2 = \Phi_2(-\boldsymbol{\xi}, \eta)$ we find that

$$\begin{aligned} &\langle \varphi \,|\, R_1(\eta)\, \mathbf{p} \cdot \mathbf{t}_\star\, R_1(\eta)\, \varphi \rangle \\ &= \int_{\mathbb{R}^3} \boldsymbol{\xi} \cdot \langle \Phi_2(\boldsymbol{\xi}, \eta) \,|\, \mathbf{t}_\star\, \Phi_2(\boldsymbol{\xi}, \eta) \rangle\, |\widehat{\varphi}_1(\boldsymbol{\xi})|^2 d^3\boldsymbol{\xi} = 0\,, \end{aligned} \tag{4.14}$$

due to the fact that φ_1 is real and, hence, $|\widehat{\varphi}_1(\boldsymbol{\xi})| = |\widehat{\varphi}_1(-\boldsymbol{\xi})|$. Furthermore,

$$|\widehat{D}_{\mathbf{p}}(\mathbf{P}_\star)| = \sqrt{(\boldsymbol{\alpha} \cdot (\mathbf{p} + \mathbf{t}_\star))^2 + 1}\,, \quad |\widehat{D}(\mathbf{P}_\star)| = \sqrt{(\boldsymbol{\alpha} \cdot \mathbf{t}_\star)^2 + 1}\,, \tag{4.15}$$

$$\sqrt{\mathbf{p}^2 + (\boldsymbol{\alpha} \cdot \mathbf{t}_\star)^2 + 1} \leqslant \sqrt{(\boldsymbol{\alpha} \cdot \mathbf{t}_\star)^2 + 1} + \sqrt{\mathbf{p}^2 + 1} - 1\,, \tag{4.16}$$

where we used $[\mathbf{p}^2, (\boldsymbol{\alpha}\cdot\mathbf{t}_\star)^2] = 0$ in the second line. Combining (4.13)–(4.16) we arrive at

$$\langle \varphi \,|\, U H_V^{\mathrm{PF}} U^* \varphi \rangle \leqslant \langle \varphi_2 \,|\, H_0^{\mathrm{PF}}(\mathbf{P}_\star)\, \varphi_2 \rangle + \langle \varphi_1 \,|\, h_V\, \varphi_1 \rangle - 1\,.$$

By a limiting argument the previous inequality extends to any real-valued $\varphi_1 \in \mathcal{Q}(h_V)$. We choose φ_1 to be he normalized, strictly positive eigenfunction of h_V corresponding to the eigenvalue at the bottom of its spectrum and $\varphi_2 = \varphi_\star$. By the choice of $\varphi_\star$ in (4.8), where $\varepsilon > 0$ is arbitrary, this proves the assertion.

Remark 4.3: As already mentioned the proof of Theorem 4.1 for the no-pair operator employs ideas similar to those described above. However, due to the more complex structure of the no-pair Hamiltonian the resulting bound (4.3) is not as satisfactory as the one for H_V^{PF}. Again, we have the representation $H_0^{\mathrm{np}} \cong \int_{\mathbb{R}^3}^{\oplus} H_0^{\mathrm{np}}(\mathbf{P})\, d^3\mathbf{P}$ with

$$H_0^{\mathrm{np}}(\mathbf{P}) = \widehat{P}(\mathbf{P})\, \big(\widehat{D}(\mathbf{P}) + H_{\mathrm{f}}\big)\, \widehat{P}(\mathbf{P})\,, \qquad \widehat{P}(\mathbf{P}) := \mathbb{1}_{[0,\infty)}(\widehat{D}(\mathbf{P}))\,.$$

Given $\varepsilon > 0$, we again find $\mathbf{P}_\star \in \mathbb{R}^3$ and some normalized $\varphi_\star = \widehat{P}(\mathbf{P})\, \varphi_\star \in \mathcal{Q}(H^{\mathrm{np}}(\mathbf{P}_\star))$ such that $\langle \varphi_\star \,|\, H_0^{\mathrm{np}}(\mathbf{P}_\star)\, \varphi_\star \rangle_{\mathbb{C}^4 \otimes \mathscr{F}_{\mathrm{b}}[\mathscr{K}]} < \Sigma^{\mathrm{np}} + \varepsilon$. Conjugating H_γ^{np} with U defined as in (4.9) we obtain

$$U\, H_\gamma^{\mathrm{np}}\, U^* = \widehat{P}_{\mathbf{p}}(\mathbf{P}_\star)\, \big(\widehat{D}_{\mathbf{p}}(\mathbf{P}_\star) + V H_{\mathrm{f}}\big)\, \widehat{P}_{\mathbf{p}}(\mathbf{P}_\star)\,, \tag{4.17}$$

where $\widehat{P}_{\mathbf{p}}(\mathbf{P}_\star) := \mathbb{1}_{[0,\infty)}(\widehat{D}_{\mathbf{p}}(\mathbf{P}_\star))$. As a test function we now choose $\Phi_R := \widehat{P}_{\mathbf{p}}(\mathbf{P}_\star)(\chi_R \otimes \varphi_\star)$, where χ_R is some normalized smooth and non-negative function supported in $\{R \leqslant |\mathbf{x}| \leqslant 2R\}$ with $R \geqslant 1$. It turns out that $\|\Phi_R\| = 1 + \mathcal{O}(1/R^2)$. We then exploit the fact that the negative contribution of the potential in (4.17) to the expectation value of Φ_R decays as $1/R^{2-\vartheta}$ whereas all other terms yield a contribution $\Sigma^{\mathrm{np}} + \varepsilon + \mathcal{O}(1/R^2)$, as R tends to infinity.

To show all this it is convenient to work on a non-projected Hilbert space by adding a suitable "positronic" no-pair operator to $H_\gamma^{\rm np}$ (similarly as in (5.1) below). For then one can again make use of the bounds (4.13)–(4.16) derived in the previous proof; see Section V of [29]. (It is entirely obvious that the Coulomb potential can be replaced by a more general one in Section V of [29].)

5. Exponential Localization

Our next aim is to discuss the exponential localization with respect to the electron coordinates of low-lying spectral subspaces of the no-pair and semi-relativistic Pauli-Fierz operators. Since the multiplication with some exponential weight function, e^F, acting on the electron coordinates does not map the projected Hilbert space $\mathscr{H}_{\mathbf{A}}^+$ into itself it is convenient to extend the no-pair operator to some continuously invertible operator on the whole Hilbert space $\mathscr{H}$ in the discussion below. Therefore, we set

$$\widehat{H}_V^{\rm np} := H_V^{\rm np} + H_0^{{\rm np},-}\,, \qquad H_0^{{\rm np},-} := P_{\mathbf{A}}^-\,(|D_{\mathbf{A}}| + H_{\rm f})\,P_{\mathbf{A}}^-\,. \tag{5.1}$$

In Lemma 8.3 below we show that $\Sigma^{\rm np} = \inf\sigma[H_0^{{\rm np},-}]$ by constructing some anti-linear map $\tau : \mathscr{H} \to \mathscr{H}$ with $H_0^{\rm np}\,\tau = \tau\,H_0^{{\rm np},-}$. Therefore,

$$\widehat{H}_0^{\rm np} \geqslant \Sigma^{\rm np}\,. \tag{5.2}$$

To unify the notation we further set $\widehat{H}_V^{\rm PF} := H_V^{\rm PF}$ and write $\widehat{H}_V^\sharp$, $\Sigma^\sharp$, etc., when we treat both $H_V^{\rm PF}$ and $\widehat{H}_V^{\rm np}$ at the same time. In the whole section we assume that $\mathbf{G}_{\mathbf{x}}$ fulfills Hypothesis 2.1.

Theorem 5.1: *(i) Let $V \in L^2_{\rm loc}(\mathbb{R}^3, \mathbb{R})$ be relatively form bounded with respect to $\sqrt{-\Delta}$ with relative form bound less than or equal to one. Moreover, assume that*

$$\exists\, r \geqslant 1 : \quad \sup_{|\mathbf{x}|\geqslant r/4} |V(\mathbf{x})| < \infty \quad \text{and} \quad V(\mathbf{x}) \xrightarrow{|\mathbf{x}|\to\infty} 0\,. \tag{5.3}$$

Define

$$\rho(a) := 1 - (1-a^2)^{1/2}, \quad a \in [0,1)\,,$$

and let $I \subset (-\infty, \Sigma^{\rm PF})$ be some compact interval. Then there exists $k \in (0,\infty)$, such that, for all $a \in (0,1)$ satisfying $\varepsilon := \Sigma^{\rm PF} - \sup I - \rho(a) \in (0,1]$, we have

$$\big\|\, e^{a|\mathbf{x}|}\, \mathbb{1}_I(H_V^{\rm PF})\,\big\| \leqslant k\,(\Sigma^{\rm PF} - E_V^{\rm PF})\,e^{k/\varepsilon}. \tag{5.4}$$

(ii) Assume that $V \in L^2_{\mathrm{loc}}(\mathbb{R}^3, \mathbb{R})$ *satisfies* $H^1(\mathbb{R}^3) \subset \mathcal{D}(V)$ *(which implies* $|V| \leqslant \mathrm{const}\,|D_{\mathbf{0}}|$*) as well as* (3.30) *with* $a < 1$ *and* (5.3)*. Let* $I \subset (-\infty, \Sigma^{\mathrm{np}})$ *be some compact interval. Then we find some* $a' > 0$ *such that* $\mathrm{Ran}(\mathbb{1}_I(\widehat{H}^{\mathrm{np}}_V)) \subset \mathcal{D}(e^{a'|\mathbf{x}|})$*. In particular,*

$$\big\|\, e^{a'|\mathbf{x}|}\, \mathbb{1}_I(H^{\mathrm{np}}_V)\,\big\|_{\mathscr{L}(\mathscr{H}^+_{\mathbf{A}},\mathscr{H})} \leqslant \mathrm{const}\,. \tag{5.5}$$

If $\mathbf{G}_{\mathbf{x}}$ *is modified, then we get uniform lower bounds on* a' *and uniform upper bounds on the constant in* (5.5)*, provided that we have uniform upper bounds on* $d_{-1}, d_1, \Sigma^{\mathrm{np}}$ *and uniform lower bounds on* $\Sigma^{\mathrm{np}} - \sup I$*.*

Proof: The proof is given in the succeeding three subsections. □

Note that the potential V is *not* assumed to be a *small* form perturbation of $\sqrt{-\Delta}$ in Part (i) of the previous theorem. In particular, the assumptions of (i) cover the Coulomb potential $-\gamma/|\mathbf{x}|$ with coupling constants $\gamma \in [0, \gamma_{\mathrm{c}}^{\mathrm{PF}}]$ including the critical one. This improves on [37] where Coulomb potentials have been treated, for subcritical γ. By a modification of the arguments of this section it is actually also possible to prove exponential localization for the no-pair operator with Coulomb potential in the critical case $\gamma = \gamma_{\mathrm{c}}^{\mathrm{np}}$, which is not covered by Part (ii) of the above theorem; see [25].

The bound on the decay rate a of Part (i) has been found first in [25] (where only the Coulomb potential is treated explicitly). It reduces to the typical relativistic decay rate known for eigenvectors of electronic Dirac or square-root operators when $\mathbf{G}_{\mathbf{x}}$ is set equal to zero.

5.1. *A general strategy to prove the localization of spectral subspaces*

The general strategy of the proof of Theorem 5.1 is essentially due to [4]. We shall present a variant of the argument used in [4] in Lemma 5.2 below. In order to apply this lemma to $H^\sharp_V$ we shall also benefit from some useful observations made in [16]. The main advantage of Lemma 5.2 and its earlier variants is that it allows to study the localization of *spectral subspaces* without any a priori knowledge on the spectrum. Its proof does not exploit eigenvalue equations as it is the case in Agmon type arguments nor does it assume discreteness of the spectrum or the presence of spectral gaps. This is important for us since the spectra of both the no-pair and the semi-relativistic Pauli-Fierz operators will be continuous up to their minima. Roughly speaking the proof of Lemma 5.2 rests on a combination of the following:

- The representation (5.11) for the spectral projection $\mathbb{1}_I(\widehat{H}_\gamma^\sharp)$. Here a *comparison operator*, Y, enters into the analysis whose resolvent stays bounded after conjugation with suitable exponential weights, for all relevant spectral parameters. (5.11) is valid since its also satisfies $\mathbb{1}_I(Y) = 0$.
- The *Helffer-Sjöstrand formula* (re-derived below for the convenience of the reader) which is used to represent smoothed versions of $\mathbb{1}_I(\widehat{H}_\gamma^\sharp)$ and $\mathbb{1}_I(Y)$ as integrals over resolvents.
- The second resolvent identity; in fact, Y will be chosen such that $\widehat{H}_\gamma^\sharp - Y$ is well-localized and is hence able to control exponential weights.

In the somewhat technical parts of this section following after Lemma 5.2 we shall verify the applicability of Lemma 5.2 to our models by defining and analyzing suitable comparison operators Y.

Let us now introduce some prerequisites for the proof of Lemma 5.2. In order to find a representation of the spectral projection which is accessible for an analysis we smooth out the projection and employ the Helffer-Sjöstrand formula. More precisely, let $I \subset (-\infty, \Sigma^\sharp)$ be some compact interval. Then we pick some slightly larger compact interval $J \subset (-\infty, \Sigma^\sharp)$, $\mathring{J} \supset I$, and some $\chi \in C_0^\infty(\mathbb{R}, [0,1])$ such that $\chi \equiv 1$ on I and $\chi \equiv 0$ outside J. We pick another cut-off function $\rho \in C_0^\infty(\mathbb{R}, [0,1])$ such that $\rho = 1$ in a neighborhood of 0 and $\rho(y) = 0$, for $|y| \geqslant 1/2$, and extend χ to a compactly supported smooth function of the complex plane setting

$$\widetilde{\chi}(x+iy) := \chi(x) + \chi'(x)\, iy\, \rho(y)\,, \qquad x, y \in \mathbb{R}\,;$$

compare, e.g., [11, 12]. Then we have

$$2\,\partial_{\bar{z}}\widetilde{\chi}(z) := (\partial_x + i\partial_y)\chi(z) = \chi'(x)\,(1 - \rho(y) - y\,\rho'(y)) + \chi''(x)\, iy\, \rho(y)\,,$$

for every $z = x + iy \in \mathbb{C}$, and the choice of ρ implies

$$|\partial_{\bar{z}}\widetilde{\chi}(z)| \leqslant C_\chi |\mathrm{Im}\, z|\,, \qquad z \in \mathbb{C}\,, \tag{5.6}$$

for some $C_\chi \in (0, \infty)$. Moreover, the following Helffer-Sjöstrand formula is valid, for every self-adjoint operator, X, on some Hilbert space,

$$\chi(X) = \int_{\mathbb{C}} (X - z)^{-1}\, d\mu(z)\,, \qquad d\mu(z) := -\frac{1}{\pi}\partial_{\bar{z}}\widetilde{\chi}(z)\, dxdy\,. \tag{5.7}$$

If X were a complex number (5.7) were just a special case of Pompeiu's formula for some path encircling the support of χ. X can, however, be inserted in that formula by means of the spectral calculus; see, e.g., [12].

To facilitate the discussion of operator domains we replace $a|\mathbf{x}|$ in (5.4) by some $F:\mathbb{R}^3_{\mathbf{x}}\to\mathbb{R}$ satisfying

$$F\in C^\infty\cap L^\infty(\mathbb{R}^3_{\mathbf{x}},\mathbb{R})\,,\quad F=a\,r\text{ on }\mathcal{B}_{r/2}(0)\,,\quad F\geqslant 0\,,\quad |\nabla F|\leqslant a\,,\tag{5.8}$$

where $r\geqslant 1$ is the parameter appearing in (5.3).

Finally, we recall our notation $\sharp\in\{\mathrm{np},\mathrm{PF}\}$.

Lemma 5.2: *Let I, J, $\widetilde{\chi}$, and C_χ be as described above and assume that Y is a self-adjoint operator in $\mathscr{H}$ with $\mathcal{D}(Y)=\mathcal{D}(\widehat{H}^\sharp_V)$ and $Y>\sup J$. Furthermore, let $a>0$ and assume there exist $C,C'\in(0,\infty)$ such that, for all F satisfying* (5.8),

$$\big\|\,e^F\,(\widehat{H}^\sharp_V-Y)\,\big\|\leqslant C\,,\qquad\sup_{z\in J+i\mathbb{R}}\big\|\,e^F\,(Y-z)^{-1}\,e^{-F}\,\big\|\leqslant C'\,.\tag{5.9}$$

Then $\mathrm{Ran}(\mathbb{1}_I(\widehat{H}^\sharp_V))\subset\mathcal{D}(e^{a|\mathbf{x}|})$ *and*

$$\big\|\,e^{a|\mathbf{x}|}\,\mathbb{1}_I(\widehat{H}^\sharp_V)\,\big\|\leqslant C\,C'\,C_\chi\,\mathcal{L}(\mathrm{supp}(\chi'))/\pi\,,\tag{5.10}$$

where $\mathcal{L}$ denotes the Lebesgue measure on $\mathbb{R}$.

Proof: Since $\mathbb{1}_I=\chi\,\mathbb{1}_I$ and $J\subset\varrho(Y)$ we have

$$\mathbb{1}_I(\widehat{H}^\sharp_V)=(\chi(\widehat{H}^\sharp_V)-\chi(Y))\,\mathbb{1}_I(\widehat{H}^\sharp_V)\,.\tag{5.11}$$

Applying the Helffer-Sjöstrand formula (5.7) and the second resolvent identity we infer that

$$e^F\,\mathbb{1}_I(\widehat{H}^\sharp_V)=\int_{\mathbb{C}}\{e^F\,(Y-z)^{-1}\,e^{-F}\}\{e^F(Y-\widehat{H}^\sharp_V)\}(\widehat{H}^\sharp_V-z)^{-1}\,d\mu(z)\,.\tag{5.12}$$

Estimating the norm of these expressions using (5.6) and $\|(\widehat{H}^\sharp_V-z)^{-1}\|\leqslant 1/|\mathrm{Im}\,z|$ we obtain (5.10) with $a|\mathbf{x}|$ replaced by F. Then (5.10) follows by choosing a suitable sequence of functions F_n satisfying (5.8) and converging monotonically on $\{|\mathbf{x}|\geqslant 2r\}$ to $a|\mathbf{x}|$ and applying the monotone convergence theorem to $e^{F_n}\,\mathbb{1}_I(\widehat{H}^\sharp_V)\,\psi$, for every $\psi\in\mathscr{H}$. □

5.2. *Choice of the comparison operator Y*

In the next step we thus have to find a suitable operator Y fulfilling the conditions of Lemma 5.2. The hardest problem is to verify the second bound in (5.9) and this is actually the main new mathematical challenge in the study of the exponential localization in our non-local models. We defer the discussion of the second bound in (5.9) to the next subsection.

In order to construct an operator Y whose spectrum differs only slightly from the spectrum of the free operator $\widehat{H}_0^\sharp$, so that $Y > \sup J$, and which is defined on the same domain as $\widehat{H}_V^\sharp$ we simply add some bounded term to $\widehat{H}_V^\sharp$ which compensates for the singularities and wells of the electrostatic potential and thus pushes the spectrum up to the ionization threshold. A similar choice of a comparison operator has been employed in the non-relativistic setting in [16]. More precisely, we first introduce a scaled partition of unity. That is, we pick $\chi_{0,R}, \chi_{1,R} \in C^\infty(\mathbb{R}^3, [0,1])$, $R \geqslant r$, such that $\chi_{0,R} \equiv 1$ on $\mathcal{B}_R(0)$, $\chi_{0,R} \equiv 0$ on $\mathbb{R}^3 \setminus \mathcal{B}_{2R}(0)$, $\chi_{0,R}^2 + \chi_{1,R}^2 = 1$, and $\|\nabla \chi_{k,R}\|_\infty \leqslant c/R$, $k = 0, 1$, for some R-independent constant $c \in (0, \infty)$. Then we define Y as follows:

The semi-relativistic Pauli-Fierz operator. We define

$$Y_V^{\mathrm{PF}} := H_V^{\mathrm{PF}} + (\Sigma^{\mathrm{PF}} - E_V^{\mathrm{PF}})\, \chi_{0,R}^2 , \tag{5.13}$$

for some $R \geqslant 1$ which shall be fixed sufficiently large later on. Of course, H_V^{PF} and Y_V^{PF} have the same domain and the first bound in (5.9) which provides the control on the exponential weights holds trivially,

$$\|e^F (H_V^{\mathrm{PF}} - Y_V^{\mathrm{PF}})\| \leqslant (\Sigma^{\mathrm{PF}} - E_V^{\mathrm{PF}})\, \|e^F \chi_{0,R}^2\|_\infty \leqslant (\Sigma^{\mathrm{PF}} - E_V^{\mathrm{PF}})\, e^{ar+2aR},$$

for every F satisfying (5.8). We shall sketch the proof of the condition $Y > \sup J$ which follows from the next lemma.

Lemma 5.3: *$Y_V^{\mathrm{PF}} \geqslant \Sigma^{\mathrm{PF}} - o(R^0)$, $R \to \infty$, in the sense of quadratic forms on $\mathscr{D}$, where the little-o-symbol depends only on V and $\chi_{0,1}$.*

Proof: We employ the localization formula (3.27) with $K = 1$, the error estimate (3.29) , and $H_V^{\mathrm{PF}} \geqslant E_V^{\mathrm{PF}}$, $H_0^{\mathrm{PF}} \geqslant \Sigma^{\mathrm{PF}}$, to get

$$\begin{aligned} Y_V^{\mathrm{PF}} &\geqslant \chi_{0,R}\, H_V^{\mathrm{PF}}\, \chi_{0,R} + \chi_{1,R}\, H_0^{\mathrm{PF}}\, \chi_{1,R} \\ &\quad + (\Sigma^{\mathrm{PF}} - E_V^{\mathrm{PF}})\, \chi_{0,R}^2 + \chi_{1,R}^2\, V - \mathcal{O}(1/R^2) \geqslant \Sigma^{\mathrm{PF}} - o(R^0) . \end{aligned}$$

We also used that $\sup_{|\mathbf{x}| \leqslant R} |V(\mathbf{x})| = o(R^0)$ and $\chi_{0,R}^2 + \chi_{1,R}^2 = 1$. □

The no-pair operator. Since $|D_{\mathbf{A}}| = P_{\mathbf{A}}^+ D_{\mathbf{A}} + P_{\mathbf{A}}^- |D_{\mathbf{A}}|$ we have

$$\widehat{H}_V^{\mathrm{np}} = |D_{\mathbf{A}}| + P_{\mathbf{A}}^+\, V\, P_{\mathbf{A}}^+ + H_{\mathrm{f}}^{\mathrm{diag}}, \quad H_{\mathrm{f}}^{\mathrm{diag}} := P_{\mathbf{A}}^+\, H_{\mathrm{f}}\, P_{\mathbf{A}}^+ + P_{\mathbf{A}}^-\, H_{\mathrm{f}}\, P_{\mathbf{A}}^- ,$$

on $\mathscr{D}$. We write $\widehat{H}_{V,R}^{\mathrm{np}} := \widehat{H}_V^{\mathrm{np}} - (1/R)\, H_{\mathrm{f}}^{\mathrm{diag}}$, for some $R > 1$, so that $E_{V,R}^{\mathrm{np}} := \inf \sigma[\widehat{H}_{V,R}^{\mathrm{np}}\, P_{\mathbf{A}}^+] > -\infty$ by (3.31), and define

$$Y_V^{\mathrm{np}} = \widehat{H}_V^{\mathrm{np}} + (\Sigma^{\mathrm{np}} - E_{V,R}^{\mathrm{np}})\, \chi_{0,R}\, P_{\mathbf{A}}^+\, \chi_{0,R} . \tag{5.14}$$

Again it is clear that $\widehat{H}_V^{\mathrm{np}}$ and Y_V^{np} are self-adjoint on the same domain. Also the first bound in (5.9) again follows trivially,

$$\|e^F(\widehat{H}_V^{\mathrm{np}} - Y_V^{\mathrm{np}})\| \leqslant (\Sigma^{\mathrm{np}} - E_{V,R}^{\mathrm{np}})\, e^{ar+2aR},$$

for every F satisfying (5.8). Besides the second bound in (5.9) it remains to derive the following lemma.

Lemma 5.4: *$Y_V^{\mathrm{np}} \geqslant \Sigma^{\mathrm{np}} - \Sigma^{\mathrm{np}}/R - o(R^0) - k\, d_1^2/R^2$ as quadratic forms on $\mathscr{D}$, where $k \in (0,\infty)$ and the little-o-symbol depend only on V and $\chi_{0,1}$.*

Proof: Again we use an IMS type localization formula to infer that

$$\begin{aligned} Y_V^{\mathrm{np}} &\geqslant \chi_{0,R}\, \widehat{H}_{V,R}^{\mathrm{np}}\, \chi_{0,R} + \chi_{1,R}\, \widehat{H}_{0,R}^{\mathrm{np}}\, \chi_{1,R} + (\Sigma^{\mathrm{np}} - E_{V,R}^{\mathrm{np}})\, \chi_{0,R}\, P_{\mathbf{A}}^{+}\, \chi_{0,R} \\ &\quad + \tfrac{1}{R}\, H_{\mathrm{f}}^{\mathrm{diag}} + \chi_{1,R} P_{\mathbf{A}}^{+}\, V\, P_{\mathbf{A}}^{+}\, \chi_{1,R} + \tfrac{1}{2} \textstyle\sum_{k=0,1} \big[\chi_{k,R}\,,\, [\chi_{k,R}, \widehat{H}_{V,R}^{\mathrm{np}}]\big]\,. \end{aligned} \tag{5.15}$$

As a consequence of (2.25), (2.26), (3.29), (9.11), (9.12), and $H_{\mathrm{f}} \leqslant 2H_{\mathrm{f}}^{\mathrm{diag}}$ the double commutator in the last line is bounded from below by $-(k/R^2)(H_{\mathrm{f}}^{\mathrm{diag}} + d_1^2 + 1)$, where $k \in (0,\infty)$ depends only on V and $\chi_{0,1}$. To control this error we use the term $\frac{1}{R}\, H_{\mathrm{f}}^{\mathrm{diag}}$ in (5.15). Furthermore, we put $\mu_R := \chi_{0,R/2}$, so that $\chi_{1,R}\, \mu_R = 0$. Then $(1-\mu_R^2)\, V = o(R^0)$ and, by (2.26),

$$\begin{aligned} -\chi_{1,R} P_{\mathbf{A}}^{+}\, V\, P_{\mathbf{A}}^{+}\, \chi_{1,R} &\leqslant -\chi_{1,R}[P_{\mathbf{A}}^{+}, \mu_R]\, V\, [\mu_R, P_{\mathbf{A}}^{+}]\, \chi_{1,R} + o(R^0)\, \chi_{1,R}^2 \\ &\leqslant o(R^0)\, \chi_{1,R}^2 + \mathcal{O}(1/R^2)\, (H_{\mathrm{f}}^{\mathrm{diag}} + d_1^2 + 1)\,, \end{aligned}$$

so that the second term in the last line of (5.15) can again be controlled by the first one. Using these remarks, $\widehat{H}_{V,R}^{\mathrm{np}} \geqslant E_{V,R}^{\mathrm{np}}\, P_{\mathbf{A}}^{+} + (1-1/R)\, \Sigma^{\mathrm{np}}\, P_{\mathbf{A}}^{-}$, and $\widehat{H}_{0,R}^{\mathrm{np}} \geqslant (1-1/R)\, \Sigma^{\mathrm{np}}$ (by (5.2)), we arrive at the assertion. □

5.3. *Conjugation of Y with exponential weights*

In order to prove Theorem 5.1 it only remains to verify the second bound in (5.9). The following lemma [37] gives a criterion for this condition to hold true.

Lemma 5.5: *Let Y be a non-negative operator in $\mathscr{H}$ which admits $\mathscr{D}$ as a form core. Set $b := \inf \sigma(Y)$ and let $J \subset (-\infty, b)$ be some compact interval. Let $a \in (0,1)$ and assume that, for all F satisfying (5.8), we have $e^{\pm F}\, \mathcal{Q}(Y) \subset \mathcal{Q}(Y)$. (Notice that $e^{\pm F}$ maps $\mathscr{D}$ into itself.) Assume further that there exist constants $c(a), f(a), g(a), h(a) \in [0,\infty)$ such that $c(a) < 1/2$*

and $\delta := b - \max J - b\,g(a) - h(a) > 0$ *and, for all* F *satisfying* (5.8) *and* $\varphi \in \mathscr{D}$,

$$\big|\langle\,\varphi\,|\,(e^F\,Y\,e^{-F} - Y)\,\varphi\,\rangle\big| \leqslant c(a)\,\langle\,\varphi\,|\,Y\,\varphi\,\rangle + f(a)\,\|\varphi\|^2, \tag{5.16}$$

$$\mathrm{Re}\,\langle\,\varphi\,|\,e^F\,Y\,e^{-F}\,\varphi\,\rangle \geqslant (1-g(a))\,\langle\,\varphi\,|\,Y\,\varphi\,\rangle - h(a)\,\|\varphi\|^2. \tag{5.17}$$

Then we have, for all F *satisfying* (5.8),

$$\sup_{z\in J+i\mathbb{R}} \big\|\,e^F\,(Y-z)^{-1}\,e^{-F}\,\big\| \leqslant \delta^{-1}. \tag{5.18}$$

Proof: We only sketch the proof and refer to Lemma 5.2 of [37] for the details. The assumptions $e^{\pm F}\,\mathcal{Q}(Y) \subset \mathcal{Q}(Y)$ and (5.16) ensure that the closure, Y_F, of $(e^F\,Y\,e^{-F})\restriction_{\mathscr{D}}$ agrees with the closed operator $e^F\,Y\,e^{-F}$. The bound (5.17) shows that the numerical range of Y_F is contained in the half space $\{z \in \mathbb{C} : \mathrm{Re}\,z \geqslant \sup J + \delta\}$. Moreover, we can argue that, for $z \in J + i\mathbb{R}$, the deficiency of $Y_F - z$ is zero and, hence, the norm of $(Y_F - z)^{-1} = e^F(Y-z)e^{-F}$ can be estimated by one over the distance of z to the numerical range of Y_F. □

The semi-relativistic Pauli-Fierz operator. Next, we apply Lemma 5.5 to H_V^{PF}. In order to verify condition (5.17) with a good bound on the exponential decay rates we apply the following technical lemma from [25]:

Lemma 5.6: *For all* $a \in (0,1)$, F *satisfying* (5.8), *and* $\varphi \in \mathscr{D}$,

$$\begin{aligned}\mathrm{Re}\,\langle\,\varphi\,|\,e^F\,|D_{\mathbf{A}}|\,e^{-F}\varphi\,\rangle &\geqslant \langle\,\varphi\,|\,(D_{\mathbf{A}}^2 - |\nabla F|^2)^{1/2}\,\varphi\,\rangle\\ &\geqslant \langle\,\varphi\,|\,(|D_{\mathbf{A}}| - \rho(a))\,\varphi\,\rangle\,.\end{aligned} \tag{5.19}$$

Proof: For every $\varphi \in \mathscr{D}$, we infer from (4.12) that

$$\langle\,\varphi\,|\,(e^F\,|D_{\mathbf{A}}|\,e^{-F} - (D_{\mathbf{A}}^2 - |\nabla F|^2)^{1/2})\,\varphi\,\rangle = \int_0^\infty J[\varphi;\eta]\,\frac{\eta^{1/2}d\eta}{\pi}\,,$$

with

$$J[\varphi;\eta] := \langle\,\varphi\,|\,(\mathscr{R}_F(\eta) - e^F\mathscr{R}_0(\eta)\,e^{-F})\,\varphi\,\rangle\,,$$

$$\mathscr{R}_G(\eta) := (D_{\mathbf{A}}^2 - |\nabla G|^2 + \eta)^{-1}, \quad G \in \{0, F\}\,.$$

Now, let $\phi := e^F(D_{\mathbf{A}}^2 + \eta)\,e^{-F}\psi$, for some $\psi \in \mathscr{D}$. Then

$$\begin{aligned}\mathrm{Re}\,\langle\,\phi\,|\,e^F\mathscr{R}_0(\eta)\,e^{-F}\phi\,\rangle &= \mathrm{Re}\,\langle\,e^F(D_{\mathbf{A}}^2+\eta)\,e^{-F}\psi\,|\,\psi\,\rangle\\ &= \langle\,(D_{\mathbf{A}}^2 - |\nabla F|^2 + \eta)\,\psi\,|\,\psi\,\rangle\\ &\geqslant (1 - a^2 + \eta)\,\|\psi\|^2 \geqslant 0\,.\end{aligned}$$

Since $D_{\mathbf{A}}^2$ is essentially self-adjoint on $\mathscr{D}$ and multiplication with e^{-F} maps $\mathscr{D}$ bijectively onto itself, we know that $(D_{\mathbf{A}}^2+\eta)\,e^{-F}\mathscr{D}$ is dense in $\mathscr{H}$. Since F is bounded we conclude that the previous estimates hold, for all ϕ in some dense domain, whence $\mathrm{Re}\,[e^F\mathscr{R}_0(\eta)\,e^{-F}] \geqslant 0$ as a quadratic form on $\mathscr{H}$. Next, we set $Q := (\boldsymbol{\alpha}\cdot\nabla F)\,D_{\mathbf{A}} + D_{\mathbf{A}}\,(\boldsymbol{\alpha}\cdot\nabla F)$ and let

$$\varphi := (D_{\mathbf{A}}^2 - |\nabla F|^2 + \eta)\,\psi = e^{\pm F}(D_{\mathbf{A}}^2+\eta)\,e^{\mp F}\psi \mp iQ\,\psi\,,$$

for $\psi \in \mathscr{D}$. Then

$$J[\varphi;\eta] = i\langle\, e^{-F}\mathscr{R}_0(\eta)\,e^{F}\varphi\,|\,Q\,\psi\,\rangle = i\langle\,\psi\,|\,Q\,\psi\,\rangle + \langle\, Q\,\psi\,|\,e^{F}\mathscr{R}_0(\eta)\,e^{-F}Q\,\psi\,\rangle\,.$$

Here $D_{\mathbf{A}}^2 - |\nabla F|^2$ is essentially self-adjoint on $\mathscr{D}$ and Q is symmetric on the same domain. Hence, $\mathrm{Re}\,J[\varphi;\eta] \geqslant 0$, for all φ in a dense set, thus for all $\varphi \in \mathscr{H}$, and we arrive at the first inequality in (5.19). Since the square root is operator monotone, $|\nabla F| \leqslant a$, and $|D_{\mathbf{A}}| \geqslant 1$, we further have

$$(D_{\mathbf{A}}^2 - |\nabla F|^2)^{1/2} \geqslant |D_{\mathbf{A}}| + (D_{\mathbf{A}}^2 - a^2)^{1/2} - |D_{\mathbf{A}}| \geqslant |D_{\mathbf{A}}| - \rho(a)\,. \qquad \square$$

In what follows we abbreviate

$$\mathcal{K}_F := [P_{\mathbf{A}}^+\,,\,e^F]\,e^{-F},$$

and recall from (2.24) that $\||D_{\mathbf{A}}|^{1/2}\,\mathcal{K}_F\| = \mathcal{O}_{a_0}(a)$, for all F satisfying (5.8) with $0 < a \leqslant a_0 < 1$.

Lemma 5.7: *For all $0 < a \leqslant a_0 < 1$ and F satisfying* (5.8),

$$\mathrm{Re}\,\langle\,\varphi\,|\,e^F\,Y_V^{\mathrm{PF}}\,e^{-F}\,\varphi\,\rangle \geqslant \langle\,\varphi\,|\,Y_V^{\mathrm{PF}}\,\varphi\,\rangle - \rho(a)\,\|\varphi\|^2, \quad \varphi \in \mathscr{D}. \tag{5.20}$$

Moreover, for all $\varepsilon > 0$ and $a_0 \in (0,1)$, there is some constant, $C(a_0,\varepsilon,V) \in (0,\infty)$, such that, for all F satisfying (5.8) *with $a \in [0,a_0]$ and $\varphi \in \mathscr{D}$,*

$$\big|\langle\,\varphi\,|\,(e^F\,Y_V^{\mathrm{PF}}\,e^{-F} - Y_V^{\mathrm{PF}})\,\varphi\,\rangle\big| \leqslant \varepsilon\,\langle\,\varphi\,|\,Y_V^{\mathrm{PF}}\,\varphi\,\rangle + C(a_0,\varepsilon,V)\,\|\varphi\|^2\,. \tag{5.21}$$

Proof: (5.20) follows immediately from (5.19). To derive (5.21) we write

$$e^F\,|D_{\mathbf{A}}|\,e^{-F} - |D_{\mathbf{A}}| = -2\,D_{\mathbf{A}}\,\mathcal{K}_F + i\boldsymbol{\alpha}\cdot(\nabla F)\,e^F\,S_{\mathbf{A}}\,e^{-F}$$

on $\mathscr{D}$ and make a little observation. Since $F \equiv a\,r$ on $\mathcal{B}_{r/2}(0)$ we find some $\mu \in C^\infty(\mathbb{R}^3_{\mathbf{x}},[0,1])$ such that $\mu = 0$ on $\overline{\mathcal{B}}_{r/4}(0)$ and $\nabla F = \mu\,\nabla F$. For $\varphi,\psi \in \mathscr{D}$, we thus have (recall (2.18) and (2.22))

$$|\langle\, D_{\mathbf{A}}\,\varphi\,|\,\mathcal{K}_F\,\psi\,\rangle| \leqslant \int_{\mathbb{R}} |\langle\,\varphi\,|\,D_{\mathbf{A}}\,R_{\mathbf{A}}(iy)\,\mu\,i\boldsymbol{\alpha}\cdot\nabla F\,R_{\mathbf{A}}^F(iy)\,\psi\,\rangle|\,\frac{dy}{2\pi}\,.$$

Here we can write $D_{\mathbf{A}}\,R_{\mathbf{A}}(iy)\,\mu = \mu\,|D_{\mathbf{A}}|\,S_{\mathbf{A}}\,R_{\mathbf{A}}(iy) + [D_{\mathbf{A}},\mu]\,R_{\mathbf{A}}(iy) + D_{\mathbf{A}}\,R_{\mathbf{A}}(iy)\,[\mu,D_{\mathbf{A}}]\,R_{\mathbf{A}}(iy)$, where $\|[D_{\mathbf{A}},\mu]\| \leqslant \|\nabla\mu\|_\infty = \mathcal{O}(1)$. On account of $\| \, |D_{\mathbf{A}}|^{1/2} R_{\mathbf{A}}(iy)\| \leqslant \mathcal{O}(1)\,(1+y^2)^{-1/4}$ and $\|D_{\mathbf{A}}\,R_{\mathbf{A}}(iy)\| = \mathcal{O}(1)$ it is now straightforward to verify that

$$\big|\big\langle\, D_{\mathbf{A}}\,\varphi \,\big|\, \mathcal{K}_F\,\psi\,\big\rangle\big| \leqslant \mathcal{O}_{a_0}(a)\,\big\{\|\,|D_{\mathbf{A}}|^{1/2}\,\mu\,\varphi\,\| + \|\varphi\|\big\}\,\|\psi\|\,.$$

Some elementary estimates using $\|\nabla F\|_\infty \leqslant a$, $\|e^F\,S_{\mathbf{A}}\,e^{-F}\| = \mathcal{O}_{a_0}(1)$, and the previous bound now show that

$$\begin{aligned} &\big|\big\langle\,\varphi\,\big|\,(e^F\,|D_{\mathbf{A}}|\,e^{-F} - |D_{\mathbf{A}}|)\,\varphi\,\big\rangle\big| \\ &\quad\leqslant \varepsilon_1\,\big\langle\,\mu\,\varphi\,\big|\,|D_{\mathbf{A}}|\,\mu\,\varphi\,\big\rangle + (\varepsilon_1^{-1}\,\mathcal{O}_{a_0}(a^2) + \mathcal{O}_{a_0}(a))\,\|\varphi\|^2 \qquad (5.22)\\ &\quad\leqslant \varepsilon_1\,\mathcal{O}(1)\,\big\langle\,\mu\,\varphi\,\big|\,Y_\gamma^{\mathrm{PF}}\,\mu\,\varphi\,\big\rangle + \mathrm{const}(a_0,\varepsilon_1)\,\|\varphi\|^2, \end{aligned}$$

for every $\varepsilon_1 \in (0,1]$. In the second step we used that $\mu\,V$ is bounded because $\mu = 0$ on $\overline{\mathcal{B}}_{r/4}(0)$. Since we may assume that there is some $\widetilde{\mu} \in C_0^\infty(\mathbb{R}^3,[0,1])$ such that $\mu^2 + \widetilde{\mu}^2 = 1$ we can employ an IMS localization formula as in the proof of Lemma 5.3 to show that $\mu\,Y_V^{\mathrm{PF}}\,\mu \leqslant Y_V^{\mathrm{PF}} + \mathcal{O}(1)$ on $\mathscr{D}$. Altogether this proves (5.21). □

Lemma 5.8: *There exist constants, $c_1, c_2 \in (0,\infty)$, such that, for all $a \in (0,1/2]$ and $\pm F$ satisfying* (5.8), *and $\varphi \in \mathscr{D}$,*

$$\big\langle\, e^F\,\varphi \,\big|\, Y_V^{\mathrm{PF}}\,e^F\,\varphi\,\big\rangle \leqslant c_1\,\|e^F\|^2\,\big\langle\,\varphi\,\big|\,Y_V^{\mathrm{PF}}\,\varphi\,\big\rangle + c_2\,\|e^F\|^2\,\|\varphi\|^2. \qquad (5.23)$$

In particular, $e^F\,\mathcal{Q}(Y_V^{\mathrm{PF}}) \subset \mathcal{Q}(Y_V^{\mathrm{PF}})$.

Proof: We pick a smooth partition of unity with respect to the electron coordinates, $\mu_0^2 + \mu_1^2 = 1$, where $\mathrm{supp}(\mu_0) \subset \mathcal{B}_{r/2}(0)$ and $\mu_0 = 1$ on $\overline{\mathcal{B}}_{r/4}(0)$. Then $\big\langle\, e^F\,\varphi \,\big|\, Y_V^{\mathrm{PF}}\,e^F\,\varphi\,\big\rangle = \sum_{i=0,1}\big\langle\,\mu_i\,e^F\,\varphi\,\big|\,Y_V^{\mathrm{PF}}\,\mu_i\,e^F\,\varphi\,\big\rangle + R_\varphi$, where $|R_\varphi| \leqslant \mathcal{O}(1)\|e^F\|^2\,\|\varphi\|^2$. (This holds in particular for $F = 0$, of course.) Therefore, it is sufficient to prove the bound (5.23) with $\varphi = \mu_i\,\psi$, $i = 0,1$, $\psi \in \mathscr{D}$. For $\varphi = \mu_0\,\psi$, the bound holds, however, true trivially, for all $c_1, c_2 \geqslant 1$, since $F = 1$ on the support of μ_0.

Let us assume that $\varphi = \mu_1\,\psi$, for some $\psi \in \mathscr{D}$, in the rest of this proof. Of course, $\|\chi_{0,R}\,e^F\,\varphi\|^2 \leqslant \|e^F\|^2\,\|\chi_{0,R}\,\varphi\|^2$ and, since H_{f} and e^F commute, $\|H_{\mathrm{f}}^{1/2}\,e^F\,\varphi\|^2 \leqslant \|e^F\|^2\,\|H_{\mathrm{f}}^{1/2}\,\varphi\|^2$. Furthermore, $|\langle\,\varphi\,|\,V\,\varphi\,\rangle| \leqslant \mathcal{O}(1)\|\varphi\|^2$, since V is bounded on $\mathrm{supp}(\mu_1)$. To conclude we write $|D_{\mathbf{A}}| = P_{\mathbf{A}}^+\,D_{\mathbf{A}} - P_{\mathbf{A}}^-\,D_{\mathbf{A}}$ and employ the following bound derived in [39],

$$\big\langle\,\varphi\,\big|\,e^F\,P_{\mathbf{A}}^{\pm}\,(\pm D_{\mathbf{A}})\,e^F\,\varphi\,\big\rangle \leqslant c_3\,\|e^F\|^2\,\big\langle\,\varphi\,\big|\,P_{\mathbf{A}}^{\pm}\,(\pm D_{\mathbf{A}})\,\varphi\,\big\rangle + c_4\,\|e^F\|^2\,\|\varphi\|^2, \qquad (5.24)$$

for every $\varphi \in \mathscr{D}$. We actually derived this bound in [39] for classical vector potentials. The proof works, however, also for the quantized vector potential without any change. Moreover, we only treated the choice of the plus sign in (5.24). But again an obvious modification of the proof in [39] shows that (5.24) is still valid when we choose the minus sign. □

The no-pair operator. The following lemma implies that the conditions (5.16) and (5.17) are fulfilled in the case of the no-pair operator, too.

Lemma 5.9: *There exist* $c \equiv c(V) \in (0,\infty)$ *and* $c' \equiv c'(V, d_{-1}, d_1, \Sigma^{\mathrm{np}} - E^{\mathrm{np}}_{V,R}) \in (0,\infty)$ *such that, for all* F *satisfying* (5.8),

$$\big|\langle\,\varphi\,|\,(e^F\,Y_V^{\mathrm{np}}\,e^{-F} - Y_V^{\mathrm{np}})\,\varphi\,\rangle\big| \leqslant \mathcal{O}(a)\,\langle\,\varphi\,|\,(c\,Y_V^{\mathrm{np}} + c')\,\varphi\,\rangle, \quad \varphi\in\mathscr{D}.$$

Proof: On account of (5.22) and $\|e^F\,\chi_{0,R}\,P^+_{\mathbf{A}}\,\chi_{0,R}\,e^{-F} - \chi_{0,R}\,P^+_{\mathbf{A}}\,\chi_{0,R}\| \leqslant \|\mathcal{K}_F\| = \mathcal{O}_{a_0}(a)$ it suffices to consider

$$\Delta^{\pm}(T) := e^F\,P^{\pm}_{\mathbf{A}}\,T\,P^{\pm}_{\mathbf{A}}\,e^{-F} - P^{\pm}_{\mathbf{A}}\,T\,P^{\pm}_{\mathbf{A}} = 2\mathrm{Re}\,\big[P^{\pm}_{\mathbf{A}}\,T\,\delta P\big] + \delta P\,T\,\delta P,$$

where $\delta P := e^F\,P^{\pm}_{\mathbf{A}}\,e^{-F} - P^{\pm}_{\mathbf{A}}$ and T is H_{f} or V. Clearly,

$$|\langle\,\varphi\,|\,\Delta^{\pm}(T)\,\varphi\,\rangle| \leqslant \varepsilon\,\langle\,\varphi\,|\,P^{\pm}_{\mathbf{A}}\,T\,P^{\pm}_{\mathbf{A}}\,\varphi\,\rangle + (1+\varepsilon^{-1})\,\big\|\,|T|^{1/2}\,\delta P\,\varphi\,\big\|^2,$$

for all $\varepsilon > 0$ and $\varphi \in \mathscr{D}$. Since, by (2.25) and (2.26), $\|\,|T|^{1/2}\,\delta P\,\varphi\|^2 \leqslant \mathcal{O}(a^2)\,\langle\,\varphi\,|\,(H_{\mathrm{f}} + (4d_1)^2 + 1)\,\varphi\,\rangle \leqslant \mathcal{O}(a^2)\,\langle\,\varphi\,|\,(H_{\mathrm{f}}^{\mathrm{diag}} + d_1^2 + 1)\,\varphi\,\rangle$, we may choose ε proportional to a and use (3.31) to conclude. □

To complete the proof of Theorem 5.1 also in the case of the no-pair operator we note that the bound (5.23) still holds true when Y_V^{PF} is replaced by Y_0^{np}. To this end we only have to observe in addition to the remarks in the proof of Lemma 5.8 that $\|\check{H}_{\mathrm{f}}^{1/2}\,P^{\pm}_{\mathbf{A}}\,e^F\,\varphi\,\| \leqslant \mathcal{O}(1)\,\|e^F\|\,\|\check{H}_{\mathrm{f}}^{1/2}\,\varphi\|$. This follows, however, immediately from (9.9) which implies $\|\check{H}_{\mathrm{f}}^{1/2}\,P^{\pm}_{\mathbf{A}}\,e^F\,\varphi\,\| \leqslant (1 + \|\mathcal{C}_{1/2}\|/2)\,\|e^F\,\check{H}_{\mathrm{f}}^{1/2}\,\varphi\|$. Thus, $e^{\pm F}\mathcal{Q}(Y_0^{\mathrm{np}}) \subset \mathcal{Q}(Y_0^{\mathrm{np}})$. By Theorem 3.6 and the assumptions on V we know that $\mathcal{Q}(Y_V^{\mathrm{np}}) = \mathcal{Q}(Y_0^{\mathrm{np}})$ and we conclude.

6. Existence of Ground States with Mass

In this section we present an intermediate step of the proof of the existence of ground states for $H_V^{\sharp}$. Namely, we prove the existence of ground state eigenvectors, $\phi_m^{\sharp}$, for modified Hamiltonians, $H^{\sharp}_{V,m}$, which are defined by means of an infra-red cut-off coupling function. The infra-red cut-off parameter, $m > 0$, is referred to as the photon mass. Later on in Section 8 we shall remove the infra-red cut-off by showing that every sequence, $\{\phi^{\sharp}_{m_j}\}_j$, $m_j \searrow 0$, contains a strongly convergent subsequence whose limit turns out to be a ground state eigenvector of $H_V^\sharp$. The compactness argument used to show this in Section 8 requires the infra-red bounds derived before in Section 7.

In the present section the existence of $\phi_m^\sharp$ is shown by discretizing the photon degrees of freedom. After the infra-red cut-off operators $H_{V,m}^\sharp$ have been defined in Subsection 6.1 we construct discretized versions of them, denoted by $H_{V,m,\varepsilon}^\sharp$, in Subsection 6.2. We collect some technical estimates needed to compare the original, infra-red cut-off, and discretized operators in Subsections 6.3 and 6.4. As another preparation we study the continuity of the ground state energy and ionization threshold with respect to the parameters m and ε in Subsection 6.5. The main result of this section, Theorem 6.11 on the existence of $\phi_m^\sharp$, is stated and proved in Subsection 6.6 and we refer the reader to that subsection for some brief remarks on its proof. Many arguments of this section (in particular those in Subsections 6.4 and 6.5) are alternatives to the corresponding ones in [28, 29].

In the whole section $\mathbf{G_x}$ is the coupling function given by (2.2). To clarify which properties of V are exploited we introduce the following hypothesis. It is fulfilled by the Coulomb potential in the subcritical cases:

Hypothesis 6.1: In the case $\sharp = \mathrm{PF}$ the potential $V \in L^2_{\mathrm{loc}}(\mathbb{R}^3, \mathbb{R})$ is relatively form-bounded with respect to $\sqrt{-\Delta}$ with relative form bound strictly less than one. In the case $\sharp = \mathrm{np}$ the potential $V \in L^2_{\mathrm{loc}}(\mathbb{R}^3, \mathbb{R})$ satisfies $H^1(\mathbb{R}^3) \subset \mathcal{D}(V)$ and (3.30) with $a < 1$.

We shall strengthen the assumptions on V later on in order to apply the localization estimates of Section 5.

6.1. *Operators with photon mass*

For every $m > 0$, the infra-red cut-off coupling function is given as

$$\mathbf{G}_{\mathbf{x},m}(k) := -e\,\frac{\mathbb{1}_{\{m \leqslant |\mathbf{k}| \leqslant \Lambda\}}}{2\pi\sqrt{|\mathbf{k}|}}\,e^{-i\mathbf{k}\cdot\mathbf{x}}\,\boldsymbol{\varepsilon}(k)\,, \tag{6.1}$$

for all $\mathbf{x} \in \mathbb{R}^3$ and almost every $k = (\mathbf{k}, \lambda) \in \mathbb{R}^3 \times \mathbb{Z}_2$. To compare $\mathbf{G}_{\mathbf{x},m}$ with $\mathbf{G_x}$ defined in (2.2) we introduce the parameter

$$\triangle^2(m) := \int \left(\omega(k) + \omega(k)^{-1}\right) \sup_{\mathbf{x}} |\mathbf{G_x}(k) - \mathbf{G}_{\mathbf{x},m}(k)|^2\,dk\,. \tag{6.2}$$

Of course, $\triangle^2(m) = (e^2/2\pi^2)\int_{|\mathbf{k}|<m}(1+|\mathbf{k}|^{-2})d^3\mathbf{k} \to 0$, as $m \searrow 0$. For $m>0$, we further define the infra-red cut-off vector potential,

$$\mathbf{A}_m := a^\dagger(\mathbf{G}_m) + a(\mathbf{G}_m)\,, \quad a^\sharp(\mathbf{G}_m) := \int_{\mathbb{R}^3}^{\oplus} \mathbb{1}_{\mathbb{C}^4} \otimes a^\sharp(\mathbf{G}_{\mathbf{x},m})\,d^3\mathbf{x}\,,$$

and the infra-red regularized Hamiltonians

$$H^{\mathrm{PF}}_{V,m} := |D_{\mathbf{A}_m}| + V + H_{\mathrm{f}}\,, \tag{6.3}$$

$$H^{\mathrm{np}}_{V,m} := P^+_{\mathbf{A}_m}\,(D_{\mathbf{A}_m} + V + H_{\mathrm{f}})\,P^+_{\mathbf{A}_m}\,. \tag{6.4}$$

We define these operators as self-adjoint Friedrichs extensions starting from $\mathscr{D}$. The ground state energies and ionization thresholds, for positive photon mass $m>0$, are denoted by

$$E^\sharp_{V,m} := \inf\sigma[H^\sharp_{V,m}]\,, \qquad \Sigma^\sharp_m := \inf\sigma[H^\sharp_{0,m}]\,.$$

As a first step we introduce a truncated Fock space where the radiation field energy H_{f} is bounded from below by $m>0$ on the orthogonal complement of the vacuum sector. Namely, we split the one-photon Hilbert space into two mutually orthogonal subspaces

$$\mathscr{K} = \mathscr{K}^>_m \oplus \mathscr{K}^<_m\,, \qquad \mathscr{K}^>_m := L^2(\mathcal{A}_m \times \mathbb{Z}_2)\,, \qquad \mathcal{A}_m := \{|\mathbf{k}| \geqslant m\}\,.$$

It is well-known that $\mathscr{F}_{\mathrm{b}}[\mathscr{K}] = \mathscr{F}_{\mathrm{b}}[\mathscr{K}^>_m] \otimes \mathscr{F}_{\mathrm{b}}[\mathscr{K}^<_m]$. We observe that $\mathbf{A}_m$ creates and annihilates photon states in $\mathscr{K}^>_m$ only and H_{f} leaves the Fock space factors associated with the subspaces $\mathscr{K}^{\lessgtr}_m$ invariant. We shall designate operators acting in the Fock space factors $\mathscr{F}_{\mathrm{b}}[\mathscr{K}^>_m]$ or $\mathscr{F}_{\mathrm{b}}[\mathscr{K}^<_m]$ by a superscript $>$ or $<$, respectively. Under the isomorphism

$$\mathscr{H} \cong \big(L^2(\mathbb{R}^3,\mathbb{C}^4) \otimes \mathscr{F}_{\mathrm{b}}[\mathscr{K}^>_m]\big) \otimes \mathscr{F}_{\mathrm{b}}[\mathscr{K}^<_m] =: \mathscr{H}^>_m \otimes \mathscr{F}_{\mathrm{b}}[\mathscr{K}^<_m], \tag{6.5}$$

we then have $D_{\mathbf{A}_m} \cong D_{\mathbf{A}^>_m} \otimes \mathbb{1}$, $|D_{\mathbf{A}_m}| \cong |D_{\mathbf{A}^>_m}| \otimes \mathbb{1}$, $P^+_{\mathbf{A}_m} \cong P^+_{\mathbf{A}^>_m} \otimes \mathbb{1}$, and $H_{\mathrm{f}} = H^>_{\mathrm{f}} \otimes \mathbb{1} + \mathbb{1} \otimes H^<_{\mathrm{f}}$ with $H^>_{f,m} := d\Gamma(\omega\restriction_{\mathcal{A}_m \times \mathbb{Z}_2})$, $H^<_{f,m} := d\Gamma(\omega\restriction_{\mathcal{A}^c_m \times \mathbb{Z}_2})$. As a consequence, the semi-relativistic Pauli-Fierz and no-pair operators decompose under the isomorphism (6.5) as

$$H^{\mathrm{PF}}_{V,m} = \overline{H^{\mathrm{PF}}_{V,m,0} \otimes \mathbb{1} + \mathbb{1} \otimes H^<_{\mathrm{f}}}\,, \tag{6.6}$$

$$H^{\mathrm{np}}_{V,m} = \overline{H^{\mathrm{np}}_{V,m,0} \otimes \mathbb{1} + P^+_{\mathbf{A}^>_m} \otimes H^<_{\mathrm{f}}}\,,$$

where

$$H^{\mathrm{PF}}_{V,m,0} := |D_{\mathbf{A}^>_m}| + V + H^>_{\mathrm{f}}\,,$$

$$H^{\mathrm{np}}_{V,m,0} := P^+_{\mathbf{A}^>_m}\,(D_{\mathbf{A}^>_m} + V + H^>_{\mathrm{f}})\,P^+_{\mathbf{A}^>_m}\,.$$

The latter operators act in the Hilbert spaces $\mathscr{H}_m^>$ and $P^+_{\mathbf{A}_m^>}\mathscr{H}_m^>$, respectively. The following lemma [28, 29] shows in particular that it suffices to prove the existence of ground states in these truncated Hilbert spaces.

Lemma 6.2: *Assume that V fulfills Hypothesis 6.1. Then, for all $m > 0$,*

$$E^\sharp_{V,m} = \inf \sigma[H^\sharp_{V,m,0}], \qquad \Sigma^\sharp_m = \inf \sigma[H^\sharp_{0,m,0}] .$$

Moreover, if $E^\sharp_{V,m}$ is an eigenvalue of $H^\sharp_{V,m,0}$, then it is also an eigenvalue of $H^\sharp_{V,m}$.

Proof: It is clear that $H^\sharp_{V,m,0} \otimes \mathbb{1} \leqslant H^\sharp_{V,m}$, whence $\inf \sigma[H^\sharp_{V,m,0}] \leqslant \inf \sigma[H^\sharp_{V,m}]$. Next, we pick a minimizing sequence of normalized vectors $\psi_n^> \in \mathcal{Q}(H^\sharp_{V,m,0})$, $\langle \psi_n^> \,|\, H^\sharp_{V,m,0}\, \psi_n^> \rangle \to \inf \sigma[H^\sharp_{V,m,0}]$. Setting $\psi_n := \psi_n^> \otimes \Omega^>$, where $\Omega^>$ is the vacuum vector in $\mathscr{F}_{\mathrm{b}}[\mathscr{K}_m^>]$, we observe that $\langle \psi_n \,|\, H^\sharp_{V,m}\, \psi_n \rangle = \langle \psi_n^> \,|\, H^\sharp_{V,m,0}\, \psi_n^> \rangle$, thus $\inf \sigma[H^\sharp_{V,m,0}] \geqslant \inf \sigma[H^\sharp_{V,m}]$. Likewise, if $\phi^\sharp_m \in \mathscr{H}_m^>$ is a ground state eigenvector of $H^\sharp_{V,m,0}$, then $\phi^\sharp_m \otimes \Omega^>$ is a ground state eigenvector of $H^\sharp_{V,m}$. □

In order to show the existence of a ground state for $H^\sharp_{V,m,0}$ in the next step it is sufficient to show that the spectrum of $H^\sharp_{V,m,0}$ is discrete in a neighborhood of $E^\sharp_{V,m}$. ($E^\sharp_{V,m}$ is contained in the essential spectrum of $H^\sharp_{V,m}$ on the contrary.) A general strategy to achieve this would be the following. We could seek for a self-adjoint operator, A, satisfying $-\infty < A \leqslant H^\sharp_{V,m,0}$ and having discrete spectrum in $(-\infty,\, E^\sharp_{V,m} + c]$, for some $c > 0$. If such an operator A exists then also $H^\sharp_{V,m,0}$ has discrete spectrum in $(-\infty, E^\sharp_{V,m} + c]$. We need, however, a modification of this strategy. Let χ denote the spectral projection of $H^\sharp_{V,m,0}$ corresponding to some half-line $(-\infty,\, E^\sharp_{V,m} + c]$, $c > 0$. Then we seek for a self-adjoint auxiliary operator A such that

$$\chi\, A\, \chi \leqslant \chi\, \{H^\sharp_{V,m,0} - E^\sharp_{V,m} - 2c\}\, \chi \leqslant -c\, \chi \tag{6.7}$$

and $\mathrm{Tr}\{\chi\, A\, \chi\} > -\infty$, where Tr denotes the trace. For in this case we have $\mathrm{Tr}\{\chi\} < \infty$. The latter strategy is advantageous since we only have to compare A and $H^\sharp_{V,m,0}$ on the range of χ whose elements are exponentially localized by Theorem 5.1, provided that $c > 0$ is appropriately chosen. A suitable comparison operator A is constructed by means of a discretization of the photon momenta in the next subsection.

6.2. *Discretization of the photon momenta*

On $\mathscr{H}_m^>$ we introduce a discretization in the photon momenta: For every $\varepsilon > 0$, we decompose $\mathcal{A}_m = \{|\mathbf{k}| \geqslant m\}$ as

$$\mathcal{A}_m = \bigcup_{\boldsymbol{\nu} \in (\varepsilon\mathbb{Z})^3} Q_m^\varepsilon(\boldsymbol{\nu}), \quad Q_m^\varepsilon(\boldsymbol{\nu}) := \big(\boldsymbol{\nu} + [-\varepsilon/2\,, \varepsilon/2)^3\big) \cap \mathcal{A}_m\,, \;\; \boldsymbol{\nu} \in (\varepsilon\mathbb{Z})^3.$$

Of course, for every $\mathbf{k} \in \mathcal{A}_m$, we find a unique vector, $\widetilde{\boldsymbol{\nu}}_\varepsilon(\mathbf{k}) \in (\varepsilon\mathbb{Z})^3$, such that $\mathbf{k} \in Q_m^\varepsilon(\widetilde{\boldsymbol{\nu}}_\varepsilon(\mathbf{k}))$. To each $\boldsymbol{\nu} \in (\varepsilon\mathbb{Z})^3$ with $Q_m^\varepsilon(\boldsymbol{\nu}) \neq \varnothing$ we further associate some $\varkappa_{m,\varepsilon}(\boldsymbol{\nu}) \in \overline{Q_m^\varepsilon(\boldsymbol{\nu})}$ such that

$$|\varkappa_{m,\varepsilon}(\boldsymbol{\nu})| = \inf_{\mathbf{k} \in Q_m^\varepsilon(\boldsymbol{\nu})} |\mathbf{k}|\,.$$

In this way we obtain a map

$$\boldsymbol{\nu}_\varepsilon \,:\, \mathcal{A}_m \times \mathbb{Z}_2 \longrightarrow \mathbb{R}^3\,, \qquad k = (\mathbf{k}, \lambda) \longmapsto \boldsymbol{\nu}_\varepsilon(k) := \varkappa_{m,\varepsilon}\big(\widetilde{\boldsymbol{\nu}}_\varepsilon(\mathbf{k})\big)\,. \tag{6.8}$$

It is evident that the vectors $\varkappa_{m,\varepsilon}(\boldsymbol{\nu})$ can be chosen such that

$$\boldsymbol{\nu}_\varepsilon(-\mathbf{k}, \lambda) = -\boldsymbol{\nu}_\varepsilon(\mathbf{k}, \lambda)\,, \qquad \text{for almost every } \mathbf{k} \in \mathcal{A}_m\,. \tag{6.9}$$

The set of Lebesgue measure zero where the identity (6.9) might not hold is contained in the union of all planes which are perpendicular to some coordinate axis and contain points of the lattice $(\varepsilon\,\mathbb{Z})^3$. We define the ε-average of a locally integrable function, f, on $\mathcal{A}_m \times \mathbb{Z}_2$ by

$$[P_\varepsilon f](k) := \frac{1}{|Q_m^\varepsilon(\widetilde{\boldsymbol{\nu}}_\varepsilon(\mathbf{k}))|} \int\limits_{Q_m^\varepsilon(\widetilde{\boldsymbol{\nu}}_\varepsilon(\mathbf{k}))} f(\mathbf{p}, \lambda)\, d^3\mathbf{p}\,, \tag{6.10}$$

and introduce the following discretized coupling function,

$$\mathbf{G}_{\mathbf{x},m,\varepsilon}(k) := -(e/2\pi)\, e^{-i\boldsymbol{\nu}_\varepsilon(k)\cdot\mathbf{x}}\, P_\varepsilon\big[\mathbb{1}_{\mathcal{A}_m}\, \omega^{-1/2}\, \boldsymbol{\varepsilon}\big](k)\,,$$

for all $\mathbf{x} \in \mathbb{R}^3$ and almost every $k = (\mathbf{k}, \lambda) \in \mathcal{A}_m \times \mathbb{Z}_2$. In order to compare $\mathbf{G}_{\mathbf{x},m}$ with $\mathbf{G}_{\mathbf{x},m,\varepsilon}$ we put

$$\triangle_*^2(a, m, \varepsilon) := \int\limits_{\mathcal{A}_m \times \mathbb{Z}_2} \big(\omega(k) + \omega^{-1}(k)\big) \sup_{\mathbf{x}} \big\{e^{-a|\mathbf{x}|}|\mathbf{G}_{\mathbf{x},m}(k) - \mathbf{G}_{\mathbf{x},m,\varepsilon}(k)|^2\big\} dk, \tag{6.11}$$

for $a, m, \varepsilon > 0$. It is elementary to verify that $\triangle_*(a, m, \varepsilon) \to 0$, as $\varepsilon \searrow 0$, for fixed $a, m > 0$. Notice that $\triangle_*(a, m, \varepsilon)$ did not converge to zero if we chose $a = 0$. The fact that we need some weight function in $\mathbf{x}$ to control the difference between $\mathbf{G}_{\mathbf{x},m}$ and $\mathbf{G}_{\mathbf{x},m,\varepsilon}$ is one of the reasons why a localization estimate for spectral subspaces is required to prove the existence of ground states. The discretized vector potential is now given as

$$\mathbf{A}_{m,\varepsilon} := a^\dagger(\mathbf{G}_{m,\varepsilon}) + a(\mathbf{G}_{m,\varepsilon}), \quad a^\sharp(\mathbf{G}_{m,\varepsilon}) := \int_{\mathbb{R}^3}^{\oplus} \mathbb{1}_{\mathbb{C}^4} \otimes a^\sharp(\mathbf{G}_{\mathbf{x},m,\varepsilon})\, d^3\mathbf{x}\,.$$

The reason why we choose vectors $\boldsymbol{\nu}_\varepsilon$ fulfilling (6.9) is its consequence $\mathbf{G}_{\mathbf{x},m,\varepsilon}(-\mathbf{k},\lambda) = \epsilon(\lambda)\,\overline{\mathbf{G}}_{\mathbf{x},m,\varepsilon}(\mathbf{k},\lambda)$ with $\epsilon(0) = -1$ and $\epsilon(1) = 1$. The latter identity ensures that different components, $\mathbf{A}^{(i)}_{m,\varepsilon}(\mathbf{x}), \mathbf{A}^{(j)}_{m,\varepsilon}(\mathbf{y})$, $\mathbf{x},\mathbf{y} \in \mathbb{R}^3$, $i,j \in \{1,2,3\}$, of the discretized vector potential still commute. We have used this property in Section 3. The dispersion relation is discretized as

$$\omega_\varepsilon(k) := \inf\{\,|\mathbf{p}| \,:\, \mathbf{p} \in Q^\varepsilon_m(\widetilde{\boldsymbol{\nu}}_\varepsilon(\mathbf{k}))\,\}\,, \qquad k = (\mathbf{k},\lambda) \in \mathcal{A}_m \times \mathbb{Z}_2\,.$$

Then $|\boldsymbol{\nu}_\varepsilon| \leqslant \omega_\varepsilon$ on $\mathcal{A}_m \times \mathbb{Z}_2$ and

$$m \leqslant \omega_\varepsilon \leqslant \omega \ \text{ on } \ \mathcal{A}_m \times \mathbb{Z}_2\,, \qquad H_{\mathrm{f},m,\varepsilon} := d\Gamma(\omega_\varepsilon) \leqslant H^>_{\mathrm{f},m}\,. \tag{6.12}$$

Here the operators $H_{\mathrm{f},m,\varepsilon}$ and $H^>_{\mathrm{f},m}$ are acting in $\mathscr{F}_{\mathrm{b}}[\mathscr{K}^>_m]$. Finally, we define discretized Hamiltonians, $H^\sharp_{V,m,\varepsilon}$, acting in $L^2(\mathbb{R}^2,\mathbb{C}^4) \otimes \mathscr{F}_{\mathrm{b}}[\mathscr{K}^>_m]$,

$$\begin{aligned} H^{\mathrm{PF}}_{V,m,\varepsilon} &:= |D_{\mathbf{A}_{m,\varepsilon}}| + V + H_{\mathrm{f},m,\varepsilon}\,,\\ H^{\mathrm{np}}_{V,m,\varepsilon} &:= P^+_{\mathbf{A}_{m,\varepsilon}}\,(D_{\mathbf{A}_{m,\varepsilon}} + V + H_{\mathrm{f},m,\varepsilon})\,P^+_{\mathbf{A}_{m,\varepsilon}}\,. \end{aligned}$$

6.3. *Comparison of operators with different coupling functions*

In order to compare the various modified operators we derive some bounds on differences of projections whose proofs are essentially consequences of the ideas collected in Subsection 2.4 and the bounds

$$\big\|\,\boldsymbol{\alpha}\cdot(\mathbf{A} - \mathbf{A}_m)\,\check{H}_{\mathrm{f}}^{-1/2}\,\big\| = \mathcal{O}(\triangle(m))\,, \tag{6.13}$$

$$\big\|\,\boldsymbol{\alpha}\cdot(\mathbf{A}^>_m - \mathbf{A}_{m,\varepsilon})\,\check{H}_{\mathrm{f}}^{-1/2}\,e^{-a|\mathbf{x}|}\,\big\| = \mathcal{O}(\triangle_*(a,m,\varepsilon))\,. \tag{6.14}$$

Here we use the notation (6.2) and (6.11).

Lemma 6.3: *Let V be a symmetric multiplication operator in $L^2(\mathbb{R}^3)$ which is relatively form bounded with respect to $\sqrt{-\Delta}$.*

(i) Set $\check{H}_{\mathrm{f}} = H_{\mathrm{f}} + E$, for some sufficiently large $E \geqslant 1$ depending on e and

Λ, and let $\nu \geqslant 0$. Then, as $m \searrow 0$,

$$\big\| \, |D_{\mathbf{A}}|^{1/2} \, (P^{\pm}_{\mathbf{A}} - P^{\pm}_{\mathbf{A}_m}) \, \check{H}_{\mathrm{f}}^{-1/2} \, \big\| \leqslant \mathcal{O}\big(\triangle(m)\big) \,, \tag{6.15}$$

$$\big\| \, |D_{\mathbf{A}_m}|^{1/2} \, (P^{\pm}_{\mathbf{A}} - P^{\pm}_{\mathbf{A}_m}) \, \check{H}_{\mathrm{f}}^{-1/2} \, \big\| \leqslant \mathcal{O}\big(\triangle(m)\big) \,, \tag{6.16}$$

$$\big\| \, \check{H}_{\mathrm{f}}^{\nu} \, (P^{\pm}_{\mathbf{A}} - P^{\pm}_{\mathbf{A}_m}) \, \check{H}_{\mathrm{f}}^{-\nu-1/2} \, \big\| \leqslant \mathcal{O}\big(\triangle(m)\big) \,, \tag{6.17}$$

$$\big\| \, |V|^{1/2} \, (P^{\pm}_{\mathbf{A}} - P^{\pm}_{\mathbf{A}_m}) \, \check{H}_{\mathrm{f}}^{-1} \, \big\| \leqslant \mathcal{O}\big(\triangle(m)\big) \,. \tag{6.18}$$

(ii) Let $\check{H}_{\mathrm{f}}$ be $H^{>}_{\mathrm{f},m} + E$ or $H_{\mathrm{f},m,\varepsilon} + E$, for some sufficiently large $E \geqslant 1$ depending on e and Λ, and let $\nu \geqslant 0$ and $a_0 \in (0,1)$. Then, for every $a \in (0, a_0]$ and $F \in C^{\infty}(\mathbb{R}^3_{\mathbf{x}}, [0,\infty))$ satisfying $|\nabla F| \leqslant a$, $F(\mathbf{x}) \geqslant a|\mathbf{x}|$, for all $\mathbf{x} \in \mathbb{R}^3$, and $F(\mathbf{x}) = a|\mathbf{x}|$, for large $|\mathbf{x}|$, and for all sufficiently small $m, \varepsilon > 0$,

$$\big\| \, |D_{\mathbf{A}^{>}_m}|^{1/2} \, (P^{\pm}_{\mathbf{A}^{>}_m} - P^{\pm}_{\mathbf{A}_{m,\varepsilon}}) \, \check{H}_{\mathrm{f}}^{-1/2} \, e^{-F} \, \big\| \leqslant \mathcal{O}\big(\triangle_*(a,m,\varepsilon)\big) \,, \tag{6.19}$$

$$\big\| \, |D_{\mathbf{A}_{m,\varepsilon}}|^{1/2} \, (P^{\pm}_{\mathbf{A}^{>}_m} - P^{\pm}_{\mathbf{A}_{m,\varepsilon}}) \, \check{H}_{\mathrm{f}}^{-1/2} \, e^{-F} \, \big\| \leqslant \mathcal{O}\big(\triangle_*(a,m,\varepsilon)\big) \,, \tag{6.20}$$

$$\big\| \, \check{H}_{\mathrm{f}}^{\nu} \, (P^{\pm}_{\mathbf{A}^{>}_m} - P^{\pm}_{\mathbf{A}_{m,\varepsilon}}) \, \check{H}_{\mathrm{f}}^{-\nu-1/2} \, e^{-F} \, \big\| \leqslant \mathcal{O}\big(\triangle_*(a,m,\varepsilon)\big) \,, \tag{6.21}$$

$$\big\| \, |V|^{1/2} \, (P^{\pm}_{\mathbf{A}^{>}_m} - P^{\pm}_{\mathbf{A}_{m,\varepsilon}}) \, \check{H}_{\mathrm{f}}^{-1} \, e^{-F} \, \big\| \leqslant \mathcal{O}\big(\triangle_*(a,m,\varepsilon)\big) \,. \tag{6.22}$$

Proof: By the assumption on V and Theorem 3.4 we have $|V| \leqslant C\,(|D_{\mathbf{A}}| + \check{H}_{\mathrm{f}})$. Therefore, the bounds (6.18) and (6.22) are consequences of (6.15) and (6.17) and (6.19) and (6.21), respectively. In order to prove the remaining estimates we pick two vector potentials, $\mathbf{A}_1$ and $\mathbf{A}_2$, such that the set $\{\mathbf{A}_1, \mathbf{A}_2\}$ equals either $\{\mathbf{A}, \mathbf{A}_m\}$ (in which case $F := 0$ in what follows) or $\{\mathbf{A}^{>}_m, \mathbf{A}_{m,\varepsilon}\}$. For $j = 1,2$ and $y \in \mathbb{R}$, we set $R_{\mathbf{A}_j}(iy) := (D_{\mathbf{A}_j} - iy)^{-1}$ and $R^F_{\mathbf{A}_j}(iy) := (D_{\mathbf{A}_j} + i\boldsymbol{\alpha} \cdot \nabla F - iy)^{-1}$. (Recall (2.20) and (2.21).) Then we have the following resolvent identity, for $\mu \geqslant 1/2$ and $y \in \mathbb{R}$,

$$\begin{aligned} &(R_{\mathbf{A}_1}(iy) - R_{\mathbf{A}_2}(iy))\, e^{-F} \, \check{H}_{\mathrm{f}}^{-\mu} \\ &\qquad = R_{\mathbf{A}_1}(iy) \, \boldsymbol{\alpha} \cdot (\mathbf{A}_2 - \mathbf{A}_1) \, \check{H}_{\mathrm{f}}^{-\mu} e^{-F} R^F_{\mathbf{A}_2}(iy) \, \Upsilon^F_{\mu}(iy) \,, \end{aligned}$$

where $\Upsilon^F_{\mu}(iy)$ is the bounded operator defined in (9.5) below (with $\mathbf{A}$ replaced by $\mathbf{A}_1$). Using (2.18) we find, for all ϕ and f in the Hilbert space,

$$\begin{aligned} &\big|\langle \, f \, | \, (P^{\pm}_{\mathbf{A}_1} - P^{\pm}_{\mathbf{A}_2}) \, \check{H}_{\mathrm{f}}^{-\mu} \, e^{-F} \, \phi \, \rangle\big| \\ &\leqslant \int_{\mathbb{R}} \big|\langle \, f \, | \, R_{\mathbf{A}_1}(iy) \, \boldsymbol{\alpha} \cdot (\mathbf{A}_2 - \mathbf{A}_1) \, \check{H}_{\mathrm{f}}^{-\mu} e^{-F} R^F_{\mathbf{A}_2}(iy) \Upsilon^F_{\mu}(iy) \, \phi \, \rangle\big| \, \frac{dy}{\pi} \,. \end{aligned} \tag{6.23}$$

In the case $\mu = 1/2$, we choose $f = |D_{\mathbf{A}_1}|^{1/2} \, \psi$, $\psi \in \mathcal{D}(|D_{\mathbf{A}_1}|^{1/2})$, and observe that the integrand in (6.23) is then bounded by

$$\mathcal{O}_* \, \big\| |D_{\mathbf{A}_1}|^{1/2} R_{\mathbf{A}_1}(iy) \big\| \, \big\| R_{\mathbf{A}_2,L}(iy) \big\| \, \big\| \Upsilon_{1/2,L}(iy) \big\| \, \|\phi\| \, \|\psi\| \,,$$

which is integrable due to (2.21), (6.13), (6.14), and the spectral theorem. Here and below $\mathcal{O}_* = \mathcal{O}(\triangle(m))$ or $\mathcal{O}_* = \mathcal{O}(\triangle_*(a,m,\varepsilon))$ depending on the choice of $\mathbf{A}_j$. This concludes the proof of (6.15), (6.16), (6.19), and (6.20).

In order to prove (6.17) and (6.21) we infer from Lemma 9.1 that the commutator $T_\nu := \check{H}_{\mathrm{f}}^{\nu}\,[\boldsymbol{\alpha}\cdot(\mathbf{A}_2-\mathbf{A}_1)\,e^{-F}\,,\,\check{H}_{\mathrm{f}}^{-\nu}]$ extends to a bounded operator with $\|T_\nu\| = \mathcal{O}_*$. Together with (6.13), (6.14), and (9.7) this implies, for $\nu \geqslant 0$ and $\psi,\phi \in \mathcal{D}(H_{\mathrm{f}}^{\nu})$,

$$\begin{aligned}
&\big|\big\langle\, \check{H}_{\mathrm{f}}^{\nu}\psi \,\big|\, R_{\mathbf{A}_1}(iy)\,\boldsymbol{\alpha}\cdot(\mathbf{A}_2-\mathbf{A}_1)\,\check{H}_{\mathrm{f}}^{-\nu-1/2}e^{-F}R^F_{\mathbf{A}_2}(iy)\,\Upsilon^F_{\nu+1/2}(iy)\,\phi\,\big\rangle\big|\\
&= \big|\big\langle\, \psi \,\big|\, R_{\mathbf{A}_1}(iy)\,\Upsilon^F_{\nu}(iy)\,\check{H}_{\mathrm{f}}^{\nu}\,\boldsymbol{\alpha}\cdot(\mathbf{A}_2-\mathbf{A}_1)\,\check{H}_{\mathrm{f}}^{-\nu-1/2}e^{-F}\\
&\qquad\qquad\qquad\qquad\qquad\qquad \times R^F_{\mathbf{A}_2}(iy)\,\Upsilon^F_{\nu+1/2}(iy)\,\phi\,\big\rangle\big|\\
&\leqslant \mathcal{O}_*\,\|R_{\mathbf{A}_1}(iy)\|\,\|\Upsilon^F_{\nu}(iy)\|\,\|R^F_{\mathbf{A}_2}(iy)\|\,\|\Upsilon^F_{\nu+1/2}(iy)\|\,\|\psi\|\,\|\phi\|\\
&\leqslant \mathcal{O}_*\,(1+y^2)^{-1}\,\|\psi\|\,\|\phi\|\,.
\end{aligned}$$

Therefore, (6.17) and (6.21) follow from (6.23) upon choosing $\mu = \nu + 1/2$ and $f = \check{H}_{\mathrm{f}}^{\nu}\psi$. □

6.4. *Higher order estimates and their consequences*

As a preparation for the proof of the existence of ground states for $H^{\sharp}_{V,m,0}$ we derive bounds on certain expectation values of the difference $H^{\sharp}_{V,m,0} - H^{\sharp}_{V,m,\varepsilon}$ in this subsection. Moreover, we compare the ground state energies and ionization thresholds of $H^{\sharp}_{V,m,0}$ with those of $H^{\sharp}_{V,m,\varepsilon}$ defined by

$$E^{\sharp}_{V,m,\varepsilon} := \inf\sigma[H^{\sharp}_{V,m,\varepsilon}]\,, \qquad \Sigma^{\sharp}_{m,\varepsilon} := \inf\sigma[H^{\sharp}_{0,m,\varepsilon}]\,,$$

for $m,\varepsilon > 0$. It is not possible to compare the no-pair operators $H^{\mathrm{np}}_{V,m,0}$ and $H^{\mathrm{np}}_{V,m,\varepsilon}$ in a quadratic form sense. For some error terms in the difference of these two operators "have the size of $|\mathbf{x}|\,H_{\mathrm{f}}^{3/2}$" and can only be controlled when we take expectations with respect to states in some low-lying exponentially localized spectral subspace. To control the higher power $H_{\mathrm{f}}^{3/2}$ of the radiation field energy we need, however, yet another non-trivial ingredient, namely the higher order estimates of the next theorem. Because of lack of space we cannot comment on the proof of Theorem 6.4 and refer to [36] instead. We remark that, for the semi-relativistic Pauli-Fierz operator, higher order estimates have been obtained earlier in [14]. Their proof given in [36] is, however, different and more model-independent so that the no-pair operator can also be treated along the same lines in [36]. In the case of the no-pair operator only the Coulomb potential is considered in [36].

An inspection of the proofs in [36] shows, however, that they immediately extend to all potentials satisfying Hypothesis 6.1.

Theorem 6.4: *Let $e \in \mathbb{R}$, $\Lambda \in (0,\infty)$, and assume that V fulfills Hypothesis 6.1. Then $\mathcal{D}((H^{\sharp}_{V,m,\varepsilon})^{n/2}) \subset \mathcal{D}(H_{\mathrm{f}}^{n/2})$, for every $n \in \mathbb{N}$, and there exist constants, $\varepsilon_0, m_0, C \in (0,\infty)$, such that, for all $\varepsilon \in [0,\varepsilon_0]$ and $m \in (0,m_0]$,*

$$\big\|H^{n/2}_{\mathrm{f},m,\varepsilon}\restriction_{P^+_{\mathbf{A}_{m,\varepsilon}}\mathscr{H}^{>}_m}\big(H^{\mathrm{np}}_{V,m,\varepsilon} - (E^{\mathrm{np}}_{V,m,\varepsilon}-1)P^+_{\mathbf{A}_{m,\varepsilon}}\big)^{-n/2}\big\| \leqslant C(1+|E^{\mathrm{np}}_{V,m,\varepsilon}|)^{2n},$$

$$\big\|H^{n/2}_{\mathrm{f},m,\varepsilon}\big(H^{\mathrm{PF}}_{V,m,\varepsilon} - (E^{\mathrm{PF}}_{V,m,\varepsilon}-1)\big)^{-n/2}\big\| \leqslant C(1+|E^{\mathrm{PF}}_{V,m,\varepsilon}|)^{2n}.$$

If $V=0$, then $E^{\sharp}_{V,m,\varepsilon}$ has to be replaced by $\Sigma^{\sharp}_{m,\varepsilon}$ in these bounds. Analogous bounds hold for $H^{\sharp}_{V}$.

Lemma 6.5: *Let $e \in \mathbb{R}$, $\Lambda \in (0,\infty)$, and assume that V fulfills Hypothesis 6.1. Then we find some $m_0 > 0$ such that the following holds:*

(i) For all $m \in (0,m_0]$ and $\psi^+ \in \mathrm{Ran}(\mathbb{1}_{(-\infty,\Sigma^{\sharp}+1]}(H^{\sharp}_V))$,

$$\big|\langle\,\psi^+\,|\,H^{\sharp}_{V,m}\,\psi^+\,\rangle - \langle\,\psi^+\,|\,H^{\sharp}_{V}\,\psi^+\,\rangle\big| \leqslant \mathrm{const}(\Sigma^{\sharp},|E^{\sharp}_V|)\,o(m^0)\,\|\psi^+\|^2.$$

(ii) For all $m \in (0,m_0]$ and $\psi^+ \in \mathrm{Ran}(\mathbb{1}_{(-\infty,\Sigma^{\sharp}_m+1]}(H^{\sharp}_{V,m}))$,

$$\big|\langle\,\psi^+\,|\,H^{\sharp}_{V,m}\,\psi^+\,\rangle - \langle\,\psi^+\,|\,H^{\sharp}_{V}\,\psi^+\,\rangle\big| \leqslant \mathrm{const}(\Sigma^{\sharp}_m,|E^{\sharp}_{V,m}|)\,o(m^0)\,\|\psi^+\|^2.$$

Proof: We treat only the no-pair operator explicitly. On account of the formula $|D_{\mathbf{A}}| = P^+_{\mathbf{A}}\,D_{\mathbf{A}} - P^-_{\mathbf{A}}\,D_{\mathbf{A}}$ it will then be clear how to obtain the result also for the semi-relativistic Pauli-Fierz operator. We remark only once that, for instance, the inclusion $P^+_{\mathbf{A}_m}\,\mathrm{Ran}(\mathbb{1}_{(-\infty,\Sigma^{\sharp}+1]}(H^{\sharp}_V)) \subset \mathcal{Q}(H^{\sharp}_{V,m})$ follows from the characterization of the form domains in Theorems 3.4 and 3.6. In the rest of this section we shall use similar remarks without further notice to simplify the exposition. Let $\delta P := P^+_{\mathbf{A}} - P^+_{\mathbf{A}_m}$ and $\mathcal{M} := \{D_{\mathbf{A}}, V, H_{\mathrm{f}}\}$. Then we have

$$H^{\mathrm{np}}_V - H^{\mathrm{np}}_{V,m} = P^+_{\mathbf{A}_m}\boldsymbol{\alpha}\cdot(\mathbf{A}-\mathbf{A}_m)P^+_{\mathbf{A}_m} + \sum_{T\in\mathcal{M}}\big\{2\mathrm{Re}\,\big[P^+_{\mathbf{A}}\,T\,\delta P\big] - \delta P\,T\,\delta P\big\} \tag{6.24}$$

in the sense of quadratic forms on $\mathcal{Q}(|D_{\mathbf{0}}|) \cap \mathcal{Q}(H_{\mathrm{f}})$. Now, let $\psi^+ \in \mathrm{Ran}(\mathbb{1}_{(-\infty,\Sigma^{\sharp}+1]}(H^{\sharp}_V))$. From (3.16), (6.13), (6.15)–(6.18), and (6.24) we readily infer that

$$\big|\langle\,\psi^+\,|\,(H^{\mathrm{np}}_V - H^{\mathrm{np}}_{V,m})\,\psi^+\,\rangle\big| \leqslant \mathcal{O}(\triangle(m))\,\big\{\|\,|D_{\mathbf{A}}|^{1/2}\,\psi^+\|^2 + \|\check{H}_{\mathrm{f}}\,\psi^+\|^2\big\}.$$

Here $\| \, |D_{\mathbf{A}}|^{1/2} \, \psi^+ \| \leqslant \mathcal{O}(1) \, \|\psi^+\|$ since ψ^+ belongs to the spectral subspace $\mathrm{Ran}(\mathbb{1}_{(-\infty,\Sigma^\sharp+1]}(H_V^\sharp))$ and the term containing V is a small form perturbation of $H_V^\sharp$ by Hypothesis 6.1 and Theorem 3.4 or Theorem 3.6, respectively. Moreover, $\|\check{H}_{\mathrm{f}} \, \psi^+\| \leqslant \mathrm{const}(\Sigma, |E_V|) \, \|\psi^+\|$ because of the higher order estimates. We can argue analogously if ψ^+ belongs to a spectral subspace of $H_{V,m}^\sharp$. In this case the right hand side of the estimate depends on Σ_m and $|E_{V,m}|$ since we apply higher order estimates for $H_{V,m}^\sharp$. □

We recall once more that some **x**-dependent weight is required to control the difference between $\mathbf{A}_m^>$ and $\mathbf{A}_{m,\varepsilon}$. If we consider only vectors ψ^+ in a spectral subspace corresponding to sufficiently low energies of $H_{V,m,\varepsilon}^\sharp$, then we can borrow this weight from the exponential localization of ψ^+. At this point we have to introduce further conditions on the potential in order to guarantee that *there are* non-trivial spectral subspaces below the ionization threshold.

Hypothesis 6.6: V satisfies (5.3) and there exist $c, m_\star, \varepsilon_\star > 0$ such that, for all $m \in (0, m_\star]$ and all $\varepsilon \in (0, \varepsilon_\star]$,

$$\Sigma^\sharp - E_V^\sharp \geqslant c\,, \quad \Sigma_m^\sharp - E_{V,m}^\sharp \geqslant c\,, \quad \Sigma_{m,\varepsilon}^\sharp - E_{V,m,\varepsilon}^\sharp \geqslant c\,. \tag{6.25}$$

Examples of potentials fulfilling the previous hypothesis have been found in Theorem 4.1. In fact, as already pointed out there, the proof of Theorem 4.1 works also for the discretized operators when the unitary transformation employed in Section 4 is modified suitably; see [28, 29] for details. There are, however, potentials fulfilling Hypotheses 6.1 and 6.6 which are not covered by Theorem 4.1, for instance, those mentioned in Remark 4.2. This is the reason why we work with the very implicit Hypothesis 6.6 in what follows.

Lemma 6.7: *Let $e \in \mathbb{R}$, $\Lambda \in (0,\infty)$, and assume that V fulfills Hypotheses 6.1 and 6.6. Then there exist $\varepsilon_0, m_0 > 0$ such that the following holds:*

(i) For all $\varepsilon \in (0, \varepsilon_0]$, $m \in (0, m_0]$, $\lambda \in (E_{V,m}^\sharp, \Sigma_m^\sharp)$, and $\psi^+ \in \mathrm{Ran}(\mathbb{1}_{(-\infty,\lambda]}(H_{V,m}^\sharp))$,

$$\begin{aligned} &\big|\langle\, \psi^+ \,|\, H_{V,m,\varepsilon}^\sharp \, \psi^+ \,\rangle - \langle\, \psi^+ \,|\, H_{V,m,0}^\sharp \, \psi^+ \,\rangle\big| \\ &\qquad \leqslant \mathrm{const}\big(\Sigma_m^\sharp, |E_{V,m}^\sharp|, (\Sigma_m^\sharp - \lambda)^{-1}\big)\, o(\varepsilon^0)\, \|\psi^+\|^2. \end{aligned} \tag{6.26}$$

(ii) For all $\varepsilon \in (0, \varepsilon_0]$, $m \in (0, m_0]$, $\lambda \in (E^{\sharp}_{V,m,\varepsilon}, \Sigma^{\sharp}_{m,\varepsilon})$, *and* $\psi^+ \in \mathrm{Ran}(\mathbb{1}_{(-\infty,\lambda]}(H^{\sharp}_{V,m,\varepsilon}))$,

$$\begin{aligned} &\big|\langle\, \psi^+ \,|\, H^{\sharp}_{V,m,\varepsilon}\, \psi^+ \,\rangle - \langle\, \psi^+ \,|\, H^{\sharp}_{V,m,0}\, \psi^+ \,\rangle\big| \\ &\qquad \leqslant \mathrm{const}\big(\Sigma^{\sharp}_{m,\varepsilon}, |E^{\sharp}_{V,m,\varepsilon}|, (\Sigma^{\sharp}_{m,\varepsilon} - \lambda)^{-1}\big)\, o(\varepsilon^0)\, \|\psi^+\|^2. \end{aligned} \tag{6.27}$$

Proof: Again we only treat the no-pair operator since the proofs for the semi-relativistic Pauli-Fierz operator will then be obvious. (i): We have a formula for $H^{\mathrm{np}}_{V,m,0} - H^{\mathrm{np}}_{V,m,\varepsilon}$ similar to (6.24) with $(\mathbf{A}, \mathbf{A}_m, H_{\mathrm{f}})$ replaced by $(\mathbf{A}^{>}_m, \mathbf{A}_{m,\varepsilon}, H^{>}_{\mathrm{f},m})$ where one additional term has to be added, namely $P^{+}_{\mathbf{A}_{m,\varepsilon}}\, d\Gamma(\omega - \omega_\varepsilon)\, P^{+}_{\mathbf{A}_{m,\varepsilon}}$. Using this formula, (6.14), (6.19)–(6.22), and $|\omega - \omega_\varepsilon| \leqslant \sqrt{3}\,\varepsilon \leqslant (\sqrt{3}\,\varepsilon/m)\,\omega_\varepsilon \leqslant (\sqrt{3}\,\varepsilon/m)\,\omega$, for $|\mathbf{k}| \geqslant m$, which yields

$$\big|\langle\, \psi^+ \,|\, P^{+}_{\mathbf{A}_{m,\varepsilon}}\, d\Gamma(\omega - \omega_\varepsilon)\, P^{+}_{\mathbf{A}_{m,\varepsilon}}\, \psi^+ \,\rangle\big| \leqslant o(\varepsilon^0)\, \big\|\check{H}_{\mathrm{f}}^{1/2}\, \psi^+\big\|^2,$$

we arrive at

$$\begin{aligned} &\big|\langle\, \psi^+ \,|\, (H^{\mathrm{np}}_{V,m,0} - H^{\mathrm{np}}_{V,m,\varepsilon})\, \psi^+ \,\rangle\big| \\ &\qquad \leqslant \big(\mathcal{O}(\triangle_*(a, m, \varepsilon)) + o(\varepsilon^0)\big)\, \big\{\|\, |D_{\mathbf{A}^{>}_m}|^{1/2}\, \psi^+\|^2 + \|e^F\, \check{H}_{\mathrm{f}}\, \psi^+\|^2\big\}, \end{aligned}$$

for $\psi^+ \in \mathrm{Ran}(\mathbb{1}_{(-\infty,\lambda]}(H^{\sharp}_{V,m}))$. Here we further have $2\|e^F\, \check{H}_{\mathrm{f}}\, \psi^+\|^2 \leqslant \|e^{2F}\, \psi^+\|^2 + \|\check{H}_{\mathrm{f}}^2\, \psi^+\|^2$, where the norms on the right side can be controlled by our exponential localization and higher order estimates, respectively. Part (ii) is derived analogously. The dependence on λ, the ionization thresholds, and the ground state energies of the constants on the right hand sides of (6.26) and (6.27) stems from the constants in the exponential localization and higher order estimates. □

6.5. *Continuity of the ionization thresholds and ground state energies*

To make use of the bounds of Lemmata 6.5 and 6.7 we still have to verify that the functions $m \mapsto \Sigma^{\sharp}_m$, $m \mapsto E^{\sharp}_{V,m}$, and $\varepsilon \mapsto \Sigma^{\sharp}_{m,\varepsilon}$, $\varepsilon \mapsto E^{\sharp}_{V,m,\varepsilon}$ are (semi-)continuous at 0. The continuity of $\varepsilon \mapsto E^{\sharp}_{V,m,\varepsilon}$ will also enter more directly into the proof of the existence of ground states in the next subsection.

Corollary 6.8: *Assume that* V *fulfills Hypothesis 6.1. Then it follows that* $\lim_{m\to 0} E^{\sharp}_{V,m} = E^{\sharp}_V$ *and* $\lim_{m\to 0} \Sigma^{\sharp}_m = \Sigma^{\sharp}$.

Proof: Using Lemma 6.5(i) it is not difficult to derive the bounds $\Sigma_m^\sharp \leqslant \Sigma^\sharp + o(m^0)$. (Given $\varepsilon > 0$ we pick some $\psi^+ \in \mathrm{Ran}(\mathbb{1}_{[\Sigma^\sharp, \Sigma^\sharp+\varepsilon)}(H_0^\sharp))$, $\|\psi^+\| = 1$, and plug it into the quadratic form of $H_{0,m}^\sharp$. In the case of the no-pair operator we also have to observe that $\|P_{\mathbf{A}_m}^+ \psi^+\| \to 1$, $m \searrow 0$.) Since this gives an upper bound on $\Sigma_m^\sharp$ which is uniform, for small m, we can then control the constants on the right hand side of the estimate in Lemma 6.5(ii) to get $\Sigma^\sharp \leqslant \Sigma_m^\sharp + o(m^0)$ by a similar argument. Since the results of Section 3 provide uniform lower bounds on $E_{V,m}$ we can now employ Lemma 6.5 in a similar fashion to show that $\lim_{m\to 0} E_{V,m}^\sharp = E_V^\sharp$. □

In order to compare the ionization thresholds $\Sigma_m^\sharp$ and $\Sigma_{m,\varepsilon}^\sharp$ we need a different argument since $\mathbf{x}$-dependent weights are required to control the difference between $\mathbf{A}_m^>$ and $\mathbf{A}_{m,\varepsilon}$ but the spectral subspaces of the free operators $H_{0,m,\varepsilon}^\sharp$ are not localized. Here the essential self-adjointness of $H_{0,m,\varepsilon}^\sharp$ asserted in Theorem 3.7 is helpful. To be able to work in one fixed Hilbert space we set

$$\widehat{H}_{0,m,\varepsilon}^{\mathrm{np}} := H_{0,m,\varepsilon}^{\mathrm{np}} + P_{\mathbf{A}_{m,\varepsilon}}^- \left(|D_{\mathbf{A}_{m,\varepsilon}}| + H_{\mathrm{f},m,\varepsilon}\right) P_{\mathbf{A}_{m,\varepsilon}}^- . \tag{6.28}$$

Then Lemma 8.3 below implies that $\Sigma_{m,\varepsilon}^{\mathrm{np}} = \inf \sigma[H_{0,m,\varepsilon}^{\mathrm{np}}] = \inf \sigma[\widehat{H}_{0,m,\varepsilon}^{\mathrm{np}}]$.

Lemma 6.9: *$H_{0,m,\varepsilon}^{\mathrm{PF}} \to H_{0,m,0}^{\mathrm{PF}}$ and $\widehat{H}_{0,m,\varepsilon}^{\mathrm{np}} \to \widehat{H}_{0,m,0}^{\mathrm{np}}$ in the strong resolvent sense, as $\varepsilon \searrow 0$. In particular, $\limsup_{\varepsilon\searrow 0} \Sigma_{m,\varepsilon}^\sharp \leqslant \Sigma_m^\sharp$.*

Proof: Since all involved operators are essentially self-adjoint on $\mathscr{D}$ it suffices to show that $H_{0,m,\varepsilon}^{\mathrm{PF}} \varphi \to H_{0,m,0}^{\mathrm{PF}} \varphi$ and $\widehat{H}_{0,m,\varepsilon}^{\mathrm{np}} \varphi \to \widehat{H}_{0,m,0}^{\mathrm{np}} \varphi$, for every fixed $\varphi \in \mathscr{D}$. Since φ has only finitely many non-vanishing Fock space components and the latter are compactly supported it is clear that $H_{\mathrm{f},m,\varepsilon}\, \varphi \to H_{\mathrm{f},m}^> \varphi$. Furthermore, we write

$$|D_{\mathbf{A}_m^>}| - |D_{\mathbf{A}_{m,\varepsilon}}| = S_{\mathbf{A}_m^>}\, \boldsymbol{\alpha} \cdot (\mathbf{A}_m^> - \mathbf{A}_{m,\varepsilon}) + (S_{\mathbf{A}_m^>} - S_{\mathbf{A}_{m,\varepsilon}})\, D_{\mathbf{A}_{m,\varepsilon}} ,$$

where $S := D\,|D|^{-1}$ denotes the sign function, which permits to get

$$\begin{aligned} \big\|\, |D_{\mathbf{A}_m^>}|\, \varphi - |D_{\mathbf{A}_{m,\varepsilon}}|\, \varphi \,\big\| &\leqslant \big\|e^{-F} \boldsymbol{\alpha} \cdot (\mathbf{A}_m^> - \mathbf{A}_{m,\varepsilon})\, \check{H}_{\mathrm{f}}^{-1/2}\big\|\, \big\|e^F\, \check{H}_{\mathrm{f}}^{1/2}\, \varphi\big\| \\ &+ \big\|(S_{\mathbf{A}_m^>} - S_{\mathbf{A}_{m,\varepsilon}})\, \check{H}_{\mathrm{f}}^{-1/2} e^{-F}\big\|\, \big\{\big\|e^F \check{H}_{\mathrm{f}}^{1/2} D_{\mathbf{0}}\, \varphi\big\| + \big\|\check{H}_{\mathrm{f}}^{1/2}\, \boldsymbol{\alpha} \cdot \mathbf{A}_{m,\varepsilon}\, e^F \varphi\big\|\big\}, \end{aligned}$$

where $\check{H}_{\mathrm{f}} := H_{\mathrm{f},m}^> + E$ and $E \equiv E(e, \Lambda)$ is sufficiently large, independently of ε. Using (6.14), (6.19), and Lemma 9.1 it is now easy to see that $H_{0,m,\varepsilon}^{\mathrm{PF}} \varphi \to H_{0,m,0}^{\mathrm{PF}} \varphi$.

To show that also $\widehat{H}^{\mathrm{np}}_{0,m,\varepsilon}\,\varphi \to \widehat{H}^{\mathrm{np}}_{0,m,0}\,\varphi$ it remains to observe that

$$\begin{aligned}
&\big\|P^{\pm}_{\mathbf{A}^{>}_{m}}\,H_{\mathrm{f}}^{>}\,P^{\pm}_{\mathbf{A}^{>}_{m}}\,\varphi - P^{\pm}_{\mathbf{A}_{m,\varepsilon}}\,H_{\mathrm{f}}^{>}\,P^{\pm}_{\mathbf{A}_{m,\varepsilon}}\,\varphi\big\| \\
&\leqslant \big\|(P^{\pm}_{\mathbf{A}^{>}_{m}} - P^{\pm}_{\mathbf{A}_{m,\varepsilon}})\,\check{H}_{\mathrm{f}}^{-1/2}\,e^{-F}\,\big\|\,\|e^{F}\,\check{H}_{\mathrm{f}}^{3/2}\,P^{\pm}_{\mathbf{A}^{>}_{m}}\,\check{H}_{\mathrm{f}}^{-3/2}\,e^{-F}\|\,\|e^{F}\,\check{H}_{\mathrm{f}}^{3/2}\,\varphi\| \\
&\quad + \big\|H_{\mathrm{f}}\,(P^{\pm}_{\mathbf{A}^{>}_{m}} - P^{\pm}_{\mathbf{A}_{m,\varepsilon}})\,\check{H}_{\mathrm{f}}^{-3/2}\,e^{-F}\,\big\|\,\|e^{F}\,\check{H}_{\mathrm{f}}^{3/2}\,\varphi\| \xrightarrow{\varepsilon\searrow 0} 0\,, \qquad (6.29)
\end{aligned}$$

and

$$\big\|P^{\pm}_{\mathbf{A}_{m,\varepsilon}}(H_{\mathrm{f}}^{>} - H_{\mathrm{f},m,\varepsilon})\,P^{\pm}_{\mathbf{A}_{m,\varepsilon}}\varphi\big\| \leqslant \frac{\sqrt{3}\,\varepsilon}{m}\,\big\|H_{\mathrm{f}}^{>}\,P^{\pm}_{\mathbf{A}_{m,\varepsilon}}\,\check{H}_{\mathrm{f}}^{-1}\big\|\,\|\check{H}_{\mathrm{f}}\,\varphi\| \xrightarrow{\varepsilon\searrow 0} 0\,. \qquad (6.30)$$

In fact, (6.29) is valid because of (6.19), (6.21), and because $e^{F}\,\check{H}_{\mathrm{f}}^{3/2}\,P^{\pm}_{\mathbf{A}^{>}_{m}}\,\check{H}_{\mathrm{f}}^{-3/2}e^{-F}$ turns out to be well-defined and bounded as a consequence of (2.24) and Lemma 9.3. Moreover, (6.30) holds true since $|\omega - \omega_\varepsilon| \leqslant \sqrt{3}\,\varepsilon$ and the photon momenta are $\geqslant m$ in modulus on $\mathscr{H}^{>}_{m}$, and since $\|H_{\mathrm{f}}^{>}\,P^{\pm}_{\mathbf{A}_{m,\varepsilon}}\,\check{H}_{\mathrm{f}}^{-1}\|$ is bounded uniformly in $\varepsilon > 0$ by Lemma 9.3. □

Corollary 6.10: *Let $m > 0$ be sufficiently small and assume that V fulfills Hypotheses 6.1 and 6.6. Then $\lim_{\varepsilon\to 0} E^{\sharp}_{V,m,\varepsilon} = E^{\sharp}_{V,m}$.*

Proof: Since Lemma 6.9 provides upper bounds on $\Sigma^{\sharp}_{m,\varepsilon}$ and we have lower bounds on the spectra of $H^{\sharp}_{V,m,\varepsilon}$ which are uniform in m and ε, we can apply Lemma 6.7 and some straightforward variational arguments to prove the assertion. We only point out one subtlety: In order to show that $E_{V,m} \leqslant E_{V,m,\varepsilon} + o(\varepsilon^0)$ we have to pick a test function in $\mathrm{Ran}(\mathbb{1}_{(-\infty,\lambda]}(H^{\sharp}_{V,m,\varepsilon}))$, for some $\lambda \in (E_{V,m,\varepsilon}\,,\,\Sigma^{\sharp}_{m,\varepsilon})$. To ensure that this is possible and in order to have an ε-independent bound on the numbers $(\Sigma^{\sharp}_{m,\varepsilon} - \lambda)^{-1}$ entering into the constant in (6.27) we need the lower bound (6.25) on the binding energy $\Sigma^{\sharp}_{m,\varepsilon} - E_{V,m,\varepsilon} \geqslant c$ which does not depend on ε. Without an estimate on the binding energy we still got $E_{V,m,\varepsilon} \leqslant E_{V,m} + o(\varepsilon^0)$, but we would not have a useful lower bound on $\Sigma^{\sharp}_{m,\varepsilon}$. □

6.6. *Proofs of the existence of ground states with mass*

The next theorem asserting compactness of spectral projections of $H^{\sharp}_{V,m,0}$ associated with sufficiently low energies is the final result of this section. As in [4] it is proved by estimating the trace of the spectral projection from above by the trace of some finite rank operator, namely the one in (6.37). This finite rank operator is constructed by means of a suitable restriction of the discretized field energy with discrete spectrum and a harmonic oscillator potential which compactifies the electronic part of the operator. In order to sneak in the harmonic oscillator potential in the proofs below we exploit the exponential localization of low-lying spectral subspaces once more. The latter idea stems from [28, 29] and replaces an argument in [4] that works only for small e and/or Λ.

Theorem 6.11: *Let $e \in \mathbb{R}$, $\Lambda \in (0, \infty)$, and assume that V fulfills Hypotheses 6.1 and 6.6. Define $\chi := \mathbb{1}_{(-\infty\,,\,E^{\sharp}_{V,m}+m/4]}(H^{\sharp}_{V,m,0})$ and assume that $m > 0$ is sufficiently small. Then $\mathrm{Tr}\{\chi\} < \infty$. In particular, $E^{\sharp}_{V,m}$ is an eigenvalue of both $H^{\sharp}_{V,m,0}$ and $H^{\sharp}_{V,m}$.*

In the proof of the preceding theorem, which is carried through separately for $\sharp = \mathrm{PF}$ and $\sharp = \mathrm{np}$ below, we shall employ an orthogonal splitting of $\mathscr{K}^{>}_{m}$ into subspaces of discrete and fluctuating photon states,

$$\mathscr{K}^{d}_{m} := P_{\varepsilon}\,\mathscr{K}^{>}_{m}\,, \qquad \mathscr{K}^{f}_{m} := \mathscr{K}^{>}_{m} \ominus \mathscr{K}^{d}_{m}\,.$$

Here P_{ε} is defined in (6.10). The splitting $\mathscr{K}^{>}_{m} = \mathscr{K}^{d}_{m} \oplus \mathscr{K}^{f}_{m}$ gives rise to an isomorphism

$$L^2(\mathbb{R}^3, \mathbb{C}^4) \otimes \mathscr{F}_{\mathrm{b}}[\mathscr{K}^{>}_{m}] \cong \big(L^2(\mathbb{R}^3, \mathbb{C}^4) \otimes \mathscr{F}_{\mathrm{b}}[\mathscr{K}^{d}_{m}]\big) \otimes \mathscr{F}_{\mathrm{b}}[\mathscr{K}^{f}_{m}]\,, \qquad (6.31)$$

and we observe that the Dirac operator and the field energy decompose under the above isomorphism as

$$D_{\mathbf{A}_{m,\varepsilon}} \cong D_{\mathbf{A}^{d}_{m,\varepsilon}} \otimes \mathbb{1}^{f}\,, \qquad H_{\mathrm{f},m,\varepsilon} = H^{d}_{\mathrm{f},m,\varepsilon} \otimes \mathbb{1}^{f} + \mathbb{1}^{d} \otimes H^{f}_{\mathrm{f},m,\varepsilon}\,. \qquad (6.32)$$

Here and in the following we designate operators acting in the Fock space factors $\mathscr{F}_{\mathrm{b}}[\mathscr{K}^{\ell}_{m}]$, $\ell \in \{d, f\}$, by the corresponding superscript d or f. In fact, the discretized vector potential $\mathbf{A}_{m,\varepsilon}$ acts on the various n-particle sectors in $\mathscr{F}_{\mathrm{b}}[\mathscr{K}^{>}_{m}]$ by tensor-multiplying or taking scalar products with elements from $\mathscr{K}^{d}_{m}$ (apart from symmetrization and a normalization constant).

For $\ell \in \{d, f\}$, we denote the identity on $\mathscr{F}_{\mathrm{b}}[\mathscr{K}^{\ell}_{m}]$ by $\mathbb{1}^{\ell}$ and the projection onto the vacuum sector in $\mathscr{F}_{\mathrm{b}}[\mathscr{K}^{\ell}_{m}]$ by $P_{\Omega^{\ell}}$ and write $P^{\perp}_{\Omega^{\ell}} := \mathbb{1}^{\ell} - P_{\Omega^{\ell}}$. The identity on $L^2(\mathbb{R}^3_{\mathbf{x}}, \mathbb{C}^4)$ is denoted as $\mathbb{1}^{\mathrm{el}}$.

The semi-relativistic Pauli-Fierz operator.

Proof: We prove Theorem 6.11 with $\sharp = \mathrm{PF}$. On account of Lemma 6.7(i) we have $\chi\, H^{\mathrm{PF}}_{V,m,0}\,\chi \geqslant \chi\, H^{\mathrm{PF}}_{V,m,\varepsilon}\,\chi - o(\varepsilon^0)\,\chi$. Using (6.32) and $H^{f}_{\mathrm{f},m,\varepsilon}\,P_{\Omega^{f}} = 0$, we then obtain

$$\begin{aligned}
&\chi\{\,H^{\mathrm{PF}}_{V,m,0} - E^{\mathrm{PF}}_{V,m} - m/2\,\}\,\chi + o(\varepsilon^0)\,\chi \\
&\quad\geqslant \chi\{\,\big[\,|D_{\mathbf{A}^d_{m,\varepsilon}}| + V + H^d_{\mathrm{f},m,\varepsilon} - E^{\mathrm{PF}}_{V,m} - m/2\,\big]\otimes P_{\Omega^f}\,\}\,\chi && (6.33)\\
&\quad+ \chi\{\,\big[\,|D_{\mathbf{A}^d_{m,\varepsilon}}| + V + H^d_{\mathrm{f},m,\varepsilon} - E^{\mathrm{PF}}_{V,m,\varepsilon}\,\big]\otimes P^{\perp}_{\Omega^f}\,\}\,\chi && (6.34)\\
&\quad+ \chi\{\,\mathbb{1}^{\mathrm{el}}\otimes\mathbb{1}^d\otimes(H^f_{\mathrm{f},m,\varepsilon} - E^{\mathrm{PF}}_{V,m} + E^{\mathrm{PF}}_{V,m,\varepsilon} - m/2)\,P^{\perp}_{\Omega^f}\,\}\,\chi\,. && (6.35)
\end{aligned}$$

In view of (6.32) we have $E^{\mathrm{PF}}_{V,m,\varepsilon} = \inf\sigma[\,|D_{\mathbf{A}^d_{m,\varepsilon}}| + V + H^d_{\mathrm{f},m,\varepsilon}]$. Therefore, the expression in (6.34) is a non-negative quadratic form. For sufficiently small $\varepsilon > 0$, the expression in (6.35) is a non-negative quadratic form, too, because of $H^f_{\mathrm{f},m,\varepsilon}\,P^{\perp}_{\Omega^f} \geqslant m\,P^{\perp}_{\Omega^f}$ and Corollary 6.10. In order to treat the remaining term in (6.33) we write

$$\begin{aligned}
&\big[\,|D_{\mathbf{A}^d_{m,\varepsilon}}| + V + H^d_{\mathrm{f},m,\varepsilon}\,\big]\otimes P_{\Omega^f}\\
&\quad= (\mathbb{1}\otimes P_{\Omega^f})\,\big\{\,|D_{\mathbf{A}_{m,\varepsilon}}| + V + H_{\mathrm{f},m,\varepsilon}\,\big\}\,(\mathbb{1}\otimes P_{\Omega^f})\\
&\quad\geqslant (\mathbb{1}\otimes P_{\Omega^f})\,\big\{\varepsilon\,|D_{\mathbf{0}}| + \varepsilon\,H_{\mathrm{f},m,\varepsilon} - \mathrm{const}(\varepsilon,m,V,e,\Lambda)\big\}\,(\mathbb{1}\otimes P_{\Omega^f})\,.
\end{aligned}$$

In the second step we assumed that $\varepsilon > 0$ is small enough. Altogether, we arrive at

$$\begin{aligned}
&\chi\{\,H^{\mathrm{PF}}_{V,m,0} - E^{\mathrm{PF}}_{V,m} - m/2\,\}\,\chi + o(\varepsilon^0)\,\chi + \varepsilon\,\chi\,|\mathbf{x}|^2\,\chi\\
&\quad\geqslant \chi\{\,\big(\,\varepsilon\,|D_{\mathbf{0}}| + \varepsilon\,|\mathbf{x}|^2 + \varepsilon\,H^d_{\mathrm{f},m,\varepsilon} - \mathrm{const}\,\big)\otimes P_{\Omega^f}\,\}\,\chi\\
&\quad\geqslant \chi\{\,\big[\,\varepsilon\,|D_{\mathbf{0}}| + \varepsilon\,|\mathbf{x}|^2 + \varepsilon\,H^d_{\mathrm{f},m,\varepsilon} - \mathrm{const}\,\big]_{-}\otimes P_{\Omega^f}\,\}\,\chi\,, && (6.36)
\end{aligned}$$

where $[\cdots]_{-} \leqslant 0$ denotes the negative part. The crucial point about the previous estimate is that both $|D_{\mathbf{0}}| + |\mathbf{x}|^2$ and $H^d_{\mathrm{f},m,\varepsilon}$ have a purely discrete spectrum as operators in the electron and discrete photon Hilbert spaces. Besides P_{Ω^f} has rank one, of course. (ω_ε has a discrete spectrum *as an operator in* $\mathscr{K}^d_m = P_\varepsilon\,\mathscr{K}$ because the eigenspace in $\mathscr{K}^d_m$ corresponding to some value attained by ω_ε is finite-dimensional. Using $\omega_\varepsilon \geqslant m > 0$ it is then easy to see that the spectrum of its second quantization, $H^d_{\mathrm{f},m,\varepsilon} = d\Gamma(\omega_\varepsilon{\restriction}_{\mathscr{K}^d_m})$, is discrete, too.) In particular, we observe that

$$W^{-}_{m,\varepsilon} := \big[\,\varepsilon\,|D_{\mathbf{0}}| + \varepsilon\,|\mathbf{x}|^2 + \varepsilon\,H^d_{\mathrm{f},m,\varepsilon} - \mathrm{const}\,\big]_{-}\otimes P_{\Omega^f} \qquad (6.37)$$

is a finite rank operator, for every sufficiently small $\varepsilon > 0$, no matter how large the value of the $(\varepsilon, m, V, e, \Lambda)$-dependent constant is. As a simple consequence of the exponential localization we further know that $\chi\,|\mathbf{x}|^2\,\chi$ is bounded. Using $\chi\{\,H^{\mathrm{PF}}_{V,m,0} - E^{\mathrm{PF}}_{V,m} - m/2\,\}\,\chi \leqslant -(m/4)\,\chi$ we obtain the bound $(o(\varepsilon^0) - m/4)\,\mathrm{Tr}\{\chi\} \geqslant \mathrm{Tr}\{\chi W^{-}_{m,\varepsilon}\chi\} > -\infty$ from (6.36). Fixing $\varepsilon > 0$ sufficiently small we conclude $\mathrm{Tr}\{\chi\} < \infty$. □

The no-pair operator.

Proof: We prove Theorem 6.11 with $\sharp=\mathrm{np}$. On account of Lemma 6.7(i) we again have $\chi\,H^{\mathrm{np}}_{V,m,0}\,\chi\geqslant\chi\,H^{\mathrm{np}}_{V,m,\varepsilon}\chi-o(\varepsilon^0)\,\chi$. In view of (6.32) we have $P^+_{\mathbf{A}_{m,\varepsilon}}=P^+_{\mathbf{A}^d_{m,\varepsilon}}\otimes\mathbb{1}^f$ with $P^+_{\mathbf{A}^d_{m,\varepsilon}}:=\mathbb{1}_{[0,\infty)}(D_{\mathbf{A}^d_{m,\varepsilon}})$ and we observe that $H^{\mathrm{np}}_{V,m,\varepsilon}$ decomposes under the isomorphism (6.31) as

$$H^{\mathrm{np}}_{V,m,\varepsilon}=\overline{X^d_\varepsilon\otimes\mathbb{1}^f+P^+_{\mathbf{A}^d_{m,\varepsilon}}\otimes H^f_{\mathrm{f},m,\varepsilon}}\,,\tag{6.38}$$
$$X^d_\varepsilon:=P^+_{\mathbf{A}^d_{m,\varepsilon}}\,(D_{\mathbf{A}^d_{m,\varepsilon}}+V+H^d_{\mathrm{f},m,\varepsilon})\,P^+_{\mathbf{A}^d_{m,\varepsilon}}\,.$$

Writing $\mathbb{1}^{\mathrm{el}}\otimes\mathbb{1}^d=P^+_{\mathbf{A}^d_{m,\varepsilon}}+P^-_{\mathbf{A}^d_{m,\varepsilon}}$ and $\mathbb{1}^f=P_{\Omega^f}+P^\perp_{\Omega^f}$ and using (6.38) and $H^f_{\mathrm{f},m,\varepsilon}\,P_{\Omega^f}=0$, we obtain

$$\begin{aligned}&\chi\,\{\,H^{\mathrm{np}}_{V,m,0}-E^{\mathrm{np}}_{V,m}-m/2\,\}\,\chi+o(\varepsilon^0)\,\chi\\
&\geqslant\chi\,\{\,(X^d_\varepsilon-(E^{\mathrm{np}}_{V,m}+m/2)\,P^+_{\mathbf{A}^d_{m,\varepsilon}})\otimes P_{\Omega^f}\,\}\,\chi &(6.39)\\
&\quad+\chi\,\{\,(X^d_\varepsilon-E^{\mathrm{np}}_{V,m,\varepsilon}\,P^+_{\mathbf{A}^d_{m,\varepsilon}})\otimes P^\perp_{\Omega^f}\,\}\,\chi &(6.40)\\
&\quad+\chi\,\{\,P^+_{\mathbf{A}^d_{m,\varepsilon}}\otimes(E^{\mathrm{np}}_{V,m,\varepsilon}-E^{\mathrm{np}}_{V,m}-m/2+H^f_{\mathrm{f},m,\varepsilon})P^\perp_{\Omega^f}\}\,\chi &(6.41)\\
&\quad-(E^{\mathrm{np}}_{V,m}+m/2)\,\chi\,\{P^-_{\mathbf{A}^d_{m,\varepsilon}}\otimes\mathbb{1}^f\}\,\chi\,. &(6.42)\end{aligned}$$

Next, we observe that $E^{\mathrm{np}}_{V,m,\varepsilon}=\inf\sigma[X^d_\varepsilon]$ and proceed as in the proof of Theorem 6.11 with $\sharp=\mathrm{PF}$: We omit the expression in (6.40) which is a non-negative quadratic form. For sufficiently small $\varepsilon>0$, the term in (6.41) is non-negative also, since $H^f_{\mathrm{f},m,\varepsilon}P^\perp_{\Omega^f}\geqslant m\,P^\perp_{\Omega^f}$ and $\lim_{\varepsilon\to0}E^{\mathrm{np}}_{V,m,\varepsilon}=E^{\mathrm{np}}_{V,m}$ by Corollary 6.10. The term in (6.42), is some $o(\varepsilon^0)$ on account of $\chi\,\{P^-_{\mathbf{A}^d_{m,\varepsilon}}\otimes\mathbb{1}^f\}\,\chi=\chi\,(P^+_{\mathbf{A}^>_m}-P^+_{\mathbf{A}_{m,\varepsilon}})\,\chi$, (6.19), and the boundedness of $e^F\,\check{H}_{\mathrm{f}}^{1/2}\,\chi$. (The latter follows from the exponential localization and higher order estimates.) Putting all these remarks together and applying (3.32) in the last step we obtain

$$\begin{aligned}&\chi\,\{H^{\mathrm{np}}_{V,m,0}-E^{\mathrm{np}}_{V,m}-m/2\}\,\chi+o(\varepsilon^0)\,\chi\\
&\geqslant\chi\,\{\,(X^d_\varepsilon-(E^{\mathrm{np}}_{V,m}+m/2)\,P^+_{\mathbf{A}^d_{m,\varepsilon}})\otimes P_{\Omega^f}\,\}\,\chi\\
&=\chi\,(\mathbb{1}\otimes P_{\Omega^f})\,\{H^{\mathrm{np}}_{V,m,\varepsilon}-(E^{\mathrm{np}}_{V,m}+m/2)\,P^+_{\mathbf{A}_{m,\varepsilon}}\}\,(\mathbb{1}\otimes P_{\Omega^f})\,\chi\\
&\geqslant\chi\,P^+_{\mathbf{A}_{m,\varepsilon}}\{\,\big[\varepsilon\,|D_{\mathbf{0}}|+\varepsilon|\mathbf{x}|^2+\varepsilon\,H^d_{f,m,\varepsilon}-\mathrm{const}\big]_-\otimes P_{\Omega^f}\}\,P^+_{\mathbf{A}_{m,\varepsilon}}\,\chi\\
&\qquad-\varepsilon\,\chi\,P^+_{\mathbf{A}_{m,\varepsilon}}\,\{|\mathbf{x}|^2\otimes P_{\Omega^f}\}\,P^+_{\mathbf{A}_{m,\varepsilon}}\,\chi\,,\end{aligned}$$

where the constant in the penultimate line again depends on ε, m, V, e, and Λ, and the operator in the last line is bounded due to the localization of χ. We may thus conclude as in the end of the proof of Theorem 6.11 with $\sharp = \mathrm{PF}$. □

7. Infra-Red Bounds

The final step in the proof of the existence of ground states is a compactness argument showing that a sequence of normalized ground state eigenfunctions, ϕ_{m_j}, $m_j \searrow 0$, of operators with photon masses m_j contains a strongly convergent subsequence whose limit turns out to be a ground state eigenfunction for the original operator. This compactness argument is explained in the subsequent Section 8. As a preparation we now discuss the infra-red bounds stated in the following proposition. They are proved in [28] for the semi-relativistic Pauli-Fierz operator and in [29] for the no-pair operator starting from a suitable representation of $a(k)\,\phi_m$. In order not to lengthen the present exposition too much we only outline the proof of the soft photon bound for the semi-relativistic Pauli-Fierz operator in Subsection 7.2. We recall the notation

$$(a(k)\,\psi)^{(n)}(k_1,\dots,k_n) \;=\; (n+1)^{1/2}\,\psi^{(n+1)}(k,k_1,\dots,k_n)\,, \qquad n \in \mathbb{N}_0\,,$$

almost everywhere, where $\psi = (\psi^{(n)})_{n=0}^{\infty} \in \mathscr{F}_{\mathrm{b}}[\mathscr{K}]$, and $a(k)\,\Omega = 0$.

Proposition 7.1: *Let $e \in \mathbb{R}$, $\Lambda \in (0,\infty)$, and assume that V fulfills Hypotheses 6.1 and 6.6. Then there is a constant, $C > 0$, such that, for all sufficiently small $m > 0$ and every normalized ground state eigenvector, $\phi_m^\sharp$, of $H_{V,m}^\sharp$, we have the following soft photon bound,*

$$\big\|\,a(k)\,\phi_m^\sharp\,\big\|^2 \;\leqslant\; \mathbb{1}_{\{m\leqslant|\mathbf{k}|\leqslant\Lambda\}}\,\frac{C}{|\mathbf{k}|}\,, \tag{7.1}$$

for almost every $k = (\mathbf{k},\lambda) \in \mathbb{R}^3 \times \mathbb{Z}_2$, as well as the following photon derivative bound,

$$\big\|\,a(k)\,\phi_m^\sharp - a(p)\,\phi_m^\sharp\,\big\| \;\leqslant\; C\,|\mathbf{k}-\mathbf{p}|\,\Big(\frac{1}{|\mathbf{k}|^{1/2}|\mathbf{k}_\perp|} + \frac{1}{|\mathbf{p}|^{1/2}|\mathbf{p}_\perp|}\Big)\,, \tag{7.2}$$

for almost every $k = (\mathbf{k},\lambda), p = (\mathbf{p},\mu) \in \mathbb{R}^3 \times \mathbb{Z}_2$ with $m < |\mathbf{k}| < \Lambda$ and $m < |\mathbf{p}| < \Lambda$.

We remark that the photon derivative bound (7.2) is actually the only place in the whole article where the special choice of the polarization vectors (2.4) enters into the analysis.

7.1. *The gauge transformed operator*

In order to derive the infra-red bounds (7.1) and (7.2) by the method outlined in Subsection 7.2 it is necessary to pass to a suitable gauge [6, 17]. For otherwise we would end up with a more singular infra-red behavior of their right hand sides. To define an appropriate operator-valued gauge transformation [17] we recall that, for $i, j \in \{1, 2, 3\}$, the components $A_m^{(i)}(\mathbf{x})$ and $A_m^{(j)}(\mathbf{y})$ of the magnetic vector potential at $\mathbf{x}, \mathbf{y} \in \mathbb{R}^3$ commute in the sense that all their spectral projections commute; see, e.g., Theorem X.43 of [43]. Therefore, it makes sense to define

$$U := \int_{\mathbb{R}^3}^{\oplus} \mathbb{1}_{\mathbb{C}^4} \otimes U_{\mathbf{x}}\, d^3\mathbf{x}, \quad U_{\mathbf{x}} := \prod_{j=1}^{3} e^{ix_j A_m^{(j)}(\mathbf{0})}, \quad \mathbf{x} = (x_1, x_2, x_3) \in \mathbb{R}^3.$$

Then the gauge transformed vector potential is given by

$$\widetilde{\mathbf{A}}_m(\mathbf{x}) := \mathbf{A}_m(\mathbf{x}) - \mathbf{A}_m(\mathbf{0}) = a^\dagger(\widetilde{\mathbf{G}}_m) + a(\widetilde{\mathbf{G}}_m),$$

where $a^\sharp(\widetilde{\mathbf{G}}_m) = \int_{\mathbb{R}^3}^{\oplus} \mathbb{1}_{\mathbb{C}^4} \otimes a^\sharp(\widetilde{\mathbf{G}}_{\mathbf{x},m})\, d^3\mathbf{x}$, and

$$\widetilde{\mathbf{G}}_{\mathbf{x},m}(k) := -e\, \frac{\mathbb{1}_{\{m \leqslant |\mathbf{k}| \leqslant \Lambda\}}}{2\pi\sqrt{|\mathbf{k}|}}\, (e^{i\mathbf{k}\cdot\mathbf{x}} - 1)\, \boldsymbol{\varepsilon}(k) = (e^{i\mathbf{k}\cdot\mathbf{x}} - 1)\mathbf{G}_{\mathbf{0},m}(k),$$

for all $\mathbf{x} \in \mathbb{R}^3$ and almost every $k = (\mathbf{k}, \lambda) \in \mathbb{R}^3 \times \mathbb{Z}_2$. Here $\mathbf{G}_{\mathbf{x},m}$ is defined in (6.1). In fact, using $[U\,,\, \boldsymbol{\alpha} \cdot \mathbf{A}_m] = 0$ we deduce that

$$U\, D_{\mathbf{A}_m}\, U^* = D_{\widetilde{\mathbf{A}}_m}, \qquad U\, P^+_{\mathbf{A}_m}\, U^* = P^+_{\widetilde{\mathbf{A}}_m}, \qquad U\, |D_{\mathbf{A}_m}|\, U^* = |D_{\widetilde{\mathbf{A}}_m}|.$$

The crucial point observed in [6] is that the transformed vector potential $\widetilde{\mathbf{A}}_m$ has a better infra-red behavior than $\mathbf{A}_m$ in view of the estimate

$$|\widetilde{\mathbf{G}}_{\mathbf{x},m}(k)| \leqslant |\mathbf{k}|\, |\mathbf{x}|\, |\mathbf{G}_{\mathbf{0},m}(k)|. \tag{7.3}$$

In particular, infra-red divergent (for $m \searrow 0$) integrals appearing in the derivation of the soft photon bound are avoided when we work with $\widetilde{\mathbf{A}}_m$ instead of $\mathbf{A}_m$. We further set

$$\begin{aligned} \widetilde{H}_{\mathrm{f}} &:= U\, H_{\mathrm{f}}\, U^* = H_{\mathrm{f}} + i\mathbf{x} \cdot \big(a(\omega\, \mathbf{G}_{\mathbf{0},m}) - a^\dagger(\omega\, \mathbf{G}_{\mathbf{0},m})\big) & (7.4)\\ &\qquad\qquad + 2\, \langle\, \omega\, \mathbf{x} \cdot \mathbf{G}_{\mathbf{0},m}\, |\, \mathbf{x} \cdot \mathbf{G}_{\mathbf{0},m}\, \rangle, &\\ \widetilde{H}^{\mathrm{PF}}_{V,m} &:= U\, H^{\mathrm{PF}}_{V,m}\, U^* = |D_{\widetilde{\mathbf{A}}_m}| + V + \widetilde{H}_{\mathrm{f}}, &\\ \widetilde{H}^{\mathrm{np}}_{V,m} &:= U\, H^{\mathrm{np}}_{V,m}\, U^* = P^+_{\widetilde{\mathbf{A}}_m}\, (D_{\widetilde{\mathbf{A}}_m} + V + \widetilde{H}_{\mathrm{f}})\, P^+_{\widetilde{\mathbf{A}}_m}, &\\ \widetilde{\phi}^\sharp_m &:= U\, \phi^\sharp_m, & (7.5) \end{aligned}$$

so that $\widetilde{\phi}^\sharp_m$ is a ground state eigenfunction of $\widetilde{H}^\sharp_{V,m}$. One can easily show that, if the infra-red bounds (7.1) and (7.2) hold true with $\phi^\sharp_m$ replaced by $\widetilde{\phi}^\sharp_m$, then they are valid for $\phi^\sharp_m$ as well (with a different constant, of course).

7.2. *Soft photon bound for the semi-relativistic Pauli-Fierz operator*

To simplify the notation we write $\widetilde{\phi}_m$ instead of $\widetilde{\phi}_m^{\mathrm{PF}}$ in this subsection. A brief calculation yields

$$\begin{aligned} 0 \leqslant \big\langle\, (\, \widetilde{H}^{\mathrm{PF}}_{V,m} - E^{\mathrm{PF}}_{V,m}\,)\, a(k)\, \widetilde{\phi}_m \,\big|\, a(k)\, \widetilde{\phi}_m \,\big\rangle &= \big\langle\, [\, \widetilde{H}^{\mathrm{PF}}_{V,m},\, a(k)\,]\, \widetilde{\phi}_m \,\big|\, a(k)\, \widetilde{\phi}_m \,\big\rangle \\ = \big\langle\, [\, S_{\widetilde{\mathbf{A}}_m},\, a(k)\,]\, D_{\widetilde{\mathbf{A}}_m}\, \widetilde{\phi}_m \,\big|\, a(k)\, \widetilde{\phi}_m \,\big\rangle &+ \big\langle\, S_{\widetilde{\mathbf{A}}_m}\, [\, D_{\widetilde{\mathbf{A}}_m},\, a(k)\,]\, \widetilde{\phi}_m \,\big|\, a(k)\, \widetilde{\phi}_m \,\big\rangle \\ + \big\langle\, [\, \widetilde{H}_{\mathrm{f}},\, a(k)\,]\, \widetilde{\phi}_m \,\big|\, a(k)\, \widetilde{\phi}_m \,\big\rangle. & \qquad (7.6) \end{aligned}$$

Combining the commutation relations

$$[\, a(k)\,,\, a^{\dagger}(f)\,] = f(k), \quad [\, a(k)\,,\, a(f)\,] = 0, \quad [\, a(k)\,,\, H_{\mathrm{f}}\,] = \omega(k)\, a(k)\,,$$

with the formula (7.4) for $\widetilde{H}_{\mathrm{f}}$ we obtain

$$[\, \widetilde{H}_{\mathrm{f}}\,,\, a(k)\,] = -\omega(k)\, a(k) + i\omega(k)\, \mathbf{x} \cdot \mathbf{G}_{\mathbf{0},m}(k)\,.$$

Furthermore, we get $[D_{\widetilde{\mathbf{A}}_m},\, a(k)] = -\boldsymbol{\alpha} \cdot \widetilde{\mathbf{G}}_{\mathbf{x},m}(k)$ and

$$[\, S_{\widetilde{\mathbf{A}}_m}\,,\, a(k)\,] = \int_{-\infty}^{\infty} R_{\widetilde{\mathbf{A}}_m}(iy)\, [\, a(k)\,,\, D_{\widetilde{\mathbf{A}}_m}\,]\, R_{\widetilde{\mathbf{A}}_m}(iy)\, \frac{dy}{\pi}, \qquad (7.7)$$

where $R_{\widetilde{\mathbf{A}}_m}(iy) = (D_{\widetilde{\mathbf{A}}_m} - iy)^{-1}$. Next, we insert (7.7) into (7.6), move the term containing $\omega(k)\, a(k)$ to the left hand side, and divide by $\omega(k)$. Furthermore, we introduce a weight function, $F(\mathbf{x}) = a\, |\mathbf{x}|^2 / \sqrt{1 + |\mathbf{x}|^2}$, for some small $a > 0$. Abbreviating

$$D^F_{\widetilde{\mathbf{A}}_m} := e^F\, D_{\widetilde{\mathbf{A}}_m}\, e^{-F} = D_{\widetilde{\mathbf{A}}_m} + i\boldsymbol{\alpha} \cdot \nabla F, \quad R^F_{\widetilde{\mathbf{A}}_m}(iy) := (D^F_{\widetilde{\mathbf{A}}_m} - iy)^{-1},$$

we obtain the following result,

$$\begin{aligned} \|\, a(k)\, \widetilde{\phi}_m\, \|^2 \leqslant \int_{-\infty}^{\infty} & \big\langle\, R_{\widetilde{\mathbf{A}}_m}(iy)\, \big\{ \boldsymbol{\alpha} \cdot \widetilde{\mathbf{G}}_{\mathbf{x},m}\, e^{-F}\, |\mathbf{k}|^{-1} \big\} \times \\ & \qquad \times\, R^F_{\widetilde{\mathbf{A}}_m}(iy)\, D^F_{\widetilde{\mathbf{A}}_m}\, e^F\, \widetilde{\phi}_m \,\big|\, a(k)\, \widetilde{\phi}_m \,\big\rangle\, \frac{dy}{\pi} \\ & - \big\langle\, S_{\widetilde{\mathbf{A}}_m}\, \big\{ \boldsymbol{\alpha} \cdot \widetilde{\mathbf{G}}_{\mathbf{x},m}\, e^{-F}\, |\mathbf{k}|^{-1} \big\}\, e^F\, \widetilde{\phi}_m \,\big|\, a(k)\, \widetilde{\phi}_m \,\big\rangle \\ & - i\mathbf{G}_{\mathbf{0},m}(k) \cdot \big\langle\, (\mathbf{x}\, e^{-F})\, e^F\, \widetilde{\phi}_m \,\big|\, a(k)\, \widetilde{\phi}_m \,\big\rangle\,. \qquad (7.8) \end{aligned}$$

The purpose of the exponentials e^{-F} introduced above is to control the factor $|\mathbf{x}|$ coming from (7.3). In fact,

$$\|\boldsymbol{\alpha} \cdot \widetilde{\mathbf{G}}_{\mathbf{x},m}\, e^{-F}\| \leqslant \sqrt{2}\, \sup_{\mathbf{x}} |\widetilde{\mathbf{G}}_{\mathbf{x},m}\, e^{-F(\mathbf{x})}| \leqslant \mathcal{O}(1)\, |\mathbf{k}|^{1/2}\, \mathbb{1}_{\{m \leqslant |\mathbf{k}| \leqslant \Lambda\}}\,.$$

Using this it is easy to see that the sum of the last two expressions in (7.8) is not greater than $\mathcal{O}(1)\,|\mathbf{k}|^{-1}\,\mathbb{1}_{\{m\leqslant|\mathbf{k}|\leqslant\Lambda\}}\,\|e^F\,\widetilde{\phi}_m\|^2+\|a(k)\,\widetilde{\phi}_m\|^2/2$. By virtue of the second resolvent identity we further get

$$R^F_{\widetilde{\mathbf{A}}_m}(iy)\,D^F_{\widetilde{\mathbf{A}}_m}=\big(1-R^F_{\widetilde{\mathbf{A}}_m}(iy)\,(i\boldsymbol{\alpha}\cdot\nabla F)\big)R_{\widetilde{\mathbf{A}}_m}(iy)\,(D_{\widetilde{\mathbf{A}}_m}+i\boldsymbol{\alpha}\cdot\nabla F)\,.$$

Since we have $\|R^F_{\widetilde{\mathbf{A}}_m}(iy)\|,\|R_{\widetilde{\mathbf{A}}_m}(iy)\|\leqslant\mathcal{O}(1)\,(1+|y|)^{-1}$ by (2.21) and $\|R_{\widetilde{\mathbf{A}}_m}(iy)\,|D_{\widetilde{\mathbf{A}}_m}|^{1/2}\|\leqslant\mathcal{O}(1)(1+|y|)^{-1/2}$, we thus obtain

$$\big\|\,R^F_{\widetilde{\mathbf{A}}_m}(iy)\,D^F_{\widetilde{\mathbf{A}}_m}\,e^F\,\widetilde{\phi}_m\,\big\|\leqslant\mathcal{O}(1)\,(1+|y|)^{-1/2}\,\|\,|D_{\widetilde{\mathbf{A}}_m}|^{1/2}\,e^F\,\widetilde{\phi}_m\|\,.$$

Therefore, the integral in (7.8) converges absolutely and we arrive at

$$\|a(k)\,\widetilde{\phi}_m\|^2\leqslant\mathcal{O}(1)\,|\mathbf{k}|^{-1}\mathbb{1}_{\{m\leqslant|\mathbf{k}|\leqslant\Lambda\}}\,\|\,|D_{\widetilde{\mathbf{A}}_m}|^{1/2}\,e^F\,\widetilde{\phi}_m\|^2+3\,\|a(k)\,\widetilde{\phi}_m\|^2/4.$$

If $a>0$ is small enough, then $\|\,|D_{\widetilde{\mathbf{A}}_m}|^{1/2}\,e^F\,\widetilde{\phi}_m\|$ is bounded uniformly in $m>0$, due to a strengthened version of our exponential localization estimates; see Lemma 5.4 of [28] whose proof works for every potential V fulfilling Hypothesis 6.1. Together with the last remark in the preceding subsection this yields the soft photon bound.

The above calculations illustrate the importance of the formal gauge invariance of our models. In fact, without the gauge invariance we could perform these calculations only for $\mathbf{G}_{\mathbf{x},m}$ instead of $\widetilde{\mathbf{G}}_{\mathbf{x},m}$. In this case we would, however, end up with an upper bound on $\|a(k)\,\phi_m\|^2$ of the form $\mathcal{O}(1)\,|\mathbf{k}|^{-3}\mathbb{1}_{\{m\leqslant|\mathbf{k}|\leqslant\Lambda\}}$, which is not integrable near zero and, hence, is not suitable for the arguments presented in the following section.

8. Existence of Ground States

We have now collected all prerequisites to show that $\inf\sigma[H^\sharp_V]$ is an eigenvalue of $H^\sharp_V$. The final compactness argument is presented in the first subsection below. The main idea behind it is borrowed from [17]: Namely, to show that $\{\phi^\sharp_{m_j}\}_j$, $m_j\searrow 0$, contains strongly convergent subsequences we may restrict our attention to finitely many Fock space sectors and to a compact subset of the $\mathbf{k}$-space. In fact, this is a consequence of the soft photon bounds. On account of the exponential localization estimates it further suffices to consider compact subsets of the $\mathbf{x}$-space. Moreover, the photon derivative bounds and the localization in energy lead to bounds on the (half-)derivatives of the vectors $\phi^\sharp_{m_j}$ w.r.t. $\mathbf{x}$ and $\mathbf{k}$ on compact sets and in the finitely many Fock space sectors. The idea proposed in [17] is to exploit such information by applying suitable compact embedding theorems. Essentially, we only have to replace the Rellich-Kondrashov theorem applied there by a suitable embedding theorem for spaces of functions with fractional derivatives. (In the non-relativistic case the ground states $\phi_m^\sharp$ possess weak derivatives with respect to the electron coordinates, whereas in our case we only have Inequality (8.5) below. For a variant of the argument that avoids the Nikol′skiĭ spaces introduced below by switching the roles of the electronic position and momentum spaces see [25].)

In the second subsection we discuss the degeneracy of the ground state energies by applying Kramers' degeneracy theorem similarly as in [40].

In the whole section the coupling function $\mathbf{G}_\mathbf{x}$ is the physical one defined in (2.2).

8.1. *Ground states without photon mass*

We shall apply the following elementary fact:

Let S be some non-negative operator acting in some Hilbert space, $\mathscr{X}$. Let $\{\eta_j\}_{j\in\mathbb{N}}$ be some sequence in $\mathscr{X}$, converging weakly to some $\eta\in\mathscr{X}$ such that $\eta_j\in\mathcal{Q}(S)$ and $\langle\,\eta_j\,|\,S\,\eta_j\,\rangle=\|S^{1/2}\,\eta_j\|\to 0$. Then η belongs to the domain of S and $S\,\eta=0$.

In fact, the linear functional $f(\phi)=\langle\,\eta\,|\,S^{1/2}\,\phi\,\rangle$, $\phi\in\mathcal{D}(S^{1/2})$, satisfies

$$f(\phi)=\lim_{j\to\infty}\langle\,\eta_j\,|\,S^{1/2}\,\phi\,\rangle=\lim_{j\to\infty}\langle\,S^{1/2}\,\eta_j\,|\,\phi\,\rangle=0\,,$$

and the self-adjointness of $S^{1/2}$ implies $\eta\in\mathcal{D}(S^{1/2})$ and $S^{1/2}\,\eta=0$.

We remind the reader of the notation $\sharp\in\{\mathrm{np},\mathrm{PF}\}$.

Theorem 8.1: *Let $e\in\mathbb{R}$, $\Lambda>0$, and assume that V fulfills Hypotheses 6.1 and 6.6. Then $E_V^\sharp$ is an eigenvalue of $H_V^\sharp$.*

Proof: For every sufficiently small $m>0$, there is some normalized ground state eigenfunction, $\phi_m^\sharp$, of $H_{V,m}^\sharp$. We thus find some sequence $m_j\searrow 0$ such that $\{\phi_{m_j}^\sharp\}_{j\in\mathbb{N}}$ converges weakly to some $\phi^\sharp\in\mathscr{H}$. According to Theorem 3.4 the form domains of H_V^{PF} and $H_{V,m}^{\mathrm{PF}}$, $m>0$, coincide whence $\phi_{m_j}^{\mathrm{PF}}\in\mathcal{Q}(H_V^{\mathrm{PF}})$. By the characterization of the form domain of the no-pair operator in Theorem 3.6 we further know that $P_\mathbf{A}^+\,\phi_{m_j}^{\mathrm{np}}\in\mathcal{Q}(H_V^{\mathrm{np}})$. It is trivial that $P_\mathbf{A}^+\,\phi_{m_j}^{\mathrm{np}}$ converges weakly to $P_\mathbf{A}^+\,\phi^{\mathrm{np}}$ in $P_\mathbf{A}^+\,\mathscr{H}$. Since $N_j:=(P_\mathbf{A}^+-P_{\mathbf{A}_{m_j}}^-)\,\check{H}_\mathrm{f}^{-1/2}$ converges to zero in norm, where $\check{H}_\mathrm{f}=H_\mathrm{f}+E$, for some sufficiently large $E\geqslant 1$, we further have $\langle\,\psi\,|\,P_\mathbf{A}^-\,\phi^{\mathrm{np}}\,\rangle=\lim_{j\to\infty}\langle\,N_j\,\check{H}_\mathrm{f}^{1/2}\,\psi\,|\,\phi_{m_j}^{\mathrm{np}}\,\rangle=0$, for every $\psi\in\mathscr{D}$, whence $P_\mathbf{A}^-\,\phi^{\mathrm{np}}=0$, or, $P_\mathbf{A}^+\,\phi^{\mathrm{np}}=\phi^{\mathrm{np}}$. Hence, by the remark preceding the statement it suffices to show that $\phi^\sharp\neq 0$ and $\langle\,\phi_{m_j}^\sharp\,|\,(H_V^\sharp-E_V^\sharp)\,\phi_{m_j}^\sharp\,\rangle\to 0$.

The latter condition immediately follows from $\lim_{m\searrow 0} E_{V,m} = E_V$ and Lemma 6.5(ii) which together imply

$$\langle\, \phi^\sharp_{m_j} \,|\, (H^\sharp_V - E^\sharp_V)\, \phi^\sharp_{m_j} \,\rangle = \langle\, \phi^\sharp_{m_j} \,|\, (H^\sharp_{V,m_j} - E^\sharp_{V,m_j})\, \phi^\sharp_{m_j} \,\rangle + o(m_j^0) = o(m_j^0)\,.$$

In order to verify that $\phi^\sharp \neq 0$ we adapt the ideas of [17] sketched in the first paragraph of this section.

We write $\phi^\sharp_m = (\phi^{(n)}_{m,\sharp})_{n=0}^\infty \in \bigoplus_{n=0}^\infty \mathscr{F}_{\mathrm{b}}^{(n)}[\mathscr{K}]$ in what follows. Let $\varepsilon > 0$. By virtue of the soft photon bound we find $n_0 \in \mathbb{N}$ and $C \in (0,\infty)$ such that, for all sufficiently small $m > 0$,

$$\sum_{n=n_0}^\infty \|\phi^{(n)}_{m,\sharp}\|^2 \leqslant \frac{1}{n_0} \sum_{n=0}^\infty n\, \|\phi^{(n)}_{m,\sharp}\|^2 = \frac{1}{n_0} \int \|a(k)\, \phi^\sharp_m\|^2\, dk \leqslant \frac{C}{n_0} < \frac{\varepsilon}{2}\,. \quad (8.1)$$

By the exponential localization estimates of Theorem 5.1, which hold uniformly for small $m > 0$, we further find some $R > 0$ such that

$$\int_{|\mathbf{x}|\geqslant R/2} \|\phi^\sharp_m\|^2_{\mathbb{C}^4\otimes\mathscr{F}_{\mathrm{b}}}(\mathbf{x})\, d^3\mathbf{x} < \frac{\varepsilon}{2}\,. \quad (8.2)$$

In addition, the soft photon bound ensures that $\phi^{(n)}_{m,\sharp}(\mathbf{x},\varsigma,k_1,\dots,k_n) = 0$, for almost every $(\mathbf{x},\varsigma,k_1,\dots,k_n) \in \mathbb{R}^3 \times \{1,2,3,4\} \times (\mathbb{R}^3 \times \mathbb{Z}_2)^n$, $k_j = (\mathbf{k}_j,\lambda_j)$, such that $|\mathbf{k}_j| > \Lambda$, for some $j \in \{1,\dots,n\}$. (Here and henceforth ς labels the four spinor components.) For $0 < n < n_0$ and some fixed $\overline{\theta} = (\varsigma,\lambda_1,\dots,\lambda_n) \in \{1,2,3,4\} \times \mathbb{Z}_2^n$ we set

$$\phi^{(n)}_{m,\overline{\theta},\sharp}(\mathbf{x},\mathbf{k}_1,\dots,\mathbf{k}_n) := \phi^{(n)}_{m,\sharp}(\mathbf{x},\varsigma,\mathbf{k}_1,\lambda_1,\dots,\mathbf{k}_n,\lambda_n)$$

and similarly for $\phi^\sharp = (\phi^{(n)}_\sharp)_{n=0}^\infty$. Moreover, we set, for every $\delta \geqslant 0$,

$$Q_{n,\delta} := \big\{\, (\mathbf{x},\mathbf{k}_1,\dots,\mathbf{k}_n) \,:\, |\mathbf{x}| < R-\delta\,,\ \delta < |\mathbf{k}_j| < \Lambda - \delta\,,\ j = 1,\dots,n \,\big\}\,.$$

Fixing some small $\delta > 0$ we pick some cut-off function $\chi \in C_0^\infty(\mathbb{R}^{3(n+1)},[0,1])$ such that $\chi \equiv 1$ on $Q_{n,2\delta}$ and $\mathrm{supp}(\chi) \subset Q_{n,\delta}$ and define $\psi^{(n)}_{m,\overline{\theta},\sharp} := \chi\, \phi^{(n)}_{m,\overline{\theta},\sharp}$. As a next step the photon derivative bound is used to show that $\{\psi^{(n)}_{m,\overline{\theta},\sharp}\}_{m\in(0,\delta]}$ is a bounded family in the anisotropic Nikol′skiĭ space[b] $H^{\mathbf{s}}_{\mathbf{q}}(\mathbb{R}^{3(n+1)})$, where $\mathbf{s} = (1/2,1/2,1/2,1,\dots,1)$ and $\mathbf{q} = (2,2,2,p,\dots,p)$ with $p \in [1,2)$. In fact, employing the Hölder inequality

[b]For $r_1,\dots,r_d \in [0,1]$, $q_1,\dots,q_d \geqslant 1$, we have $H^{(r_1,\dots,r_d)}_{q_1,\dots,q_d}(\mathbb{R}^d) := \bigcap_{i=1}^d H^{r_i}_{q_i x_i}(\mathbb{R}^d)$. For $r_i \in [0,1)$, a measurable function $f : \mathbb{R}^d \to \mathbb{C}$ belongs to the class $H^{r_i}_{q_i x_i}(\mathbb{R}^d)$, if $f \in L^{q_i}(\mathbb{R}^d)$ and there is some $M \in (0,\infty)$ such that

$$\|f(\cdot + h\,\mathrm{e}_i) - f\|_{L^{q_i}(\mathbb{R}^d)} \leqslant M\,|h|^{r_i}\,, \qquad h \in \mathbb{R}\,, \quad (8.3)$$

(w.r.t. $d^3\mathbf{x}\, d^{3(n-1)}\mathbf{K}$) and the photon derivative bound (7.2), we obtain as in [17], for $p \in [1,2)$, $m \in (0,\delta]$, and $\mathbf{h} \in \mathbb{R}^3$,

$$\int_{\substack{Q_{n,\delta}\cap\\ \{\delta<|\mathbf{k}+\mathbf{h}|<\Lambda\}}} \big|\phi^{(n)}_{m,\overline{\theta},\sharp}(\mathbf{x},\mathbf{k}+\mathbf{h},\mathbf{K}) - \phi^{(n)}_{m,\overline{\theta},\sharp}(\mathbf{x},\mathbf{k},\mathbf{K})\big|^p\, d^3\mathbf{x}\, d^3\mathbf{k}\, d^{3(n-1)}\mathbf{K}$$

$$\leqslant C \sum_{\lambda\in\mathbb{Z}_2} \int_{\substack{m<|\mathbf{k}|<\Lambda,\\ m<|\mathbf{k}+\mathbf{h}|<\Lambda}} \big\|\, a(\mathbf{k}+\mathbf{h},\lambda)\,\phi^\sharp_m - a(\mathbf{k},\lambda)\,\phi^\sharp_m\,\big\|^p d^3\mathbf{k} \leqslant C'\,|\mathbf{h}|^p,$$

where the constants $C, C' \in (0,\infty)$ do not depend on $m \in (0,\delta]$. Since $\phi^{(n)}_{m,\sharp}$ is symmetric in the photon variables the previous estimate implies [42] that the weak first order partial derivatives of $\phi^{(n)}_{m,\overline{\theta}}$ with respect to its last $3n$ variables exist on $Q_{n,\delta}$ and that

$$\|\phi^{(n)}_{m,\overline{\theta},\sharp}\|^p_{W^{\mathbf{r}}_p(Q_{n,\delta})} := \|\phi^{(n)}_{m,\overline{\theta},\sharp}\|^p_{L^p(Q_{n,\delta})} + \sum_{j=1}^{n}\sum_{i=1}^{3} \|\partial_{k^{(i)}_j}\phi^{(n)}_{m,\overline{\theta},\sharp}\|^p_{L^p(Q_{n,\delta})} \leqslant C'',$$

for $m \in (0,\delta]$ and some m-independent $C'' \in (0,\infty)$, with $\mathbf{r} := (0,0,0,1,\ldots,1)$. The previous estimate implies $\|\psi^{(n)}_{m,\overline{\theta},\sharp}\|_{W^{\mathbf{r}}_p(\mathbb{R}^{3(n+1)})} \leqslant C'''$, for some $C''' \in (0,\infty)$ which does not depend on $m \in (0,\delta]$. Moreover, the anisotropic Sobolev space $W^{\mathbf{r}}_p(\mathbb{R}^{3(n+1)})$ is continuously embedded into $H^{\mathbf{r}}_p(\mathbb{R}^{3(n+1)})$; see, e.g., [42] . By Theorems 3.4 and 3.6 we get, for $n \in \mathbb{N}$,

$$c^{-1}\,\langle\,\phi^{(n)}_{m,\sharp}\,|\,|D_{\mathbf{0}}|\,\phi^{(n)}_{m,\sharp}\,\rangle \leqslant \langle\,\phi_{m,\sharp}\,|\,H^\sharp_{V,m}\,\phi_{m,\sharp}\,\rangle + c = E^\sharp_{V,m} + c \leqslant E^\sharp_V + 2c\,, \tag{8.5}$$

for some m-independent $c \in (0,\infty)$. Therefore, $\{\phi^{(n)}_{m,\overline{\theta},\sharp}\}_{m\in(0,\delta]}$ and, hence, $\{\psi^{(n)}_{m,\overline{\theta},\sharp}\}_{m\in(0,\delta]}$ are bounded families in the Bessel potential, or, Liouville space $L^{\mathbf{r}'}_2(\mathbb{R}^{3(n+1)})$, $\mathbf{r}' := (1/2,1/2,1/2,0,\ldots,0)$, where the fractional derivatives are defined by means of the Fourier transform. The embedding $L^{\mathbf{r}'}_2(\mathbb{R}^{3(n+1)}) \to H^{\mathbf{r}'}_2(\mathbb{R}^{3(n+1)})$ is continuous, too; see §9.3 in [42]. Altogether

---

where $\mathbf{e}_i$ is the i-th canonical unit vector in $\mathbb{R}^d$. If $r_i = 1$, then (8.3) is replaced by

$$\|f(\cdot + h\,\mathbf{e}_i) - 2f + f(\cdot - h\,\mathbf{e}_i)\|_{L^{q_i}(\mathbb{R}^d)} \leqslant M\,|h|\,, \qquad h \in \mathbb{R}\,. \tag{8.4}$$

$H^{(r_1,\ldots,r_d)}_{q_1,\ldots,q_d}(\mathbb{R}^d)$ is a Banach space with norm

$$\|f\|^{(r_1,\ldots,r_d)}_{q_1,\ldots,q_d} := \max_{1\leqslant i\leqslant d}\|f\|_{L^{q_i}(\mathbb{R}^d)} + \max_{1\leqslant i\leqslant d} M_i\,,$$

where M_i is the infimum of all constants $M > 0$ satisfying (8.3) or (8.4), respectively. Finally, we abbreviate $H^{(r_1,\ldots,r_d)}_q(\mathbb{R}^d) := H^{(r_1,\ldots,r_d)}_{q,\ldots,q}(\mathbb{R}^d)$.

it follows that $\{\psi^{(n)}_{m,\overline{\theta},\sharp}\}_{m\in(0,\delta]}$ is a bounded family in $H^{\mathbf{s}}_{\mathbf{q}}(\mathbb{R}^{3(n+1)})$. Now, we may apply Theorem 3.2 in [41]. The latter ensures that $\{\psi^{(n)}_{m,\overline{\theta},\sharp}\}_{m\in(0,\delta]}$ contains a sequence which is strongly convergent in $L^2(Q_{n,2\delta})$ provided $1-3n(p^{-1}-2^{-1})>0$. Of course, we can choose $p<2$ large enough such that the latter condition is fulfilled, for all $n=1,\dots,n_0-1$. By finitely many repeated selections of subsequences we may hence assume without loss of generality that $\{\phi^{(n)}_{m_j,\overline{\theta},\sharp}\}_{j\in\mathbb{N}}$ converges strongly in $L^2(Q_{n,2\delta})$ to its weak limit $\phi^{(n)}_{\overline{\theta},\sharp}$, for $0\leqslant n<n_0$. In particular, by the choice of n_0 and R in (8.1) and (8.2),

$$\|\phi^\sharp\|^2 \geqslant \lim_{j\to\infty}\sum_{n=0}^{n_0-1}\sum_{\overline{\theta}}\|\phi^{(n)}_{m_j,\overline{\theta},\sharp}\|^2_{L^2(Q_{n,2\delta})} \geqslant \lim_{j\to\infty}\|\phi^\sharp_{m_j}\|^2-\varepsilon-o(\delta^0)\,,$$

where we use the soft photon bound to estimate

$$\begin{aligned}&\sum_{n=1}^{n_0-1}\sum_{\overline{\theta}}\big\|\,\phi^{(n)}_{m_j,\overline{\theta},\sharp}\,\mathbb{1}\{\exists i:\,|\mathbf{k}_i|\leqslant 2\delta\vee|\mathbf{k}_i|\geqslant\Lambda-2\delta\}\,\big\|^2\\&\leqslant\sum_{\lambda\in\mathbb{Z}_2}\int_{\substack{\{|\mathbf{k}|\leqslant 2\delta\}\cup\\\{|\mathbf{k}|\geqslant\Lambda-2\delta\}}}\|a(\mathbf{k},\lambda)\,\phi^\sharp_{m_j}\|^2\,d^3\mathbf{k}=o(\delta^0)\,,\qquad\delta\searrow 0\,.\end{aligned}$$

Hence, $\|\phi^\sharp\|^2\geqslant 1-\varepsilon-o(\delta^0)$, where $\delta>0$ and $\varepsilon>0$ are arbitrary, that is, $\|\phi^\sharp\|=1$. (In particular, $\phi^\sharp_{m_j}\to\phi^\sharp$ strongly in $\mathscr{H}$.) □

8.2. *Ground state degeneracy*

Suppose that $V(\mathbf{x})=V(-\mathbf{x})$. As already mentioned in the introduction it is remarked in [40] that every (speculative) eigenvalue of H^{PF}_V and, in particular, its ground state energy is evenly degenerate in this case. The authors prove this statement by constructing some anti-linear involution commuting with H^{PF}_V and applying Kramers' degeneracy theorem. We shall do the same for the no-pair operator in the next theorem which originates from [29].

Theorem 8.2: *Let $e\in\mathbb{R}$, $\Lambda>0$, assume that V fulfills Hypotheses 6.1, and assume in addition that $V(\mathbf{x})=V(-\mathbf{x})$, for almost every $\mathbf{x}$. If the ground state energy E^{np}_V is an eigenvalue of H^{np}_V, then it is evenly degenerate.*

Proof: Similarly as in [40] we introduce the anti-linear operator

$$\vartheta:=J\,\alpha_2 C R=-\alpha_2\,J\,C\,R\,,\qquad J:=\begin{pmatrix}0&\mathbb{1}_2\\-\mathbb{1}_2&0\end{pmatrix},$$

where $C : \mathscr{H} \to \mathscr{H}$ denotes complex conjugation, $C\psi := \overline{\psi}$, $\psi \in \mathscr{H}$, and $R : \mathscr{H} \to \mathscr{H}$ is the parity transformation $(R\psi)(\mathbf{x}) := \psi(-\mathbf{x})$, for almost every $\mathbf{x} \in \mathbb{R}^3$ and every $\psi \in \mathscr{H} = L^2(\mathbb{R}^3_{\mathbf{x}}, \mathbb{C}^4 \otimes \mathscr{F}_{\mathrm{b}}[\mathscr{K}])$. Obviously, $[\vartheta\,,\,-i\partial_{x_j}] = [\vartheta\,, V] = [\vartheta\,,\,H_{\mathrm{f}}] = 0$, on $\mathcal{D}(|D_{\mathbf{0}}|) \cap \mathcal{D}(H_{\mathrm{f}})$. Since α_2 squares to one and $C\,\alpha_2 = -\alpha_2\,C$, for all entries of α_2 are purely imaginary, we further get $\vartheta^2 = -\mathbb{1}$ and $[\vartheta\,,\,\alpha_2] = 0$. Moreover, the Dirac matrices α_0, α_1, and α_3 have real entries and $[J\,\alpha_2\,,\,\alpha_j] = J\,\{\alpha_2, \alpha_j\} = 0$ by (2.1), whence $[\vartheta\,,\,\alpha_j] = 0$, for $j \in \{0,1,3\}$. Finally $[\vartheta\,,\,e^{\pm i\mathbf{k}\cdot\mathbf{x}}] = 0$ implies $[\vartheta\,,\,A^{(j)}] = 0$ on $\mathcal{D}(H_{\mathrm{f}}^{1/2})$, for $j \in \{1,2,3\}$. It follows that $[\vartheta\,,\,D_{\mathbf{A}}] = 0$ on $\mathscr{D} = \vartheta\,\mathscr{D}$ and, since $D_{\mathbf{A}}$ is essentially self-adjoint on $\mathscr{D}$, we obtain $\vartheta\,\mathcal{D}(D_{\mathbf{A}}) = \mathcal{D}(D_{\mathbf{A}})$ and $[\vartheta\,,\,D_{\mathbf{A}}] = 0$ on $\mathcal{D}(D_{\mathbf{A}})$, which implies $\vartheta\,R_{\mathbf{A}}(iy) - R_{\mathbf{A}}(-iy)\,\vartheta = 0$ on $\mathscr{H}$, for every $y \in \mathbb{R}$. Using the representation (2.18) we conclude that $[\vartheta\,,\,P^+_{\mathbf{A}}] = 0$ on $\mathscr{H}$. In particular, ϑ can be considered as an operator acting on $\mathscr{H}^+_{\mathbf{A}}$. Furthermore, we obtain $H^{\mathrm{np}}_V\,\vartheta\,\varphi - \vartheta\,H^{\mathrm{np}}_V\,\varphi = 0$, for every $\varphi \in \mathscr{D}$. Since $P^+_{\mathbf{A}}\,\mathscr{D}$ is a form core for H^{np}_V we can easily extend this commutation relation and show that ϑ maps $\mathcal{D}(H^{\mathrm{np}}_V)$ into itself and $H^{\mathrm{np}}_V\,\vartheta\,\psi = \vartheta\,H^{\mathrm{np}}_V\,\psi$, for every $\psi \in \mathcal{D}(H^{\mathrm{np}}_V)$. Hence, by Kramers' degeneracy theorem every eigenvalue of H^{np}_V is evenly degenerated. (In fact, $H^{\mathrm{np}}_V\,\phi = E_V\,\phi$ implies $H^{\mathrm{np}}_V\,\vartheta\,\phi = E_V\,\vartheta\,\phi$, and $\phi \perp \vartheta\,\phi$, since $\langle\,\vartheta\,\phi\,|\,\phi\,\rangle = -\langle\,\vartheta\,\phi\,|\,\vartheta\,(\vartheta\,\phi)\,\rangle = -\langle\,C\,\phi\,|\,C\,\vartheta\,\phi\,\rangle = -\langle\,\vartheta\,\phi\,|\,\phi\,\rangle$.) □

Using similar arguments we derive the next lemma which has been referred to in Sections 5 and 6. We use the notation introduced in Subsection 6.1 and write

$$H^{\mathrm{np},-}_{0,m,\varepsilon} := P^-_{\mathbf{A}_{m,\varepsilon}}\,(|D_{\mathbf{A}_{m,\varepsilon}}| + H_{\mathrm{f},m,\varepsilon})\,P^-_{\mathbf{A}_{m,\varepsilon}}$$

for the analogs of the no-pair operators on the negative spectral subspace which appeared in (5.1) and (6.28). Then $\widehat{H}^{\mathrm{np}}_{0,m,\varepsilon} = H^{\mathrm{np}}_{0,m,\varepsilon} + H^{\mathrm{np},-}_{0,m,\varepsilon}$; see (5.1) and (6.28). We unify the notation by setting $H^{\mathrm{np}}_{0,0,0} := H^{\mathrm{np}}_0$, etc.

Lemma 8.3: *Let $e \in \mathbb{R}$, $\Lambda \in (0,\infty)$, and $m \geqslant \varepsilon \geqslant 0$. Then*

$$\Sigma^{\mathrm{np}}_{m,\varepsilon} \stackrel{\text{def.}}{=} \inf\sigma[H^{\mathrm{np}}_{0,m,\varepsilon}] = \inf\sigma[H^{\mathrm{np},-}_{0,m,\varepsilon}] = \inf\sigma[\widehat{H}^{\mathrm{np}}_{0,m,\varepsilon}]\,.$$

Proof: Let C and R be defined as in the preceding proof. We introduce the anti-linear operator $\tau : \mathscr{H}^>_m \to \mathscr{H}^>_m$, $\tau := \alpha_2\,C\,R$. Similar as in the previous proof we verify that $\tau\,D_{\mathbf{A}_{m,\varepsilon}} = -D_{\mathbf{A}_{m,\varepsilon}}\tau$ on $\mathcal{D}(D_{\mathbf{A}_{m,\varepsilon}})$, thus $R_{\mathbf{A}_{m,\varepsilon}}(iy)\,\tau = -\tau\,R_{\mathbf{A}_{m,\varepsilon}}(iy)$, thus $P^\pm_{\mathbf{A}_{m,\varepsilon}}\,\tau = \tau\,P^\mp_{\mathbf{A}_{m,\varepsilon}}$ again by (2.18). Notice that we use the property (6.9) of the discretized phase in $\mathbf{A}_{m,\varepsilon}$ to obtain these relations in the case $\varepsilon > 0$. Consequently, $H^{\mathrm{np}}_{0,m,\varepsilon}\,\tau = \tau\,H^{\mathrm{np},-}_{0,m,\varepsilon}$ on a natural dense domain (again called $\mathscr{D}$) in $\mathscr{H}^>_m$ with $\tau\mathscr{D} = \mathscr{D}$. By Theorem 3.7 we may assume that the no-pair operators in the plus or minus spaces are essentially self-adjoint on $P^{\pm}_{\mathbf{A}_{m,\varepsilon}}\mathscr{D}$, respectively, and we readily conclude. □

9. Commutator Estimates

In this final section we collect some bounds proved in [37] on the operator norm of various commutators which have been used repeatedly in the preceding sections. In the whole section we assume that $\mathbf{G_x}$ fulfills Hypothesis 2.1.

9.1. *Basic estimates*

Our estimates on commutators involving the field energy H_{f} are based on the next lemma. The following quantity appears in its statement and in various estimates below,

$$\delta_\nu^2 \equiv \delta_\nu(E)^2 := 8\int \frac{w_\nu(k,E)^2}{\omega(k)}\,\|\mathbf{G}(k)\|_\infty^2\,dk\,, \qquad E,\nu>0\,, \tag{9.1}$$

where $w_\nu(k,E) := E^{1/2-\nu}\,((E{+}\omega(k))^{\nu+1/2} - E^\nu\,(E{+}\omega(k))^{1/2})$. We observe that $w_{1/2}(k,E) \leqslant \omega(k)$ and, hence, $\delta_{1/2}(E) \leqslant 2\,d_1$, $E>0$. Moreover, $\delta_\nu(E) \leqslant \delta_\nu(1)$, for $E \geqslant 1$. We recall the identity

$$\langle\, H_{\mathrm{f}}^{1/2}\phi \,|\, H_{\mathrm{f}}^{1/2}\psi\,\rangle = \int \omega(k)\,\langle\, a(k)\phi \,|\, a(k)\psi\,\rangle dk\,, \quad \phi,\psi\in\mathcal{D}(H_{\mathrm{f}}^{1/2})\,, \tag{9.2}$$

which is a consequence of the permutation symmetry and Fubini's theorem.

Lemma 9.1: *Let ν, $E>0$, and set $\check{H}_{\mathrm{f}} := H_{\mathrm{f}} + E$. Then*

$$\big\|\,[\boldsymbol{\alpha}\cdot\mathbf{A}\,,\,\check{H}_{\mathrm{f}}^{-\nu}]\,\check{H}_{\mathrm{f}}^{\nu}\,\big\| \leqslant \delta_\nu(E)/E^{1/2}\,. \tag{9.3}$$

Proof: ([37]) We pick $\phi,\psi\in\mathscr{D}$ and write

$$\begin{aligned}&\langle\,\phi\,|\,[\boldsymbol{\alpha}\cdot\mathbf{A}\,,\,\check{H}_{\mathrm{f}}^{-\nu}]\,\check{H}_{\mathrm{f}}^{\nu}\,\psi\,\rangle\\ &= \langle\,\phi\,|\,[\boldsymbol{\alpha}\cdot a(\mathbf{G})\,,\,\check{H}_{\mathrm{f}}^{-\nu}]\,\check{H}_{\mathrm{f}}^{\nu}\,\psi\,\rangle - \langle\,[\boldsymbol{\alpha}\cdot a(\mathbf{G})\,,\,\check{H}_{\mathrm{f}}^{-\nu}]\,\phi\,|\,\check{H}_{\mathrm{f}}^{\nu}\,\psi\,\rangle\,.\end{aligned} \tag{9.4}$$

By definition of $a(k)$ and H_{f} we have the pull-through formula $a(k)\,\theta(H_{\mathrm{f}})\,\psi = \theta(H_{\mathrm{f}}+\omega(k))\,a(k)\,\psi$, for almost every k and every Borel function θ on $\mathbb{R}$, which leads to

$$\begin{aligned}&[\,a(k)\,,\,\check{H}_{\mathrm{f}}^{-\nu}\,]\,\check{H}_{\mathrm{f}}^{\nu}\,\psi\\ &\quad= \big\{((\check{H}_{\mathrm{f}}+\omega(k))^{-\nu} - \check{H}_{\mathrm{f}}^{-\nu})(\check{H}_{\mathrm{f}}+\omega(k))^{\nu+1/2}\big\}\,a(k)\,\check{H}_{\mathrm{f}}^{-1/2}\,\psi\,.\end{aligned}$$

We denote the operator $\{\cdots\}$ by $F(k)$. Then $F(k)$ is bounded and

$$\|F(k)\| \leqslant \int_0^1 \sup_{t\geqslant 0}\Big|\frac{d}{ds}\,\frac{(t+E+\omega(k))^{\nu+1/2}}{(t+E+s\,\omega(k))^{\nu}}\Big|\,ds$$
$$= -\int_0^1 \frac{d}{ds}\,\frac{(E+\omega(k))^{\nu+1/2}}{(E+s\,\omega(k))^{\nu}}\,ds \,=\, w_\nu(k,E)/E^{1/2}\,.$$

Using these remarks, the Cauchy-Schwarz inequality, and (9.2), we obtain

$$\big|\langle\,\phi\,|\,[\boldsymbol{\alpha}\cdot a(\mathbf{G})\,,\,\check{H}_{\mathrm{f}}^{-\nu}]\,\check{H}_{\mathrm{f}}^{\nu}\,\psi\,\rangle\big|$$
$$\leqslant \int \|\phi\|\,\|\boldsymbol{\alpha}\cdot\mathbf{G}(k)\|\,\|F(k)\|\,\big\|\,a(k)\,\check{H}_{\mathrm{f}}^{-1/2}\,\psi\,\big\|\,dk$$
$$\leqslant \|\phi\|\,\Big(2\int\frac{\|F(k)\|^2}{\omega(k)}\,\|\mathbf{G}(k)\|_\infty^2\,dk\Big)^{1/2}\Big(\int\omega(k)\,\big\|\,a(k)\,\check{H}_{\mathrm{f}}^{-1/2}\,\psi\,\big\|^2\,dk\Big)^{1/2}$$
$$\leqslant \frac{\delta_\nu(E)}{2E^{1/2}}\,\|\phi\|\,\big\|\,H_{\mathrm{f}}^{1/2}\,\check{H}_{\mathrm{f}}^{-1/2}\,\psi\,\big\|\,.$$

A similar argument applied to the second term in (9.4) yields

$$\big|\langle\,[\boldsymbol{\alpha}\cdot a(\mathbf{G})\,,\,\check{H}_{\mathrm{f}}^{-\nu}]\,\phi\,|\,\check{H}_{\mathrm{f}}^{\nu}\,\psi\,\rangle\big| \,\leqslant\, \frac{\widetilde{\delta}_\nu(E)}{2E^{1/2}}\,\big\|\,H_{\mathrm{f}}^{1/2}\,\check{H}_{\mathrm{f}}^{-1/2}\,\phi\,\big\|\,\|\psi\|\,,$$

where $\widetilde{\delta}_\nu(E)$ is defined by (9.1) with $w_\nu(k,E)$ replaced by $\widetilde{w}_\nu(k,E) := E^{1/2-\nu}(E^\nu\,(E+\omega(k))^{1/2} - E^{2\nu}\,(E+\omega)^{1/2-\nu})$. Evidently, $\widetilde{w}_\nu \leqslant w_\nu$, thus $\widetilde{\delta}_\nu \leqslant \delta_\nu$, which concludes the proof. □

Choosing E large enough, we can certainly make to norm in (9.3) as small as we please. This observation can be exploited to ensure that certain Neumann series converge in the next proof which yields convenient formulas allowing to interchange the field energy with resolvents of the Dirac operator.

We define $J:[0,1)\to\mathbb{R}$ by $J(0) := 1$ and $J(a) := \sqrt{6}/(1-a^2)$, for $a\in(0,1)$, so that $\|R_{\mathbf{A}}^F(iy)\| \leqslant J(a)\,(1+y^2)^{-1/2}$, where $R_{\mathbf{A}}^0(iy) := R_{\mathbf{A}}(iy)$; recall (2.20) and (2.21).

Corollary 9.2: *Let $y\in\mathbb{R}$ and $F\in C^\infty(\mathbb{R}^3_{\mathbf{x}},\mathbb{R})$ such that $|\nabla F|\leqslant a<1$. Assume that $\nu, E>0$ satisfy $\delta_\nu\,J(a)/E^{1/2}<1$, and introduce the following operators,* $T_\nu := \overline{[\check{H}_{\mathrm{f}}^{-\nu}\,,\,\boldsymbol{\alpha}\cdot\mathbf{A}]\,\check{H}_{\mathrm{f}}^{\nu}}$,

$$\Xi_\nu^F(iy) := \sum_{j=0}^{\infty}\{-R_{\mathbf{A}}^F(iy)\,T_\nu\}^j, \quad \Upsilon_\nu^F(iy) := \sum_{j=0}^{\infty}\{-T_\nu^*\,R_{\mathbf{A}}^F(iy)\}^j. \qquad (9.5)$$

Then $\|T_\nu\| \leqslant \delta_\nu/E^{1/2}$, $\|\Theta_\nu^F(iy)\| \leqslant (1-\delta_\nu J(a)/E^{1/2})^{-1}$, $\Theta \in \{\Xi, \Upsilon\}$, *and*

$$\check{H}_{\mathrm{f}}^{-\nu}\, R_{\mathbf{A}}^F(iy) \;=\; \Xi_\nu^F(iy)\, R_{\mathbf{A}}^F(iy)\, \check{H}_{\mathrm{f}}^{-\nu}\,, \tag{9.6}$$

$$R_{\mathbf{A}}^F(iy)\, \check{H}_{\mathrm{f}}^{-\nu} \;=\; \check{H}_{\mathrm{f}}^{-\nu}\, R_{\mathbf{A}}^F(iy)\, \Upsilon_\nu^F(iy)\,. \tag{9.7}$$

In particular, $R_{\mathbf{A}}^F(iy)$ *maps* $\mathcal{D}(H_{\mathrm{f}}^\nu)$ *into itself.*

Proof: The following somewhat formal computations show the simple idea behind the proof (see Section 3 of [37] for more details),

$$\begin{aligned}\check{H}_{\mathrm{f}}^{-\nu}\, R_{\mathbf{A}}^F(iy) &= R_{\mathbf{A}}^F(iy)\, \check{H}_{\mathrm{f}}^{-\nu} + R_{\mathbf{A}}^F(iy)\, \big[D_{\mathbf{A}} + i\boldsymbol{\alpha}\cdot\nabla F - iy\,,\, \check{H}_{\mathrm{f}}^{-\nu}\big]\, R_{\mathbf{A}}^F(iy)\\ &= R_{\mathbf{A}}^F(iy)\, \check{H}_{\mathrm{f}}^{-\nu} + R_{\mathbf{A}}^F(iy)\, \big[\boldsymbol{\alpha}\cdot\mathbf{A}, \check{H}_{\mathrm{f}}^{-\nu}\big]\, \check{H}_{\mathrm{f}}^{\nu}\big(\check{H}_{\mathrm{f}}^{-\nu}\, R_{\mathbf{A}}^F(iy)\big),\end{aligned}$$

which implies that $R_{\mathbf{A}}^F(iy)\, \check{H}_{\mathrm{f}}^{-\nu} = \big(1 + R_{\mathbf{A}}^F(iy)\, T_\nu\big)^{-1}\, \check{H}_{\mathrm{f}}^{-\nu}\, R_{\mathbf{A}}^F(iy)$. The operator inverse appearing here is given by the Neumann series $\Xi_\nu^F(iy)$ when we choose E in $\check{H}_{\mathrm{f}} = H_{\mathrm{f}} + E$ so large that $\delta_\nu J(a)/E^{1/2} < 1$. □

9.2. *Commuting projections with the field energy*

Lemma 9.3: *Let* $a, \kappa \in [0,1)$, *let* ν, E, *and* F *be as in Corollary 9.2, and assume that* F *is bounded. Define* $\mathcal{C}_\nu^F := e^F\,[\,S_{\mathbf{A}}\,,\, \check{H}_{\mathrm{f}}^{-\nu}\,]\, \check{H}_{\mathrm{f}}^{\nu}\, e^{-F}$ *on* $\mathcal{D}(H_{\mathrm{f}}^\nu)$. *Then*

$$\|\, |D_{\mathbf{A}}|^\kappa\, \mathcal{C}_\nu^F\,\| \;\leqslant\; (1 + a\,J(a))\, \frac{\mathrm{const}(\kappa)\, \delta_\nu\, J(a)/E^{1/2}}{1 - \delta_\nu\, J(a)/E^{1/2}}\,. \tag{9.8}$$

In particular, $S_{\mathbf{A}}$ *maps* $\mathcal{D}(H_{\mathrm{f}}^\nu)$ *into itself and, if* $E^{1/2} > \delta_\nu$, *then the following identities hold true on* $\mathcal{D}(H_{\mathrm{f}}^\nu)$, *where* $\mathcal{C}_\nu := (\mathcal{C}_\nu^0)^* \in \mathscr{L}(\mathscr{H})$,

$$\check{H}_{\mathrm{f}}^{\nu}\, S_{\mathbf{A}} = S_{\mathbf{A}}\, \check{H}_{\mathrm{f}}^{\nu} + \mathcal{C}_\nu\, \check{H}_{\mathrm{f}}^{\nu}\,, \qquad S_{\mathbf{A}}\, \check{H}_{\mathrm{f}}^{\nu} = \check{H}_{\mathrm{f}}^{\nu}\, S_{\mathbf{A}} + \check{H}_{\mathrm{f}}^{\nu}\, \mathcal{C}_\nu^*\,. \tag{9.9}$$

Proof: ([37]) Combining (9.6) with (2.18) we obtain, for all $\phi \in \mathcal{D}(|D_{\mathbf{A}}|^\kappa)$ and $\psi \in \mathcal{D}(H_{\mathrm{f}}^\nu)$,

$$\begin{aligned}\big|\langle\, |D_{\mathbf{A}}|^\kappa\, \phi\,|\, \mathcal{C}_\nu^F\, \psi\,\rangle\big| &\leqslant \int_{\mathbb{R}} \big|\langle\, |D_{\mathbf{A}}|^\kappa\, \phi\,|\, R_{\mathbf{A}}^F(iy)\, T_\nu\, \Xi_\nu^F(iy)\, R_{\mathbf{A}}^F(iy)\, \psi\,\rangle\big|\, \frac{dy}{\pi}\\ &\leqslant \|T_\nu\| \int_{\mathbb{R}} \big\|\, |D_{\mathbf{A}}|^\kappa\, R_{\mathbf{A}}^F(iy)\, \phi\,\big\|\, \|\Xi_\nu^F(iy)\|\, \|R_{\mathbf{A}}^F(iy)\|\, \frac{dy}{\pi}\, \|\phi\|\, \|\psi\|\,.\end{aligned}$$

Here we estimate $\|T_\nu\|$ by means of (9.3) and we write $|D_{\mathbf{A}}|^\kappa\, R_{\mathbf{A}}^F(iy) = |D_{\mathbf{A}}|^\kappa\, R_{\mathbf{A}}(iy)\, (\mathbb{1} - i\boldsymbol{\alpha}\cdot\nabla F\, R_{\mathbf{A}}^F(iy))$, where $\|\, |D_{\mathbf{A}}|^\kappa\, R_{\mathbf{A}}(iy)\| \leqslant \mathrm{const}(\kappa)(1 + y^2)^{-1/2+\kappa/2}$. Moreover, $\|\Xi_\nu^F(iy)\| \leqslant (1 - \delta_\nu J(a)/E^{1/2})^{-1}$, $y \in \mathbb{R}$, by Corollary 9.2. Altogether these remarks yield the asserted estimate. Now, the identity $S_{\mathbf{A}}\,\check{H}_{\mathrm{f}}^{-\nu} = \check{H}_{\mathrm{f}}^{-\nu}\,S_{\mathbf{A}} - \check{H}_{\mathrm{f}}^{-\nu}\,(C_\nu^0)^*$ in $\mathscr{L}(\mathscr{H})$ shows that $S_{\mathbf{A}}$ maps the domain of H_{f}^ν into itself and that the first identity in (9.9) is valid. Taking the adjoint of (9.9) and using $[\check{H}_{\mathrm{f}}^\nu\,S_{\mathbf{A}}]^* = S_{\mathbf{A}}\,\check{H}_{\mathrm{f}}^\nu$ (which is true since $\check{H}_{\mathrm{f}}^\nu\,S_{\mathbf{A}}$ is densely defined and $S_{\mathbf{A}} = S_{\mathbf{A}}^{-1} \in \mathscr{L}(\mathscr{H})$) we also obtain the second identity in (9.9). □

We can now prove an estimate asserted in the proof of Theorem 3.4, namely $\mathcal{T} = \mathrm{Re}\,\Theta \leqslant \varepsilon\,|D_{\mathbf{A}}| + \varepsilon^{-1}\,\mathrm{const}\,d_1/E^{1/2}$, for $\varepsilon \in (0,1]$ and $E \geqslant (4d_1)^2$, where $\Theta := [|D_{\mathbf{A}}|, \check{H}_{\mathrm{f}}^{-1/2}]\check{H}_{\mathrm{f}}^{1/2}$. In fact, this follows easily from Corollary 9.2 and Lemma 9.3 since $\Theta = |D_{\mathbf{A}}|^{1/2} S_{\mathbf{A}}\{|D_{\mathbf{A}}|^{1/2} C_{1/2}^0\} + T_{1/2}C_{1/2}^0 - T_{1/2}S_{\mathbf{A}}$.

9.3. *Double commutators*

Lemma 9.4: *Let $a \in [0,1)$ and let F satisfy (5.8). Moreover, let $\nu, E > 0$ such that $\delta_\nu\,J(a)/E^{1/2} \leqslant 1/2$. Then*

$$\big\|\,|D_{\mathbf{A}}|\,[\chi_1\,e^F,\,[P_{\mathbf{A}}^+\,,\,\chi_2\,e^{-F}]]\,\big\| \leqslant J(a)\prod_{i=1,2}(a + \|\nabla\chi_i\|_\infty)\,, \tag{9.10}$$

$$\big\|\check{H}_{\mathrm{f}}^\nu\,[\chi_1\,e^F,\,[P_{\mathbf{A}}^+\,,\,\chi_2\,e^{-F}]]\,\check{H}_{\mathrm{f}}^{-\nu}\big\| \leqslant 8\,J(a)\prod_{i=1,2}(a + \|\nabla\chi_i\|_\infty)\,. \tag{9.11}$$

Moreover, for every self-adjoint multiplication operator V in $L^2(\mathbb{R}^3)$ with $H^1(\mathbb{R}^3) \subset \mathcal{D}(V)$, we find some constant $C_V \in (0,\infty)$, depending only on V, such that

$$\big\|\,|V|\,[\chi_1\,e^F,\,[P_{\mathbf{A}}^+\,,\,\chi_2\,e^{-F}]]\,\check{H}_{\mathrm{f}}^{-1/2}\big\| \leqslant C_V\,J(a)\prod_{i=1,2}(a + \|\nabla\chi_i\|_\infty)\,. \tag{9.12}$$

In (9.12) we also assume that $E \geqslant (4d_1\,J(a))^2$ and $E \geqslant 1$.

Proof: ([37]) Let $\phi, \psi \in \mathscr{D}$, $\|\phi\| = \|\psi\| = 1$. First, we derive a bound on

$$I_{\phi,\psi} := \int_{\mathbb{R}} \Big|\Big\langle\,|D_{\mathbf{A}}|\,\phi\,\Big|\,\check{H}_{\mathrm{f}}^\nu\,\big[\chi_1\,e^F\,,\,[R_{\mathbf{A}}(iy)\,,\,\chi_2\,e^{-F}]\big]\,\check{H}_{\mathrm{f}}^{-\nu}\,\psi\Big\rangle\Big|\,\frac{dy}{2\pi}\,.$$

Expanding the double commutator we get

$$\big[\chi_1\,e^F\,,\,[R_{\mathbf{A}}(iy)\,,\,\chi_2\,e^{-F}]\big] = \eta(\chi_1,\chi_2,F\,;y) + \eta(\chi_2,\chi_1,-F\,;y)\,,$$

where

$$\begin{aligned}&\eta(\chi_1,\chi_2,F\,;y)\\ &:= R_{\mathbf{A}}(iy)\,\boldsymbol{\alpha}\cdot(\nabla\chi_1 + \chi_1\,\nabla F)\,e^F\,R_{\mathbf{A}}(iy)\,e^{-F}\,\boldsymbol{\alpha}\cdot(\nabla\chi_2 - \chi_2\,\nabla F)\,R_{\mathbf{A}}(iy)\,.\end{aligned}$$

We obtain

$$\begin{aligned}
&\int_{\mathbb{R}} \Big| \Big\langle\, |D_{\mathbf{A}}|\,\phi \,\Big|\, \check{H}_{\mathrm{f}}^{\nu}\, \eta(\chi_1,\chi_2,F\,;y)\, \check{H}_{\mathrm{f}}^{-\nu}\,\psi \Big\rangle \Big| \frac{dy}{2\pi} \\
&\leqslant \int_{\mathbb{R}} \Big| \Big\langle\, \phi \,\Big|\, |D_{\mathbf{A}}|\, R_{\mathbf{A}}(iy)\, \Upsilon_\nu^0(iy)\, \boldsymbol{\alpha}\cdot(\nabla\chi_1 + \chi_1\,\nabla F) \times \\
&\qquad \times\, R_{\mathbf{A}}^F(iy)\, \Upsilon_\nu^F(iy)\, \boldsymbol{\alpha}\cdot(\nabla\chi_2 - \chi_2\,\nabla F)\, R_{\mathbf{A}}(iy)\, \Upsilon_\nu^0(iy)\,\psi \Big\rangle \Big| \frac{dy}{2\pi} \\
&\leqslant \frac{(a+\|\nabla\chi_1\|)(a+\|\nabla\chi_2\|)}{(1-\delta_\nu/E^{1/2})^2} \cdot \frac{J(a)}{1-\delta_\nu\, J(a)/E^{1/2}} \int_{\mathbb{R}} \frac{dy}{2\pi(1+y^2)}\,. \qquad (9.13)
\end{aligned}$$

A bound analogous to (9.13) holds true when the roles of χ_1 and χ_2 are interchanged and F is replaced by $-F$. Consequently, $I_{\phi,\psi}$ is bounded by two times the right hand side of (9.13). Altogether this shows that (9.10) and (9.11) hold true. (Just ignore $|D_{\mathbf{A}}|$ or $\check{H}_{\mathrm{f}}$, respectively, in the above argument.) By the closed graph theorem we further have $\|\,|V|\,\psi\|^2 \leqslant \mathrm{const}(V)\,(\|\nabla\psi\|^2 + \|\psi\|^2)$, $\psi \in H^1(\mathbb{R}^3)$. Hence, (9.12) follows from (9.10), (9.11), and the inequality

$$\big\|\,|V|\,\varphi\,\big\|^2 \leqslant \mathrm{const}(V)\Big(\big\|\,|D_{\mathbf{A}}|\,\varphi\,\big\|^2 + \big\|\,\check{H}_{\mathrm{f}}^{1/2}\,\varphi\,\big\|^2\Big), \qquad \varphi \in \mathscr{D},$$

which holds true, for $E \geqslant d_1^2$, by (3.2), (3.23), and $\|\,|V|\,\varphi\|^2 \leqslant \mathrm{const}(V)\,(\|\nabla[\![\varphi]\!]\,\|^2 + \|\varphi\|^2)$, $\varphi \in \mathscr{D}$. □

Acknowledgments

This work has been partially supported by the DFG (SFB/TR12). O.M. and E.S. thank the Institute for Mathematical Sciences and the Center for Quantum Technologies of the National University of Singapore for their generous hospitality.

References

1. Y. Avron, I. Herbst, and B. Simon, "Schrödinger operators with magnetic fields. I. General interactions", *Duke Math. J.*, **45** (1978), 847–883.
2. V. Bach, T. Chen, J. Fröhlich, and I. M. Sigal, "Smooth Feshbach map and operator-theoretic renormalization group methods", *J. Funct. Anal.*, **203** (2003), 44–92.
3. V. Bach, J. Fröhlich, and A. Pizzo, "Infrared-finite algorithms in QED: the groundstate of an atom interacting with the quantized radiation field", *Comm. Math. Phys.*, **264** (2006), 145–165.
4. V. Bach, J. Fröhlich, and I. M. Sigal, "Quantum electrodynamics of confined nonrelativistic particles", *Adv. Math.*, **137** (1998), 299–395.

5. V. Bach, J. Fröhlich, and I. M. Sigal, "Renormalization group analysis of spectral problems in quantum field theory", *Adv. Math.*, **137** (1998), 205–298.
6. V. Bach, J. Fröhlich, and I. M. Sigal, "Spectral analysis for systems of atoms and molecules coupled to the quantized radiation field", *Comm. Math. Phys.*, **207** (1999), 249–290.
7. V. Bach and M. Könenberg, "Construction of the ground state in nonrelativistic QED by continuous flows", *J. Differential Equations*, **231** (2006), 693–713.
8. J.-M. Barbaroux, M. Dimassi, and J.-C. Guillot, "Quantum electrodynamics of relativistic bound states with cutoffs", *J. Hyperbolic Differ. Equ.*, **1** (2004), 271–314.
9. A. Berthier and V. Georgescu, "On the point spectrum of Dirac operators", *J. Funct. Anal.*, **71** (1987), 309–338.
10. K. T. Cheng, M. H. Chen, and W. R. Johnson, "Accurate relativistic calculations including QED contributions for few-electron systems", In: P. Schwerdtfeger (Editor). *Relativistic electronic structure theory. Part 2: Applications.* Theoretical and Computational Chemistry, **14**, Pages 120–187, Elsevier, 2002.
11. E. B. Davies, *Spectral theory and differential operators*, Cambridge Studies in Advanced Mathematics, **42**, Cambridge University Press, Cambridge, 1995.
12. M. Dimassi and J. Sjöstrand, *Spectral asymptotics in the semi-classical limit*, London Math. Soc. Lecture Note Series, **268**, Cambridge University Press, Cambridge, 1999.
13. W. D. Evans, P. Perry, and H. Siedentop, "The spectrum of relativistic one-electron atoms according to Bethe and Salpeter", *Comm. Math. Phys.*, **178** (1996), 733–746.
14. J. Fröhlich, M. Griesemer, and B. Schlein, "Asymptotic electromagnetic fields in models of quantum-mechanical matter interacting with the quantized radiation field", *Adv. Math.*, **164** (2001), 349–398.
15. J. Fröhlich, M. Griesemer, and I. M. Sigal, "On spectral renormalization group", *Rev. Math. Phys*, **21** (2009), 511–548.
16. M. Griesemer, "Exponential decay and ionization thresholds in non-relativistic quantum electrodynamics", *J. Funct. Anal.*, **210** (2004), 321–340.
17. M. Griesemer, E. H. Lieb, and M. Loss, "Ground states in non-relativistic quantum electrodynamics", *Invent. Math.*, **145** (2001), 557–595.
18. M. Griesemer and C. Tix, "Instability of a pseudo-relativistic model of matter with self-generated magnetic field", *J. Math. Phys.*, **40** (1999), 1780–1791.
19. F. Hiroshima, "Diamagnetic inequalities for systems of nonrelativistic particles with a quantized field", *Rev. Math. Phys.*, **8** (1996), 185–203.
20. F. Hiroshima, "Functional integral representation of a model in quantum electrodynamics", *Rev. Math. Phys.*, **9** (1997), 489–530.
21. F. Hiroshima and I. Sasaki, "On the ionization energy of the semi-relativistic Pauli-Fierz model for a single particle", *RIMS Kokyuroku Bessatsu*, **21** (2010), 25–34.
22. W. R. Johnson, "Relativistic many-body perturbation theory for highly

charged ions", In: J. J. Boyle and M. S. Pindzola (Editors). *Many-body atomic physics.* Pages 39–64, University Press, 1998.
23. T. Kato, *Perturbation theory for linear operators*, Classics in Mathematics, Springer-Verlag, Berlin, 1995. Reprint of the 1980 edition.
24. M. Könenberg, *Nichtexistenz von Grundzuständen für minimal an das quantisierte Strahlungsfeld gekoppelte, pseudorelativistische Modelle*, Diploma Thesis, Universität Mainz, 2004.
25. M. Könenberg and O. Matte, "Ground states of semi-relativistic Pauli-Fierz and no-pair Hamiltonians in QED at critical Coulomb coupling", *J. Operator Theory*, to appear.
26. M. Könenberg and O. Matte, "The mass shell in the semi-relativistic Pauli-Fierz model", *Preprint*, arXiv:1204.5123v1, 2012, 44 pages.
27. M. Könenberg and O. Matte, "On enhanced binding and related effects in the non- and semi-relativistic Pauli-Fierz models", *Comm. Math. Phys.*, to appear.
28. M. Könenberg, O. Matte, and E. Stockmeyer, "Existence of ground states of hydrogen-like atoms in relativistic quantum electrodynamics I: The semi-relativistic Pauli-Fierz operator", *Rev. Math. Phys.*, **23** (2011), 375–407.
29. M. Könenberg, O. Matte, and E. Stockmeyer, "Existence of ground states of hydrogen-like atoms in relativistic quantum electrodynamics II: The no-pair operator", *J. Math. Phys.*, **52** (2011), 123501, 34 pages.
30. E. Lieb and M. Loss, *Analysis*, Graduate Studies in Mathematics, American Mathematical Society, Providence, 2001. Second edition.
31. E. H. Lieb and M. Loss, "A bound on binding energies and mass renormalization in models of quantum electrodynamics", *J. Statist. Phys.*, **108** (2002), 1057–1069.
32. E. H. Lieb and M. Loss, "Stability of a model of relativistic quantum electrodynamics", *Comm. Math. Phys.*, **228** (2002), 561–588.
33. E. H. Lieb and M. Loss, "Existence of atoms and molecules in non-relativistic quantum electrodynamics", *Adv. Theor. Math. Phys.*, **7** (2003), 667–710.
34. E. H. Lieb, H. Siedentop, and J. P. Solovej, "Stability and instability of relativistic electrons in classical electromagnetic fields", *J. Statist. Phys.*, **89** (1997), 37–59.
35. O. Matte, *Existence of ground states for a relativistic hydrogen atom coupled to the quantized electromagnetic field*, Diploma Thesis, Universität Mainz, 2000.
36. O. Matte, "On higher order estimates in quantum electrodynamics", *Documenta Math.*, **15** (2010), 207–234.
37. O. Matte and E. Stockmeyer, "Exponential localization for a hydrogen-like atom in relativistic quantum electrodynamics", *Comm. Math. Phys.*, **295** (2010), 551–583.
38. O. Matte and E. Stockmeyer, "On the eigenfunctions of no-pair operators in classical magnetic fields", *Integr. Equ. Oper. Theory*, **65** (2009), 255–283.
39. O. Matte and E. Stockmeyer, "Spectral theory of no-pair Hamiltonians", *Rev. Math. Phys.*, **22** (2010), 1–53.
40. T. Miyao and H. Spohn, "Spectral analysis of the semi-relativistic Pauli-Fierz

Hamiltonian", *J. Funct. Anal.*, **256** (2009), 2123–2156.
41. S. M. Nikol′skiĭ, "An imbedding theorem for functions with partial derivatives considered in different metrics", *Izv. Akad. Nauk SSSR Ser. Mat.*, **22** (1958), 321–336. (Russian) English translation in: *Amer. Math. Soc. Transl. (2)*, 90:27–43, 1970.
42. S. M. Nikol′skiĭ, *Approximation of functions of several variables and imbedding theorems*, Die Grundlehren der Mathematischen Wissenschaften, Band 205, Springer-Verlag, New York, 1975.
43. M. Reed and B. Simon, *Methods of modern mathematical physics. II. Fourier analysis, self-adjointness*, Academic Press [Harcourt Brace Jovanovich Publishers], New York, 1975.
44. M. Reed and B. Simon, *Methods of modern mathematical physics. IV. Analysis of operators*, Academic Press [Harcourt Brace Jovanovich Publishers], New York, 1978.
45. M. Reed and B. Simon, *Methods of modern mathematical physics. I. Functional Analysis*, Academic Press Inc. [Harcourt Brace Jovanovich Publishers], New York, second edition, 1980.
46. M. Reiher and A. Wolf, *Relativistic quantum chemistry*, Wiley-VCH, Weinheim, 2009.
47. B. Simon, "Kato's inequality and the comparison of semigroups", *J. Funct. Anal.*, **32** (1979), 97–101.
48. E. Stockmeyer, "On the non-relativistic limit of a model in quantum electrodynamics", *Preprint*, arXiv:0905.1006v1, 2009, 13 pages.
49. C. Tix, "Strict positivity of a relativistic Hamiltonian due to Brown and Ravenhall", *Bull. London Math. Soc.*, **30** (1998), 283–290.

www.ingramcontent.com/pod-product-compliance
Lightning Source LLC
Chambersburg PA
CBHW050543190326
41458CB00007B/1896
* 9 7 8 9 8 1 4 4 6 0 1 4 9 *

www.ingramcontent.com/pod-product-compliance
Lightning Source LLC
Chambersburg PA
CBHW050543190326
41458CB00007B/1896
* 9 7 8 9 8 1 4 4 6 0 1 4 9 *